Remote Sensing of Night-time Light

Satellite images acquired at night provide a visually arresting perspective of the Earth and the human activities that light up the otherwise mostly dark Earth. These night-time light satellite images can be compiled into a geospatial time series that represent an invaluable source of information for both the natural and social sciences. Night-time light remote sensing has been shown to be particularly useful for a range of natural science and social science applications, including studies relating to urban development, demography, sociology, fishing activity, light pollution and the consequences of civil war. Key sensors for these time-series include the Defense Meteorological Satellite Program's Operational Linescan System (DMSP/OLS) and the Suomi National Polar-orbiting Partnership Satellite's Visible Infrared Imaging Radiometer Suite Day/Night Band (Suomi NPP/VIIRS DNB). An increasing number of alternative sources are also available, including high spatial resolution and multispectral sensors.

This book captures key methodological issues associated with pre-processing night-time light data, documents state of the art analysis methods, and explores a wide range of applications. Major sections focus on NPP/VIIRS DNB processing; inter-calibration between NPP/VIIRS and DMPS/OLS; applications associated with socio-economic activities, applications in monitoring urbanization; and fishing activity monitoring.

The chapters in this book were originally published as a special issue of the *International Journal of Remote Sensing*.

Christopher D. Elvidge pioneered the development of global satellite derived night-time lights and led the production of a twenty-one year time series of global DMSP night-time lights (1992–2013). Elvidge was instrumental in establishing the DMSP digital archive at NOAA and led the media migration that rescued the digital records from aging tapes. He developed the VIIRS nightfire (VNF) product, methods for discriminating flaming and smoldering combustion using night-time Landsat data, and developed a night-time boat detection product for VIIRS.

Xi Li is Associate Professor at State Key Laboratory of Information Engineering in Surveying, Mapping and Remote Sensing (LIESMARS), Wuhan University. He earned a Ph.D. in Photogrammetry and Remote Sensing at Wuhan University in 2009. His research interests include physical modeling of night-time light as well as night-time light remote sensing applications. He serves as an editorial board member of the *International Journal of Remote Sensing* and an international consultant for Asian Development Bank.

Yuyu Zhou is Associate Professor at the Department of Geological & Atmospheric Sciences, Iowa State University. His research focus has always been in quantifying spatiotemporal patterns of environmental change and developing modeling mechanisms to bridge the driving forces and consequences of environmental change so that the impacts of human activities on environment can be effectively measured, modeled and evaluated.

Changyong Cao is the Chief for the satellite calibration and data assimilation branch at the NOAA Center for Satellite Applications and Research (STAR). He oversees the calibration and validation of satellite radiometers on NOAA polar-orbiting and geostationary satellites. He is also the science team lead for the Visible Infrared Imaging Radiometer Suite (VIIRS) instrument with a research interest on VIIRS on-orbit performance for a variety of applications including night-time lights.

Timothy A. Warner is Emeritus Professor of Geology and Geography at West Virginia University. He served as editor in chief of the *International Journal of Remote Sensing* from 2014 to 2020. He is a Fellow of the American Society of Photogrammetry and Remote Sensing.

Remote Sensing of Night-time Light

Edited by
Christopher D. Elvidge, Xi Li, Yuyu Zhou, Changyong Cao and Timothy A. Warner

Routledge
Taylor & Francis Group

LONDON AND NEW YORK

First published 2022
by Routledge
2 Park Square, Milton Park, Abingdon, Oxon OX14 4RN

and by Routledge
605 Third Avenue, New York, NY 10158

Routledge is an imprint of the Taylor & Francis Group, an informa business

British Library Cataloguing in Publication Data
A catalogue record for this book is available from the British Library

ISBN: 978-0-367-76983-3 (hbk)
ISBN: 978-0-367-76988-8 (pbk)
ISBN: 978-1-003-16924-6 (ebk)

Typeset in Myriad Pro
by Newgen Publishing UK

Publisher's Note
The publisher accepts responsibility for any inconsistencies that may have arisen during the conversion of this book from journal articles to book chapters, namely the inclusion of journal terminology.

Disclaimer
Every effort has been made to contact copyright holders for their permission to reprint material in this book. The publishers would be grateful to hear from any copyright holder who is not here acknowledged and will undertake to rectify any errors or omissions in future editions of this book.

Contents

Citation Information

The chapters in this book were originally published in *International Journal of Remote Sensing*, volume 38, issue 21 (November 2017). When citing this material, please use the original page numbering for each article, as follows:

Preface

Remote sensing of night-time light
Xi Li, Christopher D. Elvidge, Yuyu Zhou, Changyong Cao and Timothy A. Warner
International Journal of Remote Sensing, volume 38, issue 21 (November 2017), pp. 5855–5859

Chapter 1

VIIRS night-time lights
Christopher D. Elvidge, Kimberly Baugh, Mikhail Zhizhin, Feng Chi Hsu and Tilottama Ghosh
International Journal of Remote Sensing, volume 38, issue 21 (November 2017), pp. 5860–5879

Chapter 2

Assessment of straylight correction performance for the VIIRS Day/Night Band using Dome-C and Greenland under lunar illumination
Shi Qiu, Xi Shao, Changyong Cao, Sirish Uprety and Wen Hui Wang
International Journal of Remote Sensing, volume 38, issue 21 (November 2017), pp. 5880–5898

Chapter 3

Improving accuracy of economic estimations with VIIRS DNB image products
Naizhuo Zhao, Feng Chi Hsu, Guofeng Cao and Eric L. Samson
International Journal of Remote Sensing, volume 38, issue 21 (November 2017), pp. 5899–5918

Chapter 10

Urban mapping using DMSP/OLS stable night-time light: a review
Xuecao Li and Yuyu Zhou
International Journal of Remote Sensing, volume 38, issue 21 (November 2017), pp. 6030–6046

Chapter 11

Analysis of urbanization dynamics in mainland China using pixel-based night-time light trajectories from 1992 to 2013
Yang Ju, Iryna Dronova, Qin Ma and Xiang Zhang
International Journal of Remote Sensing, volume 38, issue 21 (November 2017), pp. 6047–6072

Chapter 12

Assessing urban growth dynamics of major Southeast Asian cities using night-time light data
Shankar Acharya Kamarajugedda, Pradeep V. Mandapaka and Edmond Y.M Lo
International Journal of Remote Sensing, volume 38, issue 21 (November 2017), pp. 6073–6093

Chapter 13

A novel method for urban area extraction from VIIRS DNB and MODIS NDVI data: a case study of Chinese cities
Qiao Zhang, Ping Wang, Hui Chen, Qinglun Huang, Hongbing Jiang, Zijian Zhang, Yanmei Zhang, Xiang Luo and Shujuan Sun
International Journal of Remote Sensing, volume 38, issue 21 (November 2017), pp. 6094–6109

Chapter 14

Monitoring urban expansion using time series of night-time light data: a case study in Wuhan, China
Xin Xin, Bin Liu, Kaichang Di, Zhe Zhu, Zhongyuan Zhao, Jia Liu, Zongyu Yue and Guo Zhang
International Journal of Remote Sensing, volume 38, issue 21 (November 2017), pp. 6110–6128

Chapter 15

Predicting potential fishing zones of Japanese common squid (Todarodes pacificus) using remotely sensed images in coastal waters of south-western Hokkaido, Japan
Xun Zhang, Sei-Ichi Saitoh and Toru Hirawake
International Journal of Remote Sensing, volume 38, issue 21 (November 2017), pp. 6129–6146

For any permission-related enquiries please visit:
www.tandfonline.com/page/help/permissions

Notes on Contributors

Kimberly Baugh, Cooperative Institute for Research in the Environmental Sciences, University of Colorado, Boulder, CO, USA.

Mia M. Bennett, Department of Geography, University of California, Los Angeles, CA, USA.

Changyong Cao, Center for Satellite Applications and Research, NOAA/NESDIS, College Park, MD, USA.

Guofeng Cao, Department of Geosciences, Texas Tech University, Lubbock, TX, USA.

Guangping Chen, Key Laboratory for Geographical Process Analysis & Simulation, Central China Normal University, Wuhan, China; College of Urban and Environmental Science, Central China Normal University, Wuhan, PR China.

Hui Chen, Laboratory of Radar Mapping and Monitoring Technology, Sichuan Remote Sensing Geomatics Institute, Chengdu, China; Laboratory of Radar Mapping and Monitoring Technology, The Third Remote Sensing Geomatics Institute of National Administration of Surveying, Mapping and Geoinformation, Chengdu, China.

Kaichang Di, State Key Laboratory of Remote Sensing Science, Institute of Remote Sensing and Digital Earth, Chinese Academy of Sciences, Beijing, China.

Iryna Dronova, Department of Landscape Architecture and Environmental Planning, University of California Berkeley, Berkeley, CA, USA.

Christopher D. Elvidge, Earth Observation Group, Payne Institute for Public Policy, Colorado School of Mines, Golden, CO, USA.

Tilottama Ghosh, Earth Observation Group, Payne Institute for Public Policy, Colorado School of Mines, Golden, CO, USA.

Toru Hirawake, Faculty of Fisheries Sciences, Hokkaido University, Hakodate, Hokkaido, Japan.

Feng Chi Hsu, Earth Observation Group, Payne Institute for Public Policy, Colorado School of Mines, Golden, CO, USA.

Shensen Hu, College of Meteorology and Oceanography, PLA University of Science and Technology, Nanjing, China.

Qinglun Huang, Laboratory of Radar Mapping and Monitoring Technology, Sichuan Remote Sensing Geomatics Institute, Chengdu, China; Laboratory of Radar Mapping and Monitoring Technology, The Third Remote Sensing Geomatics Institute of National Administration of Surveying, Mapping and Geoinformation, Chengdu, China.

Yunxian Huang, College of Meteorology and Oceanography, PLA University of Science and Technology, Nanjing, China.

Hongbing Jiang, Laboratory of Radar Mapping and Monitoring Technology, Sichuan Remote Sensing Geomatics Institute, Chengdu, China; Laboratory of Radar Mapping and Monitoring Technology, The Third Remote Sensing Geomatics Institute of National Administration of Surveying, Mapping and Geoinformation, Chengdu, China.

Jun Jiang, Institute of Applied Meteorology, Beijing, China.

Yang Ju, Department of Landscape Architecture and Environmental Planning, University of California Berkeley, Berkeley, CA, USA.

Shankar Acharya Kamarajugedda, Interdisciplinary Graduate School, Nanyang Technological University, Singapore; Institute of Catastrophe Risk Management, Nanyang Technological University, Singapore; Future Resilient Systems, Singapore-ETH Centre, Singapore.

Chang Li, Key Laboratory for Geographical Process Analysis & Simulation, Central China Normal University, Wuhan, China; College of Urban and Environmental Science, Central China Normal University, Wuhan, PR China.

Deren Li, State Key Laboratory of Information Engineering in Surveying, Mapping and Remote Sensing, Wuhan University, Wuhan, China; Collaborative Innovation Centre of Geospatial Technology, Wuhan, China.

Shice Li, Key Laboratory for Geographical Process Analysis & Simulation, Central China Normal University, Wuhan, China; College of Urban and Environmental Science, Central China Normal University, Wuhan, PR China.

Xi Li, State Key Laboratory of Information Engineering in Surveying, Mapping and Remote Sensing (LIESMARS), Wuhan University, Wuhan, China; Collaborative Innovation Centre of Geospatial Technology, Wuhan, China.

Xuecao Li, Department of Geological and Atmospheric Sciences, Iowa State University, Ames, IA, USA.

Bin Liu, State Key Laboratory of Remote Sensing Science, Institute of Remote Sensing and Digital Earth, Chinese Academy of Sciences, Beijing, China.

Jia Liu, State Key Laboratory of Remote Sensing Science, Institute of Remote Sensing and Digital Earth, Chinese Academy of Sciences, Beijing, China.

Edmond Y.M Lo, Institute of Catastrophe Risk Management, Nanyang Technological University, Singapore; Future Resilient Systems, Singapore-ETH Centre, Singapore.

Jing Luo, Key Laboratory for Geographical Process Analysis & Simulation, Central China Normal University, Wuhan, China; College of Urban and Environmental Science, Central China Normal University, Wuhan, PR China.

Xiang Luo, Laboratory of Radar Mapping and Monitoring Technology, Sichuan Remote Sensing Geomatics Institute, Chengdu, China; Laboratory of Radar Mapping and Monitoring Technology, The Third Remote Sensing Geomatics Institute of National Administration of Surveying, Mapping and Geoinformation, Chengdu, China.

Qin Ma, Sierra Nevada Research Institute, School of Engineering, University of California Merced, Merced, CA, USA.

Shuo Ma, Academy of Ocean Science and Engineering, National University of Defense Technology, Changsha, China.

Pradeep V. Mandapaka, Institute of Catastrophe Risk Management, Nanyang Technological University, Singapore; Future Resilient Systems, Singapore-ETH Centre, Singapore.

Preeya Mohan, SALISES, University of the West Indies, St. Augustine, Trinidad and Tobago.

Boris A. Portnov, Department of Natural Resources and Environmental Management, Faculty of Management, University of Haifa, Mt. Carmel, Israel.

Shi Qiu, Department of Astronomy, University of Maryland, College Park, MD, USA.

Nataliya A. Rybnikova, Department of Natural Resources and Environmental Management, Faculty of Management, University of Haifa, Mt. Carmel, Israel.

Sei-Ichi Saitoh, Arctic Research Center, Hokkaido University, Sapporo, Hokkaido, Japan.

Eric L. Samson, Mayan Esteem Project, Farmington, CT, USA.

Xi Shao, Department of Astronomy, University of Maryland, College Park, MD, USA.

Laurence C. Smith, Department of Geography, University of California, Los Angeles, CA, USA.

Eric Strobl, Deptartment of Economics, University of Aix-Marseille, Marseille, France.

Shujuan Sun, Laboratory of Radar Mapping and Monitoring Technology, Sichuan Remote Sensing Geomatics Institute, Chengdu, China; Laboratory of Radar Mapping and Monitoring Technology, The Third Remote Sensing Geomatics Institute of National Administration of Surveying, Mapping and Geoinformation, Chengdu, China.

Sirish Uprety, Cooperative Institute for Research in the Atmosphere, Colorado State University, Fort Collins, CO, USA.

Ping Wang, Laboratory of Radar Mapping and Monitoring Technology, Sichuan Remote Sensing Geomatics Institute, Chengdu, China; Laboratory of Radar Mapping and Monitoring Technology, The Third Remote Sensing Geomatics Institute of National Administration of Surveying, Mapping and Geoinformation, Chengdu, China.

Wen Hui Wang, ERT Inc., Greenbelt, MD, USA.

Timothy A. Warner, Department of Geology and Geography, West Virginia University, Morgantown, WV, USA.

Chuanqing Wu, Economics and Management School, Wuhan University, Wuhan, China.

Xin Xin, State Key Laboratory of Remote Sensing Science, Institute of Remote Sensing and Digital Earth, Chinese Academy of Sciences, Beijing, China; University of Chinese Academy of Sciences, Beijing, China.

Huimin Xu, School of Economics, Wuhan Donghu University, Wuhan, China.

Wei Yan, College of Meteorology and Oceanography, PLA University of Science and Technology, Nanjing, China.

Jia Ye, School of design, Georgia Institute of Technology, Atlanta, GA, USA.

Zongyu Yue, State Key Laboratory of Remote Sensing Science, Institute of Remote Sensing and Digital Earth, Chinese Academy of Sciences, Beijing, China.

Guo Zhang, State Key Laboratory of Information Engineering in Surveying, Mapping and Remote Sensing, Wuhan University, Wuhan, China.

Qiao Zhang, Laboratory of Radar Mapping and Monitoring Technology, Sichuan Remote Sensing Geomatics Institute, Chengdu, China; Laboratory of Radar Mapping and Monitoring Technology, The Third Remote Sensing Geomatics Institute of National Administration of Surveying, Mapping and Geoinformation, Chengdu, China.

Xiang Zhang, School of Geographic and Oceanographic Sciences, Nanjing University, Nanjing, Peoples Republic of China.

Xun Zhang, Arctic Research Center, Hokkaido University, Sapporo, Hokkaido, Japan.

Yanmei Zhang, Laboratory of Radar Mapping and Monitoring Technology, Sichuan Remote Sensing Geomatics Institute, Chengdu, China; Laboratory of Radar Mapping and Monitoring Technology, The Third Remote Sensing Geomatics Institute of National Administration of Surveying, Mapping and Geoinformation, Chengdu, China.

Zijian Zhang, Department of Electrical Engineering and Electronics, University of Liverpool, Liverpool, UK.

Naizhuo Zhao, Center for Geospatial Technology, Texas Tech University, Lubbock, TX, USA.

Zhongyuan Zhao, Department of Surveying and Mapping, Wuhan Resources and Planning Bureau, Wuhan, China.

Mikhail Zhizhin, Earth Observation Group, Payne Institute for Public Policy, Colorado School of Mines, Golden, CO, USA. Space Dynamics and Mathematical Information Processing, Space Research Institute of the Russian Academy of Sciences, Moscow, Russia.

Yuyu Zhou, Department of Geological and Atmospheric Sciences, Iowa State University, Ames, IA, USA.

Zhe Zhu, Department of Geosciences, Texas Tech University, Lubbock, USA.

Remote sensing of night-time light

Xi Li, Christopher D. Elvidge, Yuyu Zhou ⓘ, Changyong Cao and Timothy A. Warner

The Defense Meteorological Satellite Program's Operational Linescan System (DMSP/OLS) records visible and near-infrared light from clouds and the Earth's surface at night, and in doing so provides not only visually arresting images of the Earth at night but also a geospatial time series of data that has proved to be a valuable source for both the natural and social sciences. Remote sensing of night-time light is a relatively new field in remote sensing, with origins that can be traced back to 1980s. It has become a particular focus since 2011, when a research group led by Dr Christopher Elvidge of the National Oceanic and Atmospheric Administration (NOAA) released a time series of annual composites of DMSP/OLS.

In contrast to daytime remote sensing, which is generally limited to information on physical characteristics of land cover, night-time light remote sensing provides a unique and direct perspective on human activities, specifically those activities associated with artificial light at night. Consequently, night-time light remote sensing has been widely used in the fields of human geography, demography, economy, sociology, fishery, ecology, light pollution, and human rights. Three groups of scientists in particular are working on urban lights: urban geographers studying urbanization, medical scientists documenting light pollution and its impact, and economists disaggregating or correcting gross domestic product (GDP) on a local or regional basis. It is also noteworthy that night-time light remote-sensing scientists in developing countries (e.g. China and India) have tended to focus on urbanization applications, while light-pollution studies are generally undertaken in developed countries, such as Israel, the USA, and Germany.

The future of night-time light remote sensing appears to be promising, as an increasing number of satellite-borne sensors can record night-time light. These recently launched sensors acquire greatly improved data, and thus have the potential to transform night-time light remote sensing. The original DMSP/OLS data were acquired with coarse spatial resolution, as uncalibrated data, and only in a single panchromatic band. In contrast, the data from recently launched sensors are typically radiometrically calibrated (e.g. the Suomi National Polar-orbiting Partnership satellite's Visible Infrared Imaging Radiometer Suite Day/Night Band (Suomi NPP/VIIRS DNB)). Some sensors have relatively high spatial resolution (e.g. night-time aerial photography, Israel's EROS-B, and China's Jilin-1 constellation for video and night-time light imaging). There are even sensors that incorporate multispectral bands (e.g. International Space Station imagery and China's Jilin-1 constellation for video and night-time light imaging).

Despite the great progress in the recent years, night-time light remote sensing is still in a primary stage of development. The field lacks a physical model to relate night-light imagery and light generation. Moreover, the availability of high-quality data remains limited. This special issue addresses these and other challenges, as well as focusing on recent advances in night-time light remote sensing. The articles can be grouped into five areas, as described below.

1. NPP/VIIRS DNB processing

Elvidge et al. (2017) document the procedures used for producing NOAA VIIRS-integrated global night-time light products, at both monthly and annual scales. The processing includes the suppression of background noise, solar and lunar contamination, and contamination by cloud cover and from sources unrelated to electric lighting, such as fires, flares, and volcanoes.

Although VIIRS is a substantial improvement over DMSP/OLS, the sensor is sensitive to stray light from the Sun when the satellite passes through the day–night terminator (i.e. the boundary between day and night). The problem is most evident during the solstice. Qiu et al. (2017) assess the extent of this problem and present a correction to mitigate the effects.

Zhao et al. (2017) provide an approach for dealing with another VIIRS data quality issue, namely data gaps, for example, in high latitudes in the summer months, ephemeral lights in the monthly composites, and relatively high radiance in the winter months. Their approach incorporates both interpolation of gaps and exponential smoothing. They demonstrate that the resulting time-series data set results in a better prediction of annual economic activity than the original raw data.

Hu et al. (2016) develop a new algorithm for identifying fog and low stratus in NPP/VIIRS imagery. The algorithm uses multiple channels of data, is based on a radiative transfer modelling, and incorporates consideration of the observation geometry.

2. Inter-calibration and comparison between NPP/VIIRS and DMSP/OLS

Those seeking to model trends in night-time data face the challenge of obtaining a consistent time series. For example, the monthly composites of DMSP/OLS data are only available up to February 2014, but NPP/VIIRS are only available since 2011. Li et al. (2017b) therefore develop an approach for inter-calibrating the two data sets by down-scaling the NPP/VIIRS data to the resolution of the DMSP/OLS data. They use a power function to degrade the radiometric resolution and a Gaussian low-pass filter to degrade the spatial resolution. They demonstrate the applications of this approach such as monitoring the socio-economic damage and evaluating human right violations in Syria.

As described above, NPP/VIIRS represents substantial improvements in data quality compared to DMSP/OLS. Rybnikova and Portnov (2017) evaluated the benefit of these improvements in the study of the relationship between breast cancer incidence and night lights. They found that ALAN from NPP/VIIRS is a stronger predictor of breast cancer incidence than that of the DMSP-OLS.

3. Applications associated with socio-economic activities

It is well known that increases in night-time light intensity and extent tend to be correlated with economic growth. Bennett and Smith (2017) question whether this association can also be used to monitor economic decline. In a comparison between Russia and China over the period 1992 to 2012, they find that although the correlation between GDP, population, and DMSP/OLS data is strong for China, a growing economy, the correlation is weak for Russia, which experienced considerable economic and population decline during that period. These results provide an important caution regarding the limits of night-light data for modelling human activities.

In contrast, using night-time lights, Mohan and Strobl (2017) are able to monitor the temporary economic dislocation and subsequent recovery following the passage of Hurricane Pam over the South Pacific Islands in 2015. In this case, the storm affected the physical infrastructure, which presumably was directly related to economic activity.

Li et al. (2017a) investigate the potential of night-time light data for modelling port economics for individual cities in China. Their study illustrates that even relatively fine-scale phenomena, such as the economics associated with individual cities, can be modelled using DMSP/OLS data.

4. Applications in monitoring urbanization

One of the primary focus areas of night-light remote-sensing research has been the monitoring of urban expansion and the associated increase in urban light pollution. Li and Zhou (2017), in a review of the extensive literature on this subject, emphasize methodological challenges in this work, such as dealing with saturation, blooming, inter-calibration between various sensors, and temporal pattern adjustment.

Several articles in this special issue present pixel-based analyses of night-time light trajectories for mapping urbanization. Ju et al. (2017), in a study of mainland China, identify five dominant trends in night-time lights: stable urban, high-level steady growth, accelerating growth, low-level steady growth, and a fluctuating pattern.

Kamarajugedda, Mandapaka, and Lo (2017) study the trajectories of night-time light in 15 Southeast Asian cities. They identify three major urban categories: countryside, peri-urban, and core-urban areas. Countryside areas tend to show a decline in night-time light, whereas the other two categories show strong growth.

Zhang et al. (2017) map urbanization in nine Chinese cities using VIIRS in combination with MOIDS data based on an adaptive mutation particle swarm approach.

In another fine-scale study, Xin et al. (2017) study the urban expansion of Wuhan, a major city in China. Their analysis incorporates socio-economic factors and highlights the role of construction policies.

5. Fishing activity

Zhang, Saitoh, and Hirawake (2017) identify areas where fishing vessels harvesting Japanese common squid are concentrating using night-time light data. Using derived information they then develop an environmental model for predicting potential fish-harvesting zones.

6. Concluding thoughts

This special issue has contributed in providing new insights in methodological issues associated with processing night-light data. It has also demonstrated the wide range of applications, which include:

- urban mapping (Li and Zhou 2017; Zhang et al. 2017, Xin et al. 2017; Kamarajugedda, Mandapaka, and Lo 2017; Ju et al. 2017);
- regional and economic development mapping (Bennett and Smith 2017; Ju et al. 2017; Li et al. 2017a; Zhang et al. 2017; Mohan and Strobl 2017; Kamarajugedda, Mandapaka, and Lo 2017);
- meteorology (Hu, 2016);
- light pollution and health (Rybnikova and Portnov 2017);
- fisheries (Zhang, Saitoh, and Hirawake 2017); and
- human rights (Li et al. 2017b).

The breadth of these topics is a powerful indicator of the relevance of night-time light remote sensing for a wide range of applications and ensures that this field will remain a key area of future research.

Disclosure statement

No potential conflict of interest was reported by the authors.

ORCID

Yuyu Zhou ⓘ http://orcid.org/0000-0003-1765-6789

References

Bennett, M., and L. C. Smith. 2017. "Using Multitemporal Night-Time Lights Data to Compare Regional Development in Russia and China, 1992–2012." *International Journal of Remote Sensing* 38: 5962–5991. doi:10.1080/01431161.2017.1312035.

Elvidge, C. D., K. Baugh, M. Zhizhin, F. C. Hsu, and T. Ghosh. 2017. "VIIRS Night-Time Lights." *International Journal of Remote Sensing* 38: 5860–5879. doi:10.1080/01431161.2017.1342050.

Hu, S., S. Ma, W. Yan, J. Jiang, and Y. Huang. 2016. "A New Multichannel Threshold Algorithm Based on Radiative Transfer Characteristics for Detecting Fog/low Stratus Using Night-time NPP/ VIIRS Data." *International Journal of Remote Sensing* 38: 5919–5933. doi: 10.1080/ 01431161.2016.1265691.

Ju, Y., I. Dronova, Q. Ma, and X. Zhang. 2017. "Analysis of Urbanization Dynamics in Mainland China Using Pixel-Based Night-Time Light Trajectories from 1992 to 2013." *International Journal of Remote Sensing* 38: 6047–6072. doi:10.1080/01431161.2017.1302114.

Kamarajugedda, S. A., P. V. Mandapaka, and E. Y. M. Lo. 2017. "Assessing Urban Growth Dynamics of Major Southeast Asian Cities Using Night-Time Light Data." *International Journal of Remote Sensing* 38: 6073–6093. doi:10.1080/01431161.2017.1346846.

Li, C., G. Chen, J. Luo, S. Li, and J. Ye. 2017a. "Port Economics Comprehensive Scores for Major Cities in the Yangtze Valley, China Using the DMSPOLS Night-Time Light Imagery." *International Journal of Remote Sensing* 38: 6007–6029. doi:10.1080/01431161.2017.1312034.

Li, X., D. Li, H. Xu, and C. Wu. 2017b. "Intercalibration between DMSP/OLS and VIIRS Night-Time Light Images to Evaluate City Light Dynamics O Syria's Major Human Settlement during Syrian Civil War." *International Journal of Remote Sensing* 38: 5934–5951. doi:10.1080/01431161.2017.1331476.

Li, X., and Y. Zhou. 2017. "Urban Mapping Using DMSP/OLS Stable Night-Time Light: A Review." *International Journal of Remote Sensing* 38: 6030–6046. doi:10.1080/01431161.2016.1274451.

Mohan, P., and E. Strobl. 2017. "The Short-Term Economic Impact of Tropical Cyclone Pam: An Analysis Using VIIRS Nightlight Satellite Imagery." *International Journal of Remote Sensing* 38: 5992–6006. doi:10.1080/01431161.2017.1323288.

Qiu, S., X. Shao, C. Y. Cao, S. Uprety, and W. H. Wang. 2017. "Assessment of Straylight Correction Performance for the VIIRS Day/Night Band Using Dome-C and Greenland under Lunar Illumination." *International Journal of Remote Sensing* 38: 5880–5898. doi:10.1080/01431161.2017.1338786.

Rybnikova, N. A., and B. Portnov. 2017. "Outdoor Light and Breast Cancer Incidence: A Comparative Analysis of DMSP and VIIRS-DNB Satellite Data." *International Journal of Remote Sensing* 38: 5952–5961. doi:10.1080/01431161.2016.1246778.

Xin, X., B. Liu, K. Di, Z. Zhu, Z. Zhao, J. Liu, Z. Yue, and G. Zhang. 2017. "Monitoring Urban Expansion Using Time Series of Night-Time Light Data: A Case Study in Wuhan, China." *International Journal of Remote Sensing* 38: 6110–6128. doi:10.1080/01431161.2017.1312623.

Zhang, Q., P. Wang, H. Chen, Q. Huang, H. Jiang, Z. Zhang, Y. Zhang, X. Luo, and S. Sun. 2017. "A Novel Method for Urban Area Extraction from VIIRS DNB and MODIS NDVI Data: A Case Study of Chinese Cities." *International Journal of Remote Sensing* 38: 6094–6109. doi:10.1080/01431161.2017.1339927.

Zhang, X., S.-I. Saitoh, and T. Hirawake. 2017. "Predicting Potential Fishing Zones of Japanese Common Squid (*Todarodes Pacificus*) Using Remotely Sensed Images in Coastal Waters of South-Western Hokkaido, Japan." *International Journal of Remote Sensing* 38: 6129–6146. doi:10.1080/01431161.2016.1266114.

Zhao, N., F.-C. Hsu, G. Cao, and E. L. Samson. 2017. "Improving Accuracy of Economic Estimations with VIIRS DNB Image Products." *International Journal of Remote Sensing* 38: 5899–5918. doi:10.1080/01431161.2017.1331060.

VIIRS night-time lights

Christopher D. Elvidge, Kimberly Baugh, Mikhail Zhizhin, Feng Chi Hsu and Tilottama Ghosh

ABSTRACT

The Visible Infrared Imaging Radiometer Suite (VIIRS) Day/Night Band (DNB) collects global low-light imaging data that have significant improvements over comparable data collected for 40 years by the DMSP Operational Linescan System. One of the prominent features of DNB data is the detection of electric lighting present on the Earth's surface. Most of these lights are from human settlements. VIIRS collects source data that could be used to generate monthly and annual science grade global radiance maps of human settlements with electric lighting. There are a substantial number of steps involved in producing a product that has been cleaned to exclude background noise, solar and lunar contamination, data degraded by cloud cover, and features unrelated to electric lighting (e.g. fires, flares, volcanoes). This article describes the algorithms developed for the production of high-quality global VIIRS night-time lights. There is a broad base of science users for VIIRS night-time lights products, ranging from land-use scientists, urban geographers, ecologists, carbon modellers, astronomers, demographers, economists, and social scientists.

1. Introduction

Global satellite-observed night-time lights have emerged as one of the widely used geospatial data products (Amaral et al. 2005; Small, Pozzi, and Elvidge 2005; Sutton et al. 2007; Bharti et al. 2009; Chand et al. 2009; Ghosh et al. 2010; Oda and Maksyutov 2011; Witmer and O'loughlin 2011; He et al. 2012; Mazor et al. 2013; Min et al. 2013; Falchi et al. 2016). These products show the locations where artificial lighting is present and a measure of the brightness as observed from space. From 1992 to 2013, there is a consistently processed annual time series of night-time lights processed from low-light imaging data collected by the US Air Force Defense Meteorological Satellite Program (DMSP) Operational Linescan System (OLS) (Baugh et al. 2010). The follow on to DMSP for global low-light imaging of the Earth at night is the Visible Infrared Imaging Radiometer Suite (VIIRS) Day/Night Band (DNB), flown jointly by NASA and NOAA. The

VIIRS DNB provides several key improvements over DMSP-OLS data, including a vast reduction in the pixel footprint (ground instantaneous field of view [GIFOV]), uniform GIFOV from nadir to edge of scan, lower detection limits, wider dynamic range, finer quantization, and in-flight calibration (Miller et al. 2012; Elvidge et al. 2013; Miller et al. 2013).

The fundamental purpose of both DMSP and VIIRS low-light imaging is to enable the detection of clouds using moonlight instead of sunlight as the illumination source. The requirement for this capability comes from meteorologists. However, a broad swath of natural and social sciences accrues benefits from the unintended ability of these systems to detect lighting at the Earth's surface. On a heavily moonlit night, the DNB images look like daytime images with clouds and Earth's surface features clearly visible (Figure 1(a)). The telltale sign that the image was acquired at night is the city lights present on the land surface. Under new moon conditions, the outlines of clouds and land surface

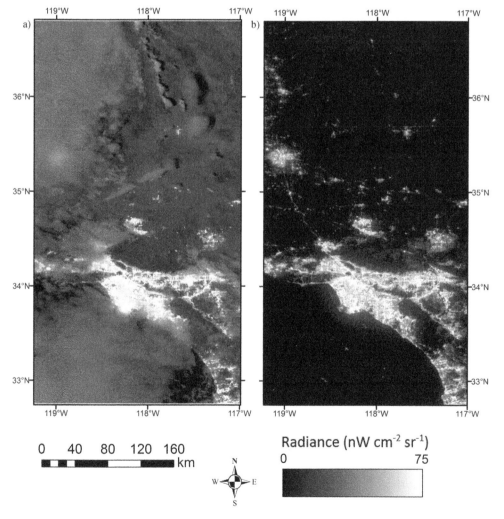

Figure 1. Full moon *versus* new moon VIIRS Day/Night Band (DNB) data for the Los Angeles region. (a) Full moon data. (b) New moon data.

features are suppressed (Figure 1(b)). However, close inspection of moonless night-time DNB data reveals the dim outline of high albedo clouds and land surface features. This remarkable capability has been linked to nocturnal airglow (Min et al. 2013). In addition, the DNB records several other types of phenomena unrelated to electric lighting, including stray light, lightning, biomass burning, gas flares, high energy particle (HEP) detections, atmospheric glow surrounding bright sources, and background noise.

To produce a night-time lights product, the low-light imaging data need to be filtered to exclude low-quality data and extraneous features through a cascading series of filtering steps prior to temporal averaging. The resulting average radiance product has null values (zero) in areas where surface lighting was not detected. Until recently, there has not been a night-time lights product available from VIIRS. The first version of a VIIRS night-time lights product is now available at https://www.ngdc.noaa.gov/eog/viirs.html. In this article, we describe the methods developed by NOAA to produce DNB night-time lights.

2. Methods

The procedures for generating VIIRS night-time lights involve a series of filtering steps to exclude the various styles of extraneous features (Table 1) followed by an averaging of the radiance values. Five styles of global grids are generated at 15 as resolution. This grid cell size is slightly finer that the 742-m DNB pixel footprints. The coverage (CVG) grid tallies the number of observations during the year that were free of sunlight, moonlight, straylight, and HEP detections. The cloud-free coverage (CF_CVG) grid tallies the number of CVG observations that were also cloud-free. Three varieties of DNB radiance grids are produced. The first DNB grid is a raw cloud-free composite (RCFC), with sunlit, moonlit, stray light, lightning, HEP filtering, and cloud filtering. This product contains lighting features, plus biomass burning, aurora, gas flares, volcanoes, and background noise. The second DNB grid is an outlier-removed cloud-free composite (ORCFC), with biomass burning and portions of the aurora are removed. The third DNB grid is VIIRS Nighttime Lights (VNL), with full filtering of all the noted extraneous features removed except lights from persistent gas flares. Below is a description of the filters and processing steps.

Table 1. Filter types and thresholds used to exclude non-lighting grid cells.

Filter name	Parameter	Threshold
Sunlit	Solar zenith angle	>101°
Moonlit	Lunar illuminance	>0.0005 lx
High energy particles	DNB radiance and SHI	Radiance > 1000 nW and SHI > 0.995
Stray light	Solar zenith angle at nadir	90–118.5°
Lightning	Scan-to-scan log-scaled relative radiance difference	>10% and >24 in a row
Outlier removal	Standard deviation	Pull high and low outliers until standard deviation stabilizes with thresholds of 1% or 0.075
Background seeds	DR and Lambda 1 (L1)	DR varies with cloud-free coverages and L1 >−0.55 over land
Surface lighting	ORCFC log-scaled radiance minus background > delta	Delta = 0.8
Gas flares	VIIRS nightfire temperature average and detection frequency	Temperature > 1200 K and frequency > 1%

SHI: Spike height index; DR: data range.

2.1. *Sunlit data*

Sunlit data are removed by discarding pixels having solar zenith angles less than 101°. At solar zenith angles above 101°, the Sun is well below the horizon and the conditions on the ground are fully free from the effects of solar illumination. Solar zenith angles are provided for each pixel from the DNB geolocation file. It should be noted that eliminating sunlit data results in high latitude black outs surrounding the summer solstice. During these periods, significant portions of the high latitude zones have extremely low numbers of dark night observations. Over a full year, it is possible to fill these gaps and make a usable product spanning 65°S–75°N.

2.2. *Moonlit data*

A lunar illumination (LI) threshold of 0.0005 lx and below is used for screening moonlit data. LI is calculated based on the per cent of LI, lunar zenith, and azimuth using a model from the US Navy (Janiczek and Deyoung 1987). The model inputs (latitude, longitude, date, and time) are extracted from DNB geolocation file.

2.3. *Stray light data*

The VIIRS DNB radiances are contaminated by stray light when the spacecraft is illuminated by sunlight while observing the areas of the Earth's surface where the sun is under the horizon. Stray light occurs at mid-to-high latitudes and the affected zone reaches its maximum latitudinal extent at the summer solstice. While there is a stray light correction (Mills, Weiss, and Liang 2013), the correction leaves residual features that would propagate into the annual product. To avoid inclusion of potentially under-or-over corrected data, we exclude stray light contaminated data from the annual composites.

2.4. *HEP detections*

Two methods are used to filter out HEP detections. An initial filtering is done using thresholds placed on the DNB radiance and spike height index (SHI) (Elvidge et al. 2015). SHI is calculated in two stages. First, by computing the relative difference between the pixel's radiance and the average of the two adjacent pixels. This calculation is done for the two adjacent pixels in the same scan line (or row), and then for the two adjacent pixels in the same sample position (or column). The final SHI is the min of these two intermediate values. Pixels with SHI values greater than .995 that also have radiance values greater than 1000 nW are filtered out. This eliminates the brightest of the HEP detections. Additional HEP detections are filtered out via the outlier removal procedure described in Section 2.7.

2.5. *Lightning*

Because DNB data are collected 16 lines at a time, lightning detections create a characteristic 16-line horizontal stripe. Each 16-line sweep across the Earth is termed a scan. A method was developed in 2014 to detect and filter out lightning (Elvidge et al.

2015). The procedure involves calculating the relative radiance difference between adjacent lines of DNB data. The analysis is only done for pairs of lines forming the edges of the 16-line scans. A threshold of 10% is placed on the log-scaled relative radiance difference. When the number of adjacent pixels passing the threshold exceeds 24, the scan segment on the higher radiance side is labelled as lightning. This eliminates most of the lightning from affecting the composites, but some very weak lightning can remain which is filtered out via the outlier removal procedure described in Section 2.7.

2.6. Cloud screening

Clouds tend to obscure and scatter the radiance from lighting. Opaque clouds can fully block the observation of lighting. Thinner clouds reduce the observed radiance and scatter the light, creating a fuzzy appearance. We use the VIIRS cloud mask (VCM) (Godin and Vicente 2015) to screen out pixels deemed to be clouds. Kopp et al. (2014) analysed the accuracy of the VCM and report that the night-time VCM has an 86.4% accuracy, with 4.4% false alarms and 7.3% missed clouds.

2.7. Outlier removal

The primary purpose of the outlier removal is to exclude pixels with biomass burning, which are expressed as anomalously high radiance pixels but occur infrequently over an entire year. The high side outlier removal has the additional benefit of filtering out some of the aurora affected pixels. Filtering out the low radiance outliers has a more subtle effect on the average, by filtering out observations that may be affected by clouds (errors in the cloud mask) or power outages.

   We adapted the outlier removal algorithm developed for DMSP-OLS night-time lights (Baugh et al. 2010) to work on the DNB data. The outlier removal is conducted on histograms generated for each of the 15 as grid cells in the composite. Due to the large dynamic range of the DNB, a logarithmic scaling is done on the radiance values used in the RCFC to transform the individual pixel radiances from nW to log-scaled radiance (LSR) values ranging from 0 to 255. First, a constant of 2 is added to the DNB values so that the logarithm is not applied to negative numbers. LSR values greater than 255 are forced to 255.

$$DNB_{LSR} = \min(100 \times x\log(DNB_{nW} + 2), 255)$$

The outlier removal algorithm excludes pixels one at a time from the high and low end of the histogram. After each removal, the standard deviation is recomputed and compared to the standard deviation before that pixel removal. The outlier removal stops when the standard deviation stabilizes by changing either less than 1%, or by 0.075 (Figure 2). Following outlier removal, the LSRs for the remaining pixels are averaged to generate a log-scaled ORCFC.

2.8. Defining the local background

Background is the residual radiance in areas that lack detectable surface lighting. In general, background radiances are quite small, with the vast majority under 1 nW.

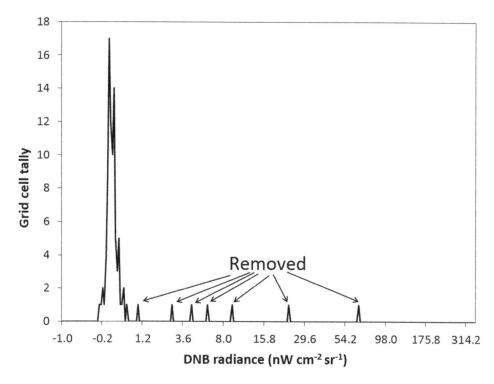

Figure 2. Outlier removal is applied to each 15 as grid cell. This histogram of 2015 data is for the grid cell at (10.43125 N, 1.31875 W) in Ghana, Africa.

However, there are sufficient variations in background that it is impossible to set a single threshold to separate lights from background. There is also a noticeable rise in background, termed glow, that surrounds large cities and gas flares.

We developed a three-step process for defining local background. In the first step, we use an image texture measure known as data range. The data range is a convolution filter, calculated for each grid cell using a kernel size of 3, or the three-by-three block of surrounding cells. The data range for the centre pixel is calculated as the maximum LSR minus the minimum in the kernel. An example, covering a portion of Myanmar is shown in Figure 3(a,b). The initial pixel set for use in defining the background pixel set is selected using the data range and CF_CVG grids. The data range threshold rises as CF_CVGs decline (Figure 4) to accommodate the higher variability present in regions having low numbers of CF_CVGs. The data range is low in areas with no surface lighting.

From this initial background pixel set, we remove remaining small faint light sources on land based on a concavity measure known as lambda1. Lambda1 is computed from the log-scaled ORCFC image and is the smallest magnitude eigenvalue of the Hessian matrix (Frangi et al. 1998; Lindeberg 1998). Lambda1 will be negative in areas where the log-scaled ORCFC image is concave down, which occurs in the centre of small lighting features. Lambda 1 values less than −0.55 were removed from the background pixel set over land only (Figure 3). The result is a global grid of background radiances with null values (zeros) in grid cells with lighting or adjacent to lighting (Figure 5(a)). We refer to

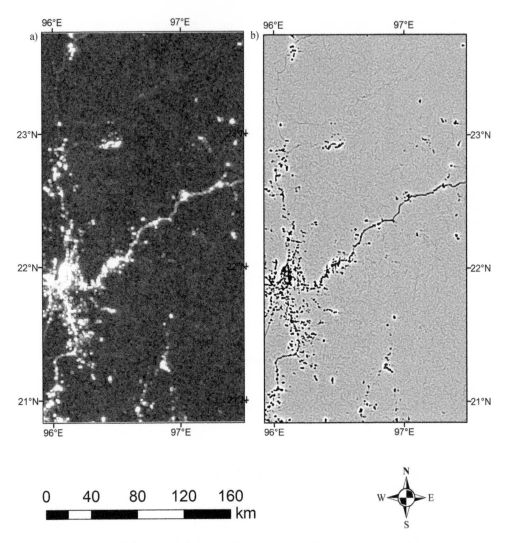

Figure 3. Two key grids used to define background pixel sets: (a) Data range calculated using a 3 × 3 kernel. Background areas have low data ranges; (b) lambda1 is used to define small dim lighting features over land that are excluded from the background pixel set.

this pixel set as the background seeds. The third step is to fill the null values with the radiance of the nearest background grid cell (Figure 5(b)).

2.9. Defining areas with surface lighting

The next step is to define areas with surface lighting by subtracting the ORCFC LSR and the log-scaled background radiance. A delta of 0.8 is added to the background to trim off some of the glow that surrounds lighting. This threshold was derived empirically based on visual inspection of difference values in the glow surrounding large cities

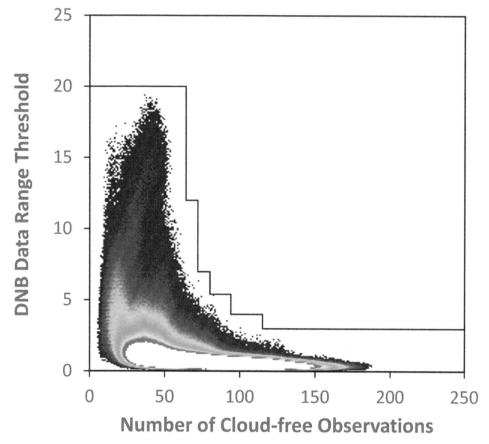

Figure 4. The threshold used to select the initial background pixel set rises as the number of cloud-free observations decreases. Thresholds are log-scaled radiance values.

where no indication of surface lighting was evident. Positive values in the difference image indicate the presence of surface lighting.

2.10. *Manual editing*

After subtracting the background radiance from the ORCFC, several types of non-lighting features are set to zero using hand drawn masks. This includes grid cells affected by aurora, high mountain tops, volcanoes, and grid cells surrounding extremely low CF_CVG zones, with numbers of cloud-free observations less than 2.

2.11. *Restoring the radiance values*

The difference image created by subtracting the local background works quite well for identifying areas with surface lighting. However, close inspection reveals that the boundaries between regions filled by different seed radiances propagate into the difference. While we are interested to develop a radiance product with a

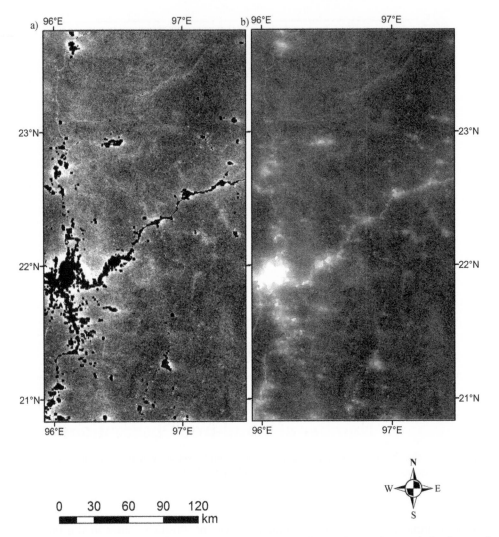

Figure 5. The final steps in defining the background: (a) After defining the seeds background pixel set, the average radiances are restored from the outlier-removed cloud-free composite, leaving zeros in the areas with data ranges that are above the selection threshold or lambda1 values over land less than −0.55; (b) the holes are filled using the radiance from the nearest background pixel. Note that the effects of glow surrounding bright lights can be seen in both the seeds grid and the background grid. The seeds grid contains part of the glow adjacent to bright lighting. The glow pixels are subsequently propagated to fill the holes in the background grid.

background correction, the current method of filling the holes in the seeds image will need to be improved for this. For this reason, the ORCFC radiance is restored for all grid cells determined to have surface lighting to produce the night-time lights product (VNL).

2.12. *Identification of gas flares*

Ideally, gas flares and the glow surrounding gas flares would be filtered out of the night-time lights product leaving only electric lighting sources. While small gas flares have no obvious glow, large gas flares have spatially extensive glow which starts out with high radiance values and declines with distance out from the source. An algorithm for filtering out the glow surrounding gas flares is under development. In the meantime, a gas flaring vector set is available, with more than 12,000 flaring sites identified in 2015. The identification of flaring sites is based on annual composites of VIIRS nightfire (Elvidge et al. 2016) temperatures and detection frequencies. Flaring sites are indicated for grid cells with average temperature exceeding 1200 K and per cent detection frequencies of 1% or more.

3. Results

A total of five global grids are produced as a VIIRS night-time lights suite. The global grids are huge, 86,400 grid cells wide and 33,600 high. This exceeds the size limitation for the standard geotiff format, so the global grids are cut into six tiles. The first grid tallies the number of usable CVGs or observations that pass through the primary filtering to remove pixels contaminated by sunlight, moonlight, stray light, lightning, and HEP (Figure 6). The histogram (Figure 7) of the CVG image shows that 0.003% of grid cells have zero CVGs and 97% of the grids cells have more than 100 CVGs in 2015. The histogram shows a series of peaks and troughs that appear to be latitude related variations in the usable CVGs.

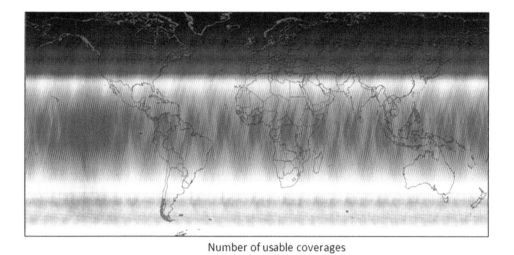

Number of usable coverages

0   246

Figure 6. Grid tallying the number of usable coverages (CVG) for 2015. The tally includes pixels that passed filtering steps for sunlight, moonlight, stray light, and HEP detections.

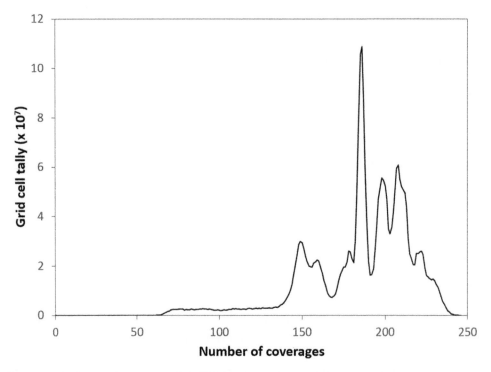

Figure 7. Histogram of coverage tallies for 2015.

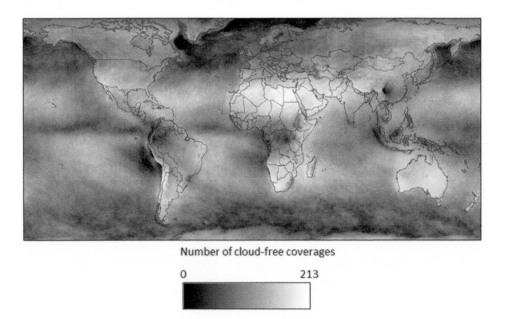

Figure 8. Grid tallying the number of cloud-free coverages (CF_CVG) for 2015. The tally includes the CVG pixels minus those pixels deemed to be cloudy by the VIIRS cloud mask.

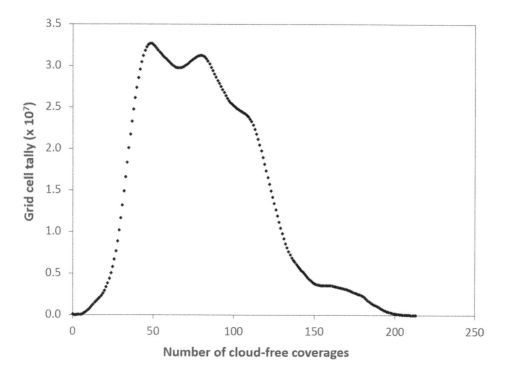

Figure 9. Histogram of the cloud-free coverages for 2015.

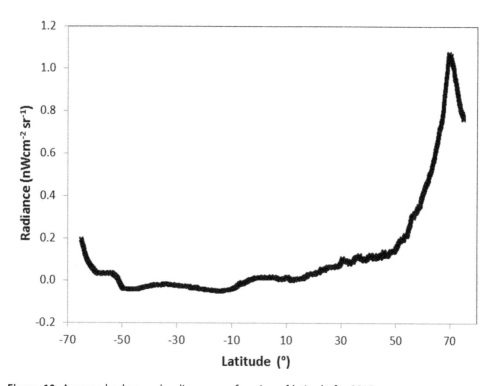

Figure 10. Average background radiance as a function of latitude for 2015.

The second grid, referred to as CF_CVG, tallies the number of CVG pixels that passed the cloud filter (Figure 8). The histogram of the CF-CVG grid (Figure 9) indicates that 99% of grid cells had 20 or more CF_CVGs and only 0.003% had zero CF_CVGs. This indicates that outside of a few low CVG areas, the vast majority of areas have sufficient numbers of cloud-free observations to generate a high-quality night-time lights product.

Visual inspection indicates that the overall background rises as latitude increases. This was also the case for the DMSP night-time lights. Figure 10 shows the average background radiance as a function of latitude. The average background varies from −0.05 to 1.075 nW, a range of approximately 1 nW. The background is lowest between 50°S and 50°N, with average radiances under 0.2 nW. The average background radiance drops slightly below zero between 10°S and 50°S latitude. This is attributed to the use of northern hemisphere dark night ocean data in establishing the dark current level for night-time DNB calibration. The background then rises in two stages from 51° to 65°S, probably due to the influence of the aurora Australis. The background rises rapidly from 50°N to 70°N and then drops at about the same rate out to 75°N. The rise and fall of the background north of 50° is believed to be a reflection of auroral activity.

Three varieties of annual VIIRS DNB cloud-free composites were produced for 2015. Examples of the three are shown in Figures 11(a–c). The first is the RCFC, which retains biomass burning, aurora, and background. The second is the ORCFC, with biomass burning and portions of the aurora filtered out. The third composite, night-time lights (VNL), is generated by removing background from the ORCFC. The data for the three are floating point values and the units are average radiances in nW cm^{-2} sr^{-1}.

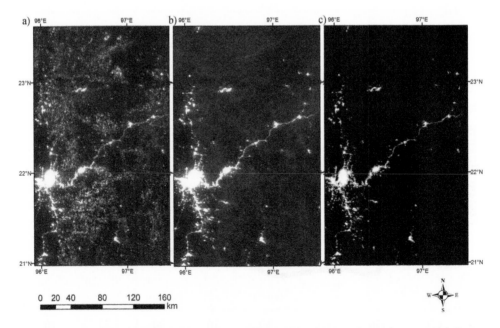

Figure 11. Three styles of VIIRS DNB average radiance grids were produced for 2015: (a) the raw cloud-free composite (RCFC), which has lights, biomass burning, and aurora present; (b) the outlier-removed cloud-free composite (ORCFC) that has been filtered to remove fires and some aurora; (c) night-time lights (VNL) with the background and all extraneous features removed except gas flaring.

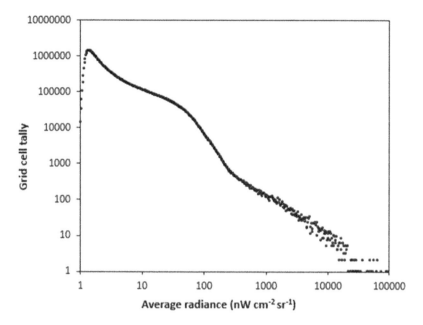

Figure 12. Histogram of the 2015 VIIRS night-time lights. There are vastly larger numbers of low radiance grid cells *versus* high radiance cells.

Figure 12 shows a log-scaled histogram of the distribution of radiance present in the VIIRS night-time lights. The histogram peaks at 1.4 nW with more than 1.3 million grid cells per bin. The numbers fall off rapidly as radiance increases and at the extreme high end, the numbers of grid cells per bin drop to single digits. Above 13,000 nW, the number of grid cells per bin is under 10, and above 22,000 nW, there are one or two grid cells per bin. Las Vegas, Nevada peaks out at 2800 nW, leaving 1475 grid cells brighter than Las Vegas. It is believed that the majority of these are gas flares. Below 1.4 nW, the number of grid cells per bin also drops precipitously to 14,000 grid cells at 1 nW. This decline is likely an expression of the VIIRS DNB detection limits.

We wanted to determine whether population centres were being removed from the VNL product as a result of outlier and background removal. Our procedures intentionally seek to remove biomass burning and the glow that surrounds urban centres. To investigate the effectiveness of the outlier and background removal, we extracted Landscan 2015 (Bright, Rose, and Urban 2016) population counts for three varieties of background. Australia was selected for the analysis because it has annual biomass burning and vast areas having scant population.

The first background set consists of the grid cells where the average of high end outliers was 5 nW or greater. These are grid cells with biomass burning activity. The remaining background grid cells were divided into two classes using a radiance threshold of 0.1 nW. Those background grid cells with radiance 0.1 nW or higher are concentrated around lighting features, indicating that these are associated with atmospheric scatter, commonly referred to as glow. The third background class is the largest in terms of grid cell numbers, representing normal background largely free of biomass burning and glow.

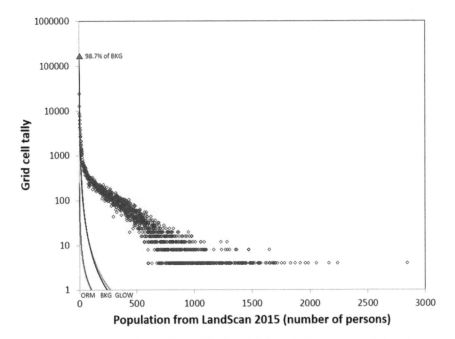

Figure 13. Comparison of Landscan population counts from four sets of grid cells in Australia. The data points are from the VIIRS night-time lights. The three trend lines show results from three varieties of background. ORM refers to grid cells where the average of high end outliers was 5 nW or more. GLOW are background grid cells where the average radiance was 0.1 nW and higher, BKG is the remainder of the background grid cells after removal of ORM and GLOW.

Figure 13 shows the results from the Landscan analysis, with Landscan population count on the x axis and grid cell count on the y axis. Note that a log scale is used on the y axis to accommodate the wide range of values. The large data cloud is from the VNL product, showing population counts where lighting was detected. The curves on the left are the fit lines for the three varieties of background: outlier removed, glow, and normal background. The triangle marks the grid cell counts for normal background having population count of zero. This group represents 98.7% of the normal background. The outlier removed grid cells had the lowest population counts of the three varieties of background and the glow has only a slight increase in population count when compared to the normal background. These results indicate that the outlier removal process is not removing population centres from the night-time lights product. Similarly, the glow assigned to background falls outside of population centres and is nearly identical to the normal background lacking glow. Our conclusion is that the outlier, glow, and background removal are working as intended.

4. Discussion

The VNL product indicates the locations and brightness of human settlements, from large cities down to small towns and many exurban housing clusters. Many bright linear road features propagate into the VNL product, but dimmer road features are frequently missed. There is a narrow ring of grid cells surrounding each feature that probably has no detectable surface lighting but exceeds the data range threshold used to select the local background.

The quality of the VIIRS night-time lights product banks on two principals. The first is that the quality of the data product can be improved by excluding data featuring signal from phenomena other than electric lighting. This includes sunlit, moonlit, and cloudy pixels. The other principal that enhances the quality of the product is the 'power of the mean.' That is to say, the stability of the average night-time lights radiance improves through the inclusion of larger numbers of observations. For example, there are scan angle effects on the brightness of lights. These effects tend to cancel out through the averaging of radiances from all available scan angles. Another example is the seasonal effects on the observed radiances, which tends to cancel out through the inclusion of observations across multiple seasons.

The night-time VCM is rated to have a global cloud omission rate of 7.3% (Kopp et al. 2014). Thus, the cloud-free composite is corrupted by the inclusion of approximately 7% cloudy observations that are incorrectly marked as clear. Opaque clouds will block the detection of lights at the Earth's surface. Thinner clouds will reduce the observed radiance and blur the spatial features. It appears that the outlier removal process works to remove pixels whose radiances have dimmed as a result of undetected clouds. For those cloud impacted pixels which pass through the outlier removal, the product again banks on the 'power of the mean' to arrive at a stable average through the inclusion of large numbers of observations.

The 2015 VNL product can be analysed to identify the brightest spot on Earth. This is often thought to be Las Vegas, Nevada, home of extraordinary casino lighting,

Figure 14. Glow surrounding the brightest cluster of gas flares in 2015, located near Madrid, Venezuela. The red spots indicate average radiances over 10,000 nW cm^{-2} sr^{-1}. These are the gas flare locations.

including the 273,000 W sky beacon shining up from the Luxor hotel casino. Indeed, the brightest spot in Las Vegas has an average radiance of 2800 nW, marking it as the brightest urban area in the world. However, the Las Vegas radiance is dwarfed by the average radiance recorded at many gas flares. Close inspection reveals the brightest spot on Earth to be a gas flare in Venezuela, tipping out at 97,892 nW. Worldwide, large gas flares clearly exceed the brightness of human settlements. While most electric lights have shielding and adhere to local regulations on lighting intensity, gas flares are under no such constraints and represent the worst offenders to dark skies. Figure 14 shows the glow extending around the bright gas flares in Venezuela.

5. Conclusion

Satellite-observed global night-time lights indicate locations and brightness of light that escapes to space. Only specialized low-light imaging sensors are capable of acquiring the source data for night-time lights. Having mid-to-longwave infrared data collected simultaneously is an important asset, enabling cloud screening. For global product generation, there needs to be dozens of observations per year to have sufficient observations to filter out extraneous features unrelated to surface lighting. From the mid-1970s to 2011, the only instrument meeting these criteria was the US Air Force DMSP OLS, which collected a complete set of night-time images of the Earth every 24 h. NOAA made a time series of annual global night-time lights product with OLS data spanning 1992–2013. While the OLS products have been widely utilized by scientists and economists, the OLS data have several flaws, including a dynamic range limited by six-bit quantization, coarse spatial resolution, and no in-flight calibration. Under normal operating conditions, the OLS saturates on bright urban cores. In 2011, NASA and NOAA launched the SNPP satellite, carrying the VIIRS, which collects low-light imaging data with 14 bit quantization, lower detection limits, wider dynamic range and 45 times smaller pixel footprints when compared to the OLS. As with the OLS, VIIRS collects a complete set of night-time images of the Earth every 24 h. In this article, we have outlined the basic set of steps involved in making an annual night-time lights product from VIIRS data.

Any instrument that has detection limits low enough to detect electric lighting present at the Earth's surface will also detect a set of phenomena that are unrelated to surface lighting, including traces of reflected sunlight, stray light, lightning, aurora, biomass burning, and HEPs. Under moonlit conditions, the array of extraneous features expands to clouds, snow, and lunar glint. To make a research quality night-time lights product from VIIRS data requires a cascading series of filtering steps to strip out data contaminated by extraneous features prior to temporal averaging. This filtering is essential to producing a highly uniform standardized night-time lights product. Solar zenith angle thresholds are used to filter out sunlit and stray light contaminated data. The satellite zenith angle for each pixel is provided in the VIIRS geolocation file. A lunar illuminance threshold is used to filter out moonlit data. We developed specialized algorithms to identify and exclude pixels contaminated by lightning and HEP detections. More than a decade ago, we developed an outlier removal algorithm to filter fires from OLS composites. This was revised to work with VIIRS data and effectively removes biomass burning and other temporal lights to produce an ORCFC. With the OLS data, background was identified with manually drawn vectors. For VIIRS, we developed a set of procedures to identify background pixel sets based on low data range (maximum minus minimum) and lack of expression for dim human settlements. The result is a grid of background seed radiances where areas with or adjacent to surface lighting have null (zero) values. The surface lighting holes are filled with the radiance of the nearest seed radiance. A delta is applied to the background to reduce false detections of surface lighting in areas having low numbers of cloud-free observations. Grid cells with positive radiances following the subtraction of the background (plus delta) from the ORCFC are deemed to have surface lighting present. The night-time lights product reports the ORCFC average radiance for the grid cells deemed to have surface lighting.

The current VNL product represents a major advance over the previous monthly DNB RCFCs. In addition to electric lighting, the RCFC products record lighting from biomass burning, aurora, and high albedo surfaces such as high-altitude mountain tops. In addition, the monthly products invariably have data outages in heavily clouded areas and at high latitudes in the months surrounding the summer solstice. These effects have been fully removed from the 2015 VIIRS night-time lights product.

The primary remaining sets of features that are unrelated to electric lighting are gas flares, which for large flares create vast quantities of glow surrounding the flare sites. Gas flares are at the top of the list in terms of the brightest spots on Earth. For artificial sky brightness studies, the inclusion of the gas flares and surrounding glow is of great value. However, for urban studies, the flares and glow are distractions. We provide a 2015 gas flare vector set as a guide to researchers.

There are several reasons for processing night-time lights using full years of data. Global night-time lights products require data from winter seasons to obtain usable CVGs at mid-to-high latitudes. Processing a full year of data ensures that the winter seasons are covered for both northern and southern hemispheres. Observations obtained during the dry season are particularly valuable for night-time lights. Dry seasons typically occur over several months, but the specific months are variable. Inclusion of a full year of data helps ensure that the local dry season is covered. Having large numbers of cloud-free observations is key to successful outlier removal, key to removal of biomass burning. All of these reasons point towards annual increments for the production of research quality night-time lights.

Future plans for improvement of the VNL grids include algorithm development to better filter out the glow surround flares and large cities. Research is ongoing on methods for extraction of dimly lit roadways that can be observed visually but are lost in the final step going from ORCFC to VNL. Other areas for improvement of the product include a background subtraction and an atmospheric correction.

Disclosure statement

No potential conflict of interest was reported by the authors.

Funding

This work was supported by the National Aeronautics and Space Administration: [Grant Number NNH15AZ01I].

References

Amaral, S., G. Camara, A. M. V. Monteiro, J. A. Quintanilha, and C. D. Elvidge. 2005. "Estimating Population and Energy Consumption in Brazilian Amazonia Using DMSP Night-Time Satellite Data." *Computers, Environment and Urban Systems* 29 (2): 179–195. doi:10.1016/j.compenvurbsys.2003.09.004.

Baugh, K., C. D. Elvidge, T. Ghosh, and D. Ziskin. 2010. "Development of a 2009 Stable Lights Product Using DMSP/OLS Data." *Proceedings of the Asia Pacific Advanced Network* 30: 114–130. doi:10.7125/APAN.30.17.

Bharti, N., A. J. Tatem, M. J. Ferrari, R. F. Grais, A. Djibo, and B. T. Grenfell. 2009. "Explaining Seasonal Fluctuations of Measles in Niger Using Nighttime Lights Imagery." *Science* 334 (6061): 1424–1427. doi:10.1126/science.1210554.

Bright, E. A., A. N. Rose, and M. L. Urban, 2016. Landscan 2015 High-Resolution Global Population Data Set. Report published by Oak Ridge National Laboratory, Oak Ridge, TN, USA.

Chand, K., K. V. S. Badarinath, C. D. Elvidge, and B. T. Tuttle. 2009. "Spatial Characterization of Electrical Power Consumption Patterns over India Using Temporal DMSP-OLS Night-Time Satellite Data." *International Journal of Remote Sensing* 30: 647–661. doi:10.1080/01431160802345685.

Elvidge, C. D., K. Baugh, M. Zhizhin, and F. C. Hsu. 2013 January 13-18. "Why VIIRS Data are Superior to DMSP for Mapping Nighttime Lights. *Proceedings of the Asia-Pacific Advanced Network*." In *Asia-Pacific Advanced Network*, edited by C. Elvidge, 62–69. USA: East-West Center, University of Hawai'i - Manoa.

Elvidge, C. D., M. Zhizhin, K. Baugh, and F. C. Hsu. 2015. "Automatic Boat Identification System for VIIRS Low Light Imaging Data." *Remote Sensing* 7: 3020–3036. doi:10.3390/rs70303020.

Elvidge, C. D., M. Zhizhin, K. Baugh, F. C. Hsu, and T. Ghosh. 2016. "Methods for Global Survey of Natural Gas Flaring from Visible Infrared Imaging Radiometer Suite Data." *Energies* 9 (1): 14. doi:10.3390/en9010014.

Falchi, F., P. Cinzano, D. Duriscoe, C. C. M. Kyba, C. D. Elvidge, and K. Baugh. 2016. "The New World Atlas of Artificial Night Sky Brightness." *Science Advances* 2 (6): e1600377. doi:10.1126/sciadv.1600377.

Frangi, A. F., W. Niessen, K. Vincken, and M. Viergever. 1998. "Multiscale Vessel Enhancement Filtering." *Medical Image Computing and Computer-Assisted Intervention — MICCAI'98 Lecture Notes in Computer Science* 130–137. doi:10.1007/bfb0056195.

Ghosh, H. T., R. Powell, C. D. Elvidge, K. E. Baugh, P. C. Sutton, and S. Anderson. 2010. "Shedding Light on the Global Distribution of Economic Activity." *Open Geography Journal* 3: 147–160. doi:10.2174/1874923201003010147.

Godin, R., and G. Vicente. 2015. *Joint Polar Satellite System (JPSS) Operational Algorithm Description (OAD) Document for VIIRS Cloud Mask (VCM) Intermediate Product (IP) Software, National Aeronautics and Space Administration (NASA)*. Greenbelt, Maryland: Goddard Space Flight Center. Doi:accessed on 12th April 2017. Available online at: https://jointmission.gsfc.nasa.gov/sciencedocs/2015-08/474-00062_OAD-VIIRS-Cloud-Mask-IP_I.pdf

He, C., Q. Ma, T. Li, Y. Yang, and Z. Liu. 2012. "Spatiotemporal Dynamics of Electric Power Consumption in Chinese Mainland from 1995 to 2008 Modeled Using DMSP/OLS Stable Nighttime Lights Data." *Journal of Geophysical Sciences* 22 (1): 125–136.

Janiczek, P. M., and J. A. Deyoung. 1987. *Computer Programs for Sun and Moon Illuminance with Contingent Tables and Diagrams*. U.S. Naval Observatory Circular 171. Doi:accessed on 12th April 2017. Washington, DC: U.S. Naval Observatory. Available online at: http://aa.usno.navy.mil/publications/docs/Circular_171.pdf

Kopp, T. J., W. M. Thomas, A. K. Heidinger, D. Botambekov, R. A. Frey, K. D. Hutchinson, B. D. Lisager, K. Brueske, and B. Reed. 2014. "The VIIRS Cloud Mask: Progress in the First Year of S-NPP toward a Common Cloud Detection Scheme." *Journal of Geophysical Research – Atmospheres* 119: 2441–2456. doi:10.1002/2013JD020458.

Lindeberg, T. 1998. "Feature Detection with Automatic Scale Selection." *International Journal of Computer Vision* 30 (2): 79–116. doi:10.1023/A:1008045108935.

Mazor, T., N. Levin, H. P. Possingham, Y. Levy, D. Rocchini, A. J. Richardson, and S. Kark. 2013. "Can Satellite-Based Night Lights Be Used for Conservation? The Case of Nesting Sea Turtles in the Mediterranean." *Biological Conservation* 159: 63–72. doi:10.1016/j.biocon.2012.11.004.

Miller, S. D., S. P. Mills, C. D. Elvidge, D. T. Lindsey, T. F. Lee, and S. J. D. Hawkin. 2012. "Suomi Satellite Brings to Light a Unique Frontier of Environmental Sensing Capabilities." *Proceedings of the National Academy of Sciences of the United States of America* 109 (39): 15706–15711. doi:10.1073/pnas.1207034109.

Miller, S. D., W. Straka III, S. P. Mills, C. D. Elvidge, T. F. Lee, J. Solbrig, A. Walther, A. K. Heidinger, and S. C. Weiss. 2013. "Illuminating the Capabilities of the Suomi NPP VIIRS Day/Night Band." *Remote Sensing* 5 (12): 6717–6766. doi:10.3390/rs5126717.

Mills, S., S. Weiss, and C. Liang. 2013. "VIIRS Day/Night Band (DNB) Stray Light Characterization and Correction." *Proceedings SPIE 8866, Earth Observing Systems XVIII,* 88661P.

Min, B., K. Gaba, O. F. Sarr, and A. Agalassou. 2013. "Detection of Rural Electrification in Africa Using DMSP-OLS Night Lights Imagery." *International Journal of Remote Sensing* 34 (22): 8118–8141. doi:10.1080/01431161.2013.833358.

Oda, T., and S. Maksyutov. 2011. "A Very High-Resolution (1 Km×1 Km) Global Fossil Fuel CO2 Emission Inventory Derived Using a Point Source Database and Satellite Observations of Nighttime Lights." *Atmospheric Chemistry and Physics* 11: 543–556. doi:10.5194/acp-11-543-2011.

Small, C., F. Pozzi, and C. D. Elvidge. 2005. "Spatial Analysis of Global Urban Extent from DMSP-OLS Night Lights." *Remote Sensing of Environment* 96: 277–291. doi:10.1016/j.rse.2005.02.002.

Sutton, P. C., D. Roberts, C. D. Elvidge, and K. E. Baugh. 2007. "Census from Heaven: An Estimate of Global Human Population Using Nighttime Satellite Imagery." *International Journal of Remote Sensing* 22 (16): 3061–3076. doi:10.1080/01431160010007015.

Witmer, F. D. W., and J. O'loughlin. 2011. "Detecting the Effects of Wars in the Caucasus Regions of Russia and Georgia Using Radiometrically Normalized DMSP-OLS Nighttime Lights Imagery." *Giscience and Remote Sensing* 48 (4): 478–500. doi:10.2747/1548-1603.48.4.478.

Assessment of straylight correction performance for the VIIRS Day/Night Band using Dome-C and Greenland under lunar illumination

Shi Qiu, Xi Shao, Changyong Cao, Sirish Uprety and Wen Hui Wang

ABSTRACT

The day/night band (DNB) of the Visible Infrared Imaging Radiometer Suite (VIIRS) on board Suomi National Polar-orbiting Partnership (Suomi-NPP) represents a major advancement in night-time imaging capabilities. However, the DNB is sensitive to noise introduced from straylight, which appears as a grey haze in radiance images. This effect on the DNB is caused by solar illumination entering the optical path when the satellite passes through the day-night terminator projected on the Earth's surface. It results in an overall increase in the recorded radiance values. This effect is more significant during solstice. Straylight correction techniques have been implemented to remove this unwanted effect. This study presents an effective method to assess straylight correction performance for VIIRS DNB using DNB observations over Dome C in the Antarctic and Greenland under lunar illumination. Nadir observations of these high-latitude regions by VIIRS are selected during the perpetual night season over various lunar phases. Through cross-comparison between the lunar-phase dependence of DNB observations of events with straylight correction and those without straylight, the quality of straylight correction can be assessed. Using this method, DNB radiance data from two different sources, i.e. the National Oceanic and Atmospheric Administration (NOAA) Operational Interface and Data Processing Segment (IDPS) and the National Aeronautics and Space Administration (NASA) Land Product Evaluation and Algorithm Test Element (PEATE), were compared for their performance in straylight correction.

1. Introduction

The Suomi National Polar-orbiting Partnership (Suomi-NPP) satellite, successfully launched on 28 October 2011, is the first next-generation polar-orbiting satellite in the Joint Polar Satellite System (JPSS) series, and is considered a bridge between the National Aeronautics and Space Administration (NASA)'s legacy Earth-observing missions and the National Oceanic and Atmospheric Administration (NOAA)'s JPSS constellation. The Visible Infrared Imaging Radiometer Suite (VIIRS) is one of the key instruments on board the Suomi-NPP satellite. The VIIRS nadir door was opened on 21 November 2011, enabling a new generation of operational moderate-resolution-imaging capabilities following the legacy of the Advanced Very High Resolution Radiometer (AVHRR) on NOAA satellites and the Moderate Resolution Imaging Spectroradiometer (MODIS) on the NASA Terra and Aqua satellites.

VIIRS allows operational environmental monitoring and numerical weather forecasting. With 22 imaging and radiometric bands including 14 reflective solar bands (RSBs), seven thermal emissive bands (TEBs), and one day and night band (DNB) covering wavelengths from 0.41 to 12.5 μm, VIIRS provides sensor data records (SDRs) for more than 20 environmental data records (EDRs) including clouds, sea-surface temperature, ocean colour, polar wind, vegetation fraction, aerosol, fire, snow and ice, vegetation, and other applications (Cao, Shao, and Uprety 2013; Cao et al. 2014; Letu et al. 2010).

The DNB of VIIRS detects radiance in a panchromatic Visible/near-infrared (Vis/NIR) band and is capable of detecting radiance from the brightest daytime scenes down to very dim night-time scenes illuminated by only a quarter moon. It shares the same optical path with the Vis/NIR Focal Plane Array (FPA), but uses a unique detector technology (Shao et al. 2014; Li and Li 2014; Mills, Weiss, and Liang 2013; Geis et al. 2012). The primary objective of the DNB is to provide imagery of clouds and other Earth features over illumination levels ranging from full sunlight to quarter moon. Other applications for the DNB, e.g. light outage detection during major storms (Cao, Shao, and Uprety 2013), have also been demonstrated.

The VIIRS DNB is also affected by straylight. Straylight is unwanted light into an optical system and is caused by two phenomena known as ghosting and scattering. Ghosting occurs when excess light enters an observation system through optical or electrical leaks, limiting the signal-to-noise ratio, or contrast ratio, by limiting how dark the system can be. This light will often set a working limit on the dynamic range of the system. On the other hand, scattering occurs when light rays are diverted from their straight trajectories due to the properties of the medium that they travel through. Excess light that is scattered towards the observation system limits the contrast ratio by adding noise to the system. This makes it more difficult to extract a clear signal. Straylight correction, which is the process of correcting for the effects of these phenomena, is the main focus of this article and is discussed in more detail in Section 2.

After the launch of Suomi-NPP, a grey haze in radiance images with offsets up to 5×10^{-9}W cm^{-2} sr^{-1} was observed by the DNB sensor due to straylight (Mills, Weiss, and Liang 2013). Three possible mechanisms have been considered: mirror scattering, which is defined as the scattering of light from the reflective optics in the observation system due to surface roughness or contamination of the optics; baffle scattering, which is defined as the scattering of excess light off of the baffles, sometimes after multiple reflections, and into the optical path; and light leaks, which is defined as light entering through holes and gaps in the structure including the baffles (Mills, Weiss, and Liang 2013). It has been found that one major source for straylight is solar illumination entering the optical path through any of the three mechanisms described above as the satellite passes through the day-night terminator projected on the Earth's surface. Straylight seems to appear on the night side of the terminator and occurs for both northern and southern terminator crossings, but it affects different segments of the orbit depending on the hemisphere. It has also been observed that straylight has detector dependence. The magnitude of straylight changes with scan angle, but extends across the entire scan. This effect is more significant during solstice.

To perform image processing, the DNB SDR is generated from the raw data record (RDR), which requires accurate knowledge of the dark offsets and gain coefficients for each DNB stage (Geis et al. 2012). These are measured on-orbit and stored in Lookup Tables (LUTs) that are used during ground processing. These tables have values for each detector and gain stage. The official DNB data are processed by the JPSS Interface and Data Processing Segment (IDPS) and are made available along with all other VIIRS data products in near-real time for all users (Liao et al. 2013). At the same time, the NASA Land Product Evaluation and Algorithm Test Element (PEATE) has a different implementation of the DNB calibration algorithm (Lee et al. 2015) and provides an alternative source to obtain mission-wide DNB SDR products calibrated based on the latest NASA DNB calibration methodology.

In this article, vicarious assessment of VIIRS DNB straylight correction performance using Dome C in Antarctica and Greenland in the Arctic under lunar illumination is demonstrated, and the performance of this correction is benchmarked. Because straylight affects different segments of the orbit depending on the hemisphere, two regions, one in the northern hemisphere and one in the southern hemisphere, were studied. We selected nadir observations of Dome C during perpetual night seasons (from April to July) over the year 2014 and Greenland during perpetual night seasons (from November to February) over the years 2013–2014 to investigate the radiometric accuracy of DNB observations with straylight correction. These observations were made under greater than quarter moon illumination. We quantified the dependence of the observed radiance over Dome C and Greenland on lunar phases and performed cross-comparisons between the DNB observations for events without straylight correction and with straylight correction. Depending on the dependence of radiance over Dome C w.r.t. the lunar phases, we found that we could assess the performance of straylight correction for the VIIRS DNB band in both hemispheres. In addition, after comparing the VIIRS DNB data from two sources, IDPS and NASA Land PEATE (NPP level 1), we further found that differences in the data between these two sources may be attributed to the difference in calibration algorithm implementations.

2. Data

2.1. *VIIRS DNB*

The spatial resolution of the VIIRS DNB sensor is about 750 m over a 3060 km-wide swath, which provides global coverage with a 12 h revisit time for cloud imagery under both solar and lunar illumination (down to quarter moon). The sensor utilizes a backside-illuminated charge-couple device (CCD) FPA for sensing radiances spanning seven orders of magnitude in one panchromatic (0.5–0.9μm) RSB. To cover this extremely broad measurement range, the DNB sensor employs four imaging arrays that comprise three gain stages: the low gain stage (LGS), the medium gain stage (MGS), and the two redundant high gain stage (HGS). The LGS produces imagery of the brightest Earth scenes such as sunlit clouds; the MGS produces imagery of scenes near the terminator or at twilight; the HGS produces imagery of night scenes. The HGS is made up of two

Table 1. DNB design specifications (data with symbol '*' cite from Liao et al. 2013).

DNB characteristics	Specification
Spectral passband centre (nm) *	700 ± 14
Spectral passband bandwidth (nm)*	400 ± 20
Dynamic range (W cm^{-2} sr^{-1})*	3×10^{-9} to 0.02
Calibration uncertainty (HGS)	30%
SNR <53°/SNR @≥53°*	≥ 6 @L_{min}/ ≥ 5@L_{min}
Number of bits in Analogue-to-Digital (A/D)	14 bits (16,384 levels) for HGS; 13 bits (8,192 levels) for MGS and LGS
Relative radiometric gains	119,000: 477: 1 (HGS: MGS: LGS)
Aggregation	32 aggregation zones
Time Delay Integration (TDI)	1, 3, and 250 pixels for LGS, MGS, and HGS, respectively
Number of samples per scan	4064

separate arrays of identical design: high gain stage A (HGA) and high gain stage B (HGB) (Geis et al. 2012; Baker 2013). Whereas the RSBs of VIIRS are calibrated regularly during daytime using solar light reflected from the on-board solar diffuser, vicarious calibration of the HGS of the DNB requires the usage of light sources at night. Because the three gain stages together cover the full radiance range of the DNB with overlap, the individual gains are correlated. The LGS values are determined by using data collected from the solar diffuser, whereas the MGS and HGS gain values are determined by multiplying the LGS gains by the MGS/LGS and HGS/MGS gain ratios, respectively (Mills, Weiss, and Liang 2013; Liao et al. 2013; Thuillier et al. 2003). The basic design specifications of the DNB are listed in Table 1. More details about the DNB can be found in Liao et al. (2013).

The spatial resolution of the DNB sensor is about 750 m over the entire 3060 km-wide swath, which provides global coverage with a 12 h revisit time for cloud imagery under both solar and lunar illumination (down to quarter moon). This makes calibrating the DNB band challenging.

2.2. The IDPS and NASA land PEATE DNB data and straylight correction methods used

It has been shown that straylight is dependent on the satellite solar zenith and azimuth angles, as well as the detector number and scan sample number in addition to other factors. Liao et al. (2013) describe the correction algorithm used by IDPS, which takes into account these factors. Briefly, correction is performed by subtracting the effective straylight value from the calibrated DNB radiance when the satellite solar zenith angle is within the straylight bounds. The effective straylight value is obtained from a LUT, which is generated by an offline process that analyses calibrated Earth view scenes during the new moon, including scenes over dark areas of the Earth. The LUT provides values based on factors such as the hemisphere, satellite solar zenith angle, the detector number, the scan sample number, and the half-angle mirror side. Satellite solar azimuth angle, which varies over the course of a year, is also accounted for by using 12 tables, one for each month (Liao et al. 2013).

IDPS routinely processes DNB observations into SDRs using this method and makes them available to the public (Mills, Weiss, and Liang 2013). The IDPS SDR team estimated the straylight at each new moon, and the LUT for straylight correction was generated to be incorporated in the DNB SDR calibration algorithm to remove excess straylight. The

IDPS started to apply straylight correction LUTs generated for the same month from the prior year after August 2013, as the straylight pattern repeats for the yearly Earth–Sun–satellite geometry cycle.

On the other hand, NASA employs a different method for DNB straylight correction. Although both methods assume that signals observed over dark Earth scenes during a new moon are affected by straylight, there are a few key differences. First, IDPS removes unwanted night light by collecting radiance data for one full day, then uses this data to mask out nightlight. NASA approaches this problem by treating pixels containing night-light as outliers and removing them from the data set. Second, NASA smooths out the computed dark scene signals over the satellite zenith and scan angles, except for the region with high contrast. There is no such smoothing procedure in the method used by IDPS. Third, NASA identifies 'airglow' in dark scenes from regions without straylight effects and subtracts it globally from dark scene signals that may have straylight effects to generate the final straylight estimates. More details of the DNB straylight correction algorithm implemented in NASA Land PEATE can be found in Lee et al. (2014).

Figure 1 shows the DNB data from January 2014 with straylight over the southern hemisphere (left) and the same data corrected for straylight using the IDPS correction method (right). A grey haze is clearly visible in the left plot while it is clearly absent from the right plot. Figure 1 therefore illustrates the necessity for straylight correction for VIIRS DNB products.

3. Methodology and regions of interest (ROIs)

The method is as follows. First, we model the dependence of local radiance on lunar phase angle using data collected over an ROI during periods where straylight effects are absent. Next, we use this model to assess the performance of straylight correction methods under

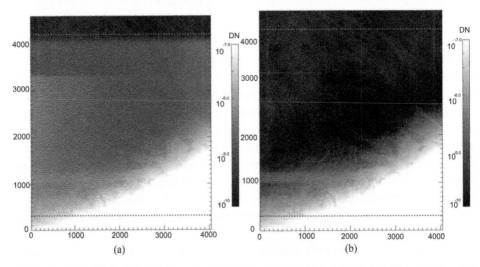

Figure 1. (a) DNB data with straylight in Southern Hemisphere and (b) same data with straylight correction on 1 January 2014. The data with straylight correction is IDPS. These two plots are pixel space figure, one pixel is about 750 m, x-axis is scan direction and y-axis is track direction.

the principle that radiance data collected during periods where straylight was present but subsequently corrected for straylight effects should fit the model well. In other words, the ideal straylight correction algorithm should remove the effects of straylight from the radiance data completely. In this study, we applied our method to data obtained from IDPS and NASA Land PEATE for two ROIs, one in the northern hemisphere and one in the southern hemisphere, in order to evaluate its feasibility and effectiveness in assessing the performance of the two straylight correction algorithms.

3.1. Regions of interest (ROIs)

The Antarctic Dome C site (75.10° S, 123.35° E) and the Greenland site (74.00° N, 41.00° W) in the southern and northern hemispheres, respectively, were selected as ROIs and used to characterize the DNB radiance measurement. The radius of each ROI is 50 km.

In searching for the vicarious calibration sites for satellite observations, many factors need to be considered, including atmospheric turbulence, cloud cover, precipitable water vapour, thermal emission from the atmosphere, auroral activity, aerosol/dust pollution, average and maximum wind speed, seismic activity, rates of snow/rain fall, light pollution, accessibility, infrastructure, and cost of operation (Cao et al. 2010; Uprety and Cao 2012).

The Dome C site is a large snow flat located in Antarctica at a high altitude of 3.25 km. The site is extremely cold and dry with temperatures ranging from −23°C to −83°C. The region has very low atmospheric absorption and other desirable characteristics that make it a good candidate for a vicarious calibration site, such as high uniformity, high reflectance, low water vapour content, and low aerosol and dust content (Cao et al. 2010; Uprety and Cao 2012). Additionally, Dome C site shows a very significant bidirectional reflectance distribution function (BRDF) effect due to the very high lunar zenith angles and is one of the Committee on Earth Observation Satellites (CEOS)-endorsed vicarious calibration sites (Cao et al. 2010; Uprety and Cao 2010; 2011; 2012; Lawrence et al. 2004). Greenland site, which is located in another hemisphere, shows similar geographical and atmospheric conditions as Dome C.

Vicarious calibration of DNB HGS requires lunar illumination over the calibration site, which occurs at night. Both Dome C and Greenland sites are located at high latitude, and consequently experience perpetual nights during half of the year. For Antarctic Dome C, perpetual night occurs from May to August each year (Cao et al. 2010; Uprety and Cao 2012). For Greenland, perpetual night occurs during the boreal winter season from November to January. The DNB observation of Greenland can be used to complement the data gap. Both of these sites have a thin and relatively constant atmosphere with clear skies most of the time. These two sites have been used in numerous past studies for sensor calibration and validation (Uprety and Cao 2010; Hagelin et al. 2008; Koç, Jansen, and Haflidason 1993).

3.2. Selection of observations

The guideline for event selection is to identify the time events corresponding to satellite overpass over the ROIs under sufficient lunar illumination. In selecting observations, attention needs to be paid to the terminator straylight effect on the instrument due to solar illumination after the satellite passes through the day-night terminator projected onto the Earth's surface. This effect is more significant during a solstice. The selection criteria are summarized as follows. For each VIIRS product generation system (IDPS and NASA Land PEATE),

- Suomi-NPP overpasses the ROIs, i.e. the nadir distance to the centre of the ROIs is less than 10 km. For the observations we have identified, satellite view angle is less than 4°. This reduces the uncertainty due to the variations in the satellite-view angle towards the ROIs.
- The absolute value of the lunar phase angle is less than 90°, i.e. the lunar phase angle is larger than the quarter moon. This condition ensures that there is adequate lunar illumination.
- Lunar zenith angle is less than 90°.
- For each ROI, select into two groups
  o Group 1 criteria: solar zenith angle is larger than 95.0° and less than 118.4° (10 cases for Dome C; 9 cases for Greenland). This ensures that the overpass of Suomi-NPP occurs at night and there are influences of straylight effects present at the ROIs during the observations. These events should have straylight correction applied.
  o Group 2 criteria: solar zenith angle is larger than 120° (eight cases for Dome C; seven cases for Greenland). At these angles, there are no influences of straylight effects present at the ROIs during the observations. These events should not have straylight correction applied.

Comparison of these two groups of data can help assess the performance of straylight correction methods. Particularly, Group 1 events can be used to analyse straylight correction differences between IDPS and NASA Land PEATE.

As an example, Figure 2 shows two DNB observations corresponding to two Suomi-NPP Dome C overpasses. The radiances of the two plots are at the same scale due to similar lunar phases. The left plot is the DNB observation of Dome C for the overpass that occurred on 18 May 2014 without straylight effect, and the right plot is the DNB observation of Dome C for the overpass that occurred on 8 July 2014 with straylight but corrected. Both are at nearly the same lunar phase (approximately 55°). Ideally, both figures should be equal but clearly they are not, demonstrating drawback in the straylight correction algorithm that was used to generate the image on the right.

Regarding event selection, Figure 3 shows the lunar zenith angle, solar zenith angle, and lunar phase angle over time for the cases we identified that satisfied our selection criteria. In total, there were 19 cases identified for Dome C for the period April–July 2014 and 16 cases identified for Greenland for the period November to February from 2013 to 2014.

The solid line at 118.4° in Figure 3 is used as a threshold to distinguish between Group 1 and Group 2 observations. Observations where the solar zenith angle is below the threshold belong to Group 1 whereas observations where the solar zenith angle is above the threshold belong to Group 2. The observations in Group 2 are selected with

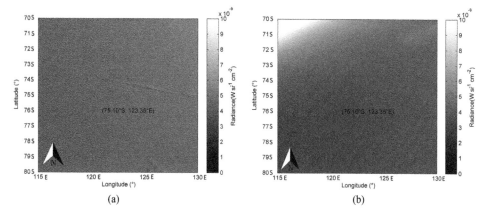

(a) (b)

Figure 2. (a) DNB observation of Dome C on 18 May 2014, the solar zenith angle is 124.32°, lunar phase angle is 51.00° and the lunar zenith angle is 67.70°; (b) DNB observation of Dome C on 8 July 2014, the solar zenith angle is 103.50°, the lunar phase angle is −56.00° and the lunar zenith angle is 69.60°. Red '+' marks the centre of the site Dome C used in this study. Data for both figures were extracted from IDPS.

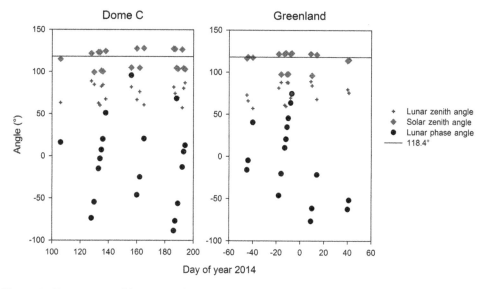

Figure 3. Time series of lunar zenith angle, solar zenith angle, and lunar phase angle for the observations satisfying the selection criteria over Dome C and Greenland. The horizontal line marks 118.4° solar zenith angle threshold. Day 0 is 1 January 2014. Positive days are the number of days after 1 January 2014 and negative days are the number of days before 1 January 2014.

solar zenith angles larger than 120° in order to nullify the influences of straylight effects for comparison.

Figure 4 shows the DNB radiance of Dome C observations with log-scales under different lunar phases, obtained from IDPS. Natural lunar illumination was the only light source for these images. Each observation occurred when straylight effects were absent, or in other words, the solar zenith angle at the time of the observation was greater than

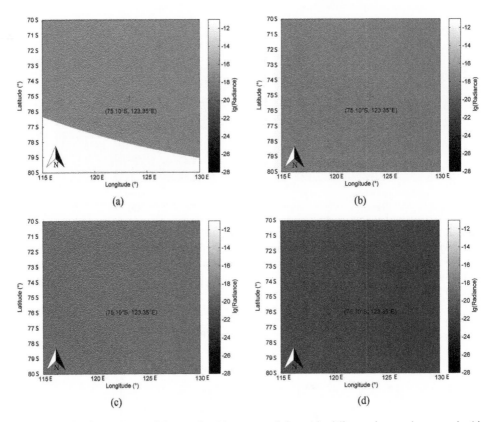

Figure 4. DNB observations of Dome C without straylight with different lunar phase angle (the colour scale is log-scale), red '+' marks the site location. (a) 14 May 2014, lunar phase angle = 3.08°, solar zenith angle = 123.02°, lunar zenith angle = 60.27°; (b) 6 June 2014, lunar phase angle = 20.62°, solar zenith angle = 127.80°, lunar zenith angle = 80.31°; (c) 18 May 2014, lunar phase angle = 51.10°, solar zenith angle = 124.32°, lunar zenith angle = 67.69°; (d) 8 May 2014, lunar phase angle = 73.60°, solar zenith angle = 121.70°, lunar zenith angle = 89.02°.

120°. Here, we can observe that lunar illumination is not only sufficient but also predictable for DNB HGS night-time calibration, as the images demonstrate clear variations from bright to dark with respect to lunar phase changes from nearly full moon to close to quarter moon.

3.3. DNB data-processing procedure

Radiance data over ROIs were extracted for analysis using the following steps.

(1) Collect radiance data from the DNB observations from both IDPS and NASA Land PEATE SDR products.
(2) For each radiance data set, collect all available pixels of an ROI within a 10 km radius centred at the calibration site.
(3) Calculate the mean radiance L_{DNB} across all of the collected pixels.
(4) Correct L_{DNB} for the distance factor to derive the radiance L_{DC}:

$$L_{DC} = L_{DNB} \times \left(\frac{d_{ME} \times d_{MS}}{\overline{d_{ME}} \times \overline{d_{MS}}}\right)^2 \times \frac{1}{\cos \theta_L}, \tag{1}$$

where d_{ME} is the distance between the moon and the Earth, d_{MS} is the distance between the moon and the Sun, $\overline{d_{ME}}$ is the mean distance between the moon and the Earth, $\overline{d_{MS}}$ is the mean distance between the moon and the Sun, and θ_L is the lunar zenith angle. L_{DC} is regarded as the characteristic radiance for the ROI.

(5) Compare the radiance over days of year obtained for data sets with straylight correction (Group 1) and without straylight (Group 2).
(6) Compare the radiance over lunar phase angle curves obtained for data sets with straylight correction (Group 1) and without straylight (Group 2).

4. Results and discussion

For the discussion of the results of the method mentioned above, this section is split into two parts. The first part compares the IDPS and NASA Land PEATE data products to demonstrate and quantify the differences between them, highlighting the differences between the results of their respective straylight correction algorithms. The second part shows, for each data source, the radiance data of the ROIs versus the lunar phase angles to assess their respective straylight corrected products.

4.1 Comparison of IDPS and NASA Land PEATE data product

The top two plots in Figure 5(a) show the distance-corrected DNB radiance data (with/without straylight correction) over Dome C extracted from IDPS and NASA Land PEATE. Similarly, the bottom two plots in Figure 5(a) show the radiance difference between them. Figure 5(b) shows the same plots but for the Greenland site. Figure 5 shows that most radiance values extracted from NASA Land PEATE are larger than the radiance values extracted from IDPS for the two ROIs.

The mean value computed using the differences between IDPS and NASA Land PEATE radiance over Dome C for cases without straylight is about $(1.88 \pm 0.50) \times 10^{-9}$ W sr^{-1} cm^{-2}, and with straylight correction it is about $(4.93 \pm 3.55) \times 10^{-9}$ W sr^{-1} cm^{-2}, which is much larger than the average of the differences without straylight. The average of the differences between them over Greenland for cases without straylight is about $(4.70 \pm 6.69) \times 10^{-9}$ W sr^{-1} cm^{-2}, and with straylight correction it is about $(46.41 \pm 55.30) \times 10^{-9}$ W sr^{-1} cm^{-2}, which is an order of magnitude larger than the average of the differences without straylight.

The differences between IDPS and NASA Land PEATE over Greenland are much larger (more than double) than over Dome C, and the standard deviations over Greenland are also much larger than the standard deviation over Dome C. This might be caused by the different methods of straylight correction for the two hemispheres (northern and southern). For example, the corrections are LUT based, which has dependencies on the satellite solar zenith angle, detector, frame, half-angle mirror side, and hemisphere.

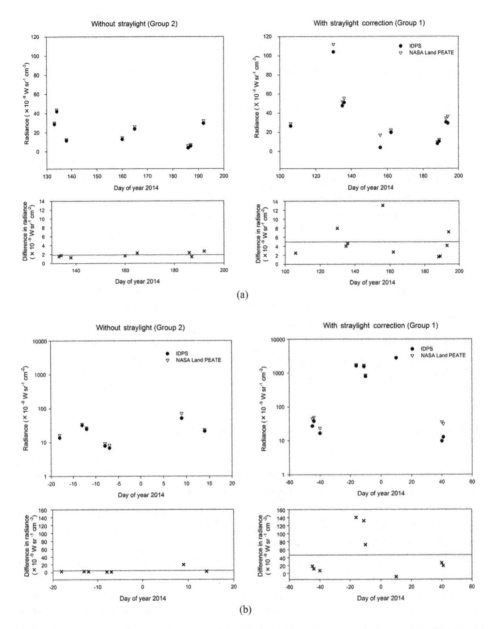

Figure 5. IDPS and NASA Land PEATE DNB data over Dome C (a) and Greenland (b) with distance correction, and the difference between them over Dome C for observations from April to July of the year 2014, and the difference between them over Greenland for observations from November 2013 to February 2014.The horizontal lines in the difference between IDPS and NASA Land PEATE DNB data plots over ROIs are the average values of them.

The percentage of difference between data extracted from IDPS and NASA Land PEATE is calculated as follows:

$$D_{\text{IDPS}-\text{LPEATE}} = (L_{\text{IDPS}} - L_{\text{LPEATE}}) \times 100\%/L_{\text{LPEATE}}, \qquad (2)$$

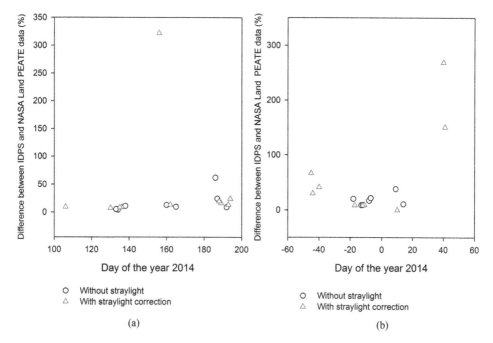

Figure 6. (a) The percentage of difference between data extracted from IDPS and NASA Land PEATE over Dome C; (b) the percentage of difference between data extracted from IDPS and NASA Land PEATE over Greenland. Distance correction has been applied.

where L_{IDPS} is the mean radiance extracted from IDPS and L_{LPEATE} is the mean radiance extracted from NASA Land PEATE. $D_{IDPS-LPEATE}$ is shown in Figure 6, which is separated to Dome C (*a*) and Greenland (*b*). For Dome C (southern hemisphere), the percentage differences between data from these two sources without straylight correction are less than 25.00%, with the sole exception being the event on 5 July 2014, which is 61.99%, and the percentage difference between data with straylight correction is less than 25.00%, except for the event on 5 June 2014, which is 322.00%. The average of the percentage differences without straylight correction is 10.96% (neglecting one anomalous point on 5 July 2014); the average of the percentage differences with straylight correction is about 13.52% (neglecting one anomalous point on 5 June 2014). To investigate the 5 June 2014 anomaly further, we show the radiance data from both sources obtained on that day in Figure 7.

For Greenland (northern hemisphere), the percentage difference between data extracted from IDPS and NASA Land PEATE without straylight correction is less than 40%, and the percentage differences between data with straylight correction are less than 70% except for the events on 9 and 10 February 2014. The percentage differences between data extracted from IDPS and NASA Land PEATE on 9 February 2014 are 268.72%, and on 10 February 2014 it is 150.86%. The average of the percentage without straylight correction is about 18.13%; the average of the percentage with straylight correction is about 23.44% (neglecting the two anomalous points on 9 and 10 February 2014). Again, to investigate further the anomalies, we show the radiance data from both sources obtained on 9 February 2014 in Figure 7.

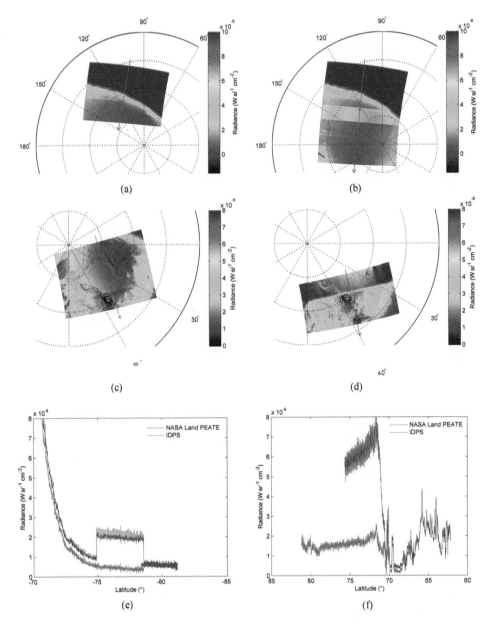

Figure 7. Radiance data on 5 June 2014 from IDPS (a) and NASA Land PEATE (b); Radiance data on 9 February 2014 from IDPS (c) and NASA Land PEATE (d); (e) and (f) show the comparison of line profile from the two sources for the two events, respectively.

The comparisons between IDPS and NASA Land PEATE DNB data on 5 June and 9 February 2014 in Figure 7 show that there are clear sharp changes on NASA Land PEATE plots, which might have been the cause for the large difference of radiance. Plots (a), (b), and (c) in the first row show the DNB observation on 5 June 2014 over Dome C, and plots (d), (e), and (f) in the second row show the DNB observation on 9 February 2014 over Greenland. The red curve shown in Figure 7(e) is the radiance data along the red arrow (along the scan direction) shown in Figure 7(a) and the blue curve is the radiance data along the red arrow shown in Figure 7(b). The same applies for Figure 7(f) with regard to plots (c) and (d). There are clear differences between the red and blue curves in both Figure 7(e) and Figure 7(f). The red curves in Figure 7(e,f) are smooth, whereas the blue curves have large gaps. These anomalies might be caused by missing straylight correction during NASA processing for one granule for some unknown reason. Otherwise, the values appear similar between the overlapping regions of the blue and red curves in (e) and (f). Next, to further investigate the cause for the difference between IDPS and NASA Land PEATE straylight correction results, the lunar phase-dependence of the DNB radiances over the two ROIs is investigated.

4.2 *The performance of straylight correction*

The radiance data from DNB observations for the selected events over the years 2013 and 2014 were collected and analysed following the procedures outlined in section 3 to derive the characteristic radiance for the ROIs. Figure 8 shows the plots of the characteristic radiance of Dome C versus the lunar phase angles. The radiance without straylight depends strongly on the lunar phase angle. As the lunar phase angle increases, the radiance decreases. This radiance-lunar phase dependence is fitted with polynomial function $f_{noSL\_DomeC}(\varphi)$ as the blue curve, where φ is the lunar phase angle. The right panel in Figure 8 shows the observed radiance over Dome C versus the lunar phase angle for cases with straylight correction (Group 1) together with the radiance-lunar phase function $f_{noSL\_DomeC}(\varphi)$ derived from the left panel (cases without straylight). In this way, the performance of straylight correction can be assessed. In general, the radiance-lunar phase dependence derived from data in Group 1 agrees with the trend derived from cases in Group 2 except those cases in May 2014 (markers with a red circle in Figure 8(b), includes 10 May 2014, 15 May 2014, and 16 May 2014). The outliers may be caused by the April 2014 solar eclipse event.

Figure 9 shows the plots of the characteristic radiance of Greenland versus the lunar phase angles. The radiance that is not affected by straylight depends strongly on the

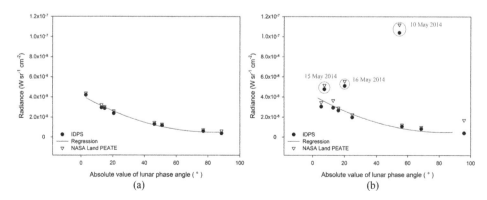

Figure 8. Radiance extracted from IDPS and NASA Land PEATE data for cases without (a) and with (b) straylight correction over Dome C with respect to the lunar phase angle.

lunar phase angle (plot a). As the lunar phase angle increases, the radiance decreases. This radiance-lunar phase dependence is fitted with the polynomial function $f_{noSL\_Greenland}(\varphi)$ as a green curve, where φ is the lunar phase angle. In addition, the blue curve that represents $f_{noSL\_DomeC}(\varphi)$ (shown in Figure 8) is also plotted as a reference. Figure 9(b) shows the observed radiance over Greenland versus the lunar phase angle for cases with straylight correction (Group 1 in Figure 9(b) neglecting the cases in December 2013 and on 10 January 2014, because the radiance on these days exceeds the scale; they are shown in the log-linear scales in Figure 9(c) together with the radiance-lunar phase function $f_{noSL\_Greenland}(\varphi)$ derived from cases without straylight (Group 2). Figure 9(b) shows that the radiance values extracted from NASA Land PEATE are much larger than the ones extracted from IDPS, which is consistent with section 4.1. This is mainly due to the different straylight correction algorithms used in IDPS and Land PEATE, and the different fitting functions they used. Figure 9(c) shows all the radiances including the ones ignored in Figure 9(b), which do not match the curve at log-scales.

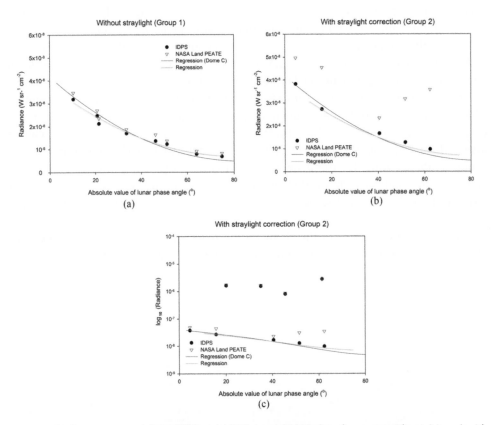

Figure 9. Radiance extracted from IDPS and NASA Land PEATE data for cases without (a) and with (b, c) straylight correction over Greenland with respect to lunar phase angle; plot (b) shows the radiance excluded the cases in December 2013 and on 10 January 2014; plot (c) shows the radiances in the log-linear scale. Plot (b) neglects the cases on December 2013 and on 10 January 2014, because the radiance on these days exceeds the scale; they are shown in the log-linear scales in plot (c).

In this way, the performance of straylight correction algorithms can be assessed. In general, the radiance-lunar phase dependence derived from data in Group 1 agrees with the trend derived from the cases in Group 2 except for those cases extracted from NASA Land PEATE and the cases on December 2013 and 10 January 2014.

For the southern hemisphere, for both IDPS and NASA Land PEATE, the radiance data from Dome C with straylight correction applied mostly agree with the radiance-lunar phase angle curve derived from the radiance data without straylight effects, except for the data in May 2014 as the circles mark in Figure 8(b). This may be because of two reasons. First, there were less cases in this month when the solar zenith angle was smaller than 102° for deriving the straylight correction LUT in the southern hemisphere, and second, some of the May 2014 IDPS DNB data were affected by the solar eclipse event in April 2014, causing the LGS gain LUT to be incorrect. The difference in the cases from the two hemispheres might be caused by the different straylight correction methods. In addition, the difference between the radiance data of Greenland with straylight correction extracted from NASA Land PEATE and the radiance data from the trend function $f_{noSL\_Greenland}(\varphi)$ at the same lunar phase angle depends strongly on the solar zenith angle. This might be due to the influence of solar (straylight).

Furthermore, the difference between the radiance extracted from the two sources of Dome C with straylight correction and the trend function $f_{noSL\_DomeC}(\varphi)$ at the same lunar phase angle is shown in Figure 10(a). The vertical line marks where the solar zenith angle is equal to 102°. The dependence of radiance versus lunar phases for cases with straylight correction agrees well with the trend derived from Group 2 (the cases without straylight) when the solar zenith angle is larger than 102°. However, the radiance with straylight correction deviates largely from the trend when the solar zenith angle is smaller than 102°; these cases are the May data as shown in the circles marked in Figure 8(b). This might be due to the less-populated DNB data in deriving the straylight correction LUT in the northern hemisphere for solar zenith angle smaller than 102°. Unlike for the northern hemisphere, there are always plenty of data to derive straylight LUTs in the southern hemisphere. The region from 95° to 102° is the twilight region. It is

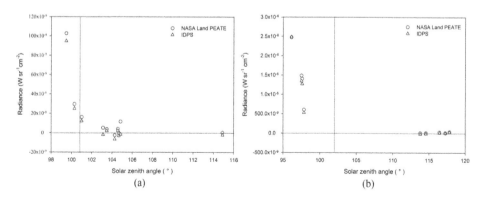

Figure 10. (a) Difference between the radiance extracted from IDPS, NASA Land PEATE with straylight correction, and the trend function $f_{noSL\_DomeC}(\varphi)$ derived from cases in Group 1 versus solar zenith angle over Dome C. The vertical line marks 102°. (b) Difference between the radiance extracted from IDPS, NASA Land PEATE with straylight correction, and the trend function $f_{noSL\_Greenland}(\varphi)$ derived from cases in Group 1 versus solar zenith angle over Greenland.

extremely difficult to estimate straylight directly from this region because it is hard to separate straylight and sunlight. IDPS and NASA Land PEATE use different extrapolation methods. There are uncharacterized straylight issues in the southern hemisphere in the methods of both IDPS and NASA Land PEATE, which may be one of the causes for the large difference between the model and the observation. Additionally, the difference between the radiance extracted from the two sources from Greenland with straylight correction and the trend function $f_{noSL\_Greenland}(\varphi)$ at the same lunar phase angle is shown in Figure 10(b). In Figure 10(b), the radiance with straylight correction deviates largely from the trend when the solar zenith angle is smaller than 102°. We can see that the difference in radiance depends strongly on the solar zenith angle.

Moreover, for the southern hemisphere (Dome C) with solar zenith angle larger than 120°, the mean difference between the radiance extracted from IDPS and the trend function $f_{noSL\_DomeC}(\varphi)$ as shown in Figure 10(a) is about 2.82×10^{-9}W sr^{-1} cm^{-2}, and it is about 3.97×10^{-9}W sr^{-1} cm^{-2} for NASA Land PEATE. Both show a good match to the model function with the solar zenith angle larger than 102°. Besides, for the northern hemisphere (Greenland), with the solar zenith angle larger than 120°, the mean difference between the radiance extracted from IDPS and the trend function $f_{noSL\_Greenland}(\varphi)$ as shown in Figure 10(a) is about 8.80×10^{-9}W sr^{-1} cm^{-2}, and it is about 2.42×10^{-8}W sr^{-1} cm^{-2} for NASA Land PEATE. The result of NASA Land PEATE is three times larger than the result of IDPS for the northern hemisphere. The observation using the IDPS method shows a good match with the model for both hemispheres, but the radiance extracted from NASA Land PEATE shows a big difference with the model for the northern hemisphere. Generally, the radiance extracted from NASA Land PEATE cannot match the model as well as the IDPS when the solar zenith angle is greater than 102°.

5. Conclusion

This study presents a method for assessing the performance of straylight correction. The method provides a simple way of evaluating the accuracy of VIIRS data products with straylight correction without having to use lunar irradiance models for each hemisphere (southern and northern). It presents comparisons of the quality of straylight correction between different methods, not only the different methods between the two hemispheres but also the different methods between two data sources (IDPS and NASA Land PEATE).

For both hemispheres, the study shows that the DNB radiance data with and without straylight extracted from NASA Land PEATE are generally larger in magnitude than the radiance data extracted from IDPS for the same ROIs. NASA Land PEATE is larger than that of IDPS by more than 10%. The difference is greater for the data with straylight correction applied than the data without straylight correction. Despite this difference, both data sets demonstrate the same general curve between radiance and lunar phase angle. Because the difference in magnitude was larger for the data with straylight correction, it might be caused by the different straylight correction methods between the two resources.

For the northern hemisphere, the radiance data of Greenland with straylight correction applied and extracted from IDPS mostly agree with the radiance-lunar phase angle curve derived from the IDPS radiance data without straylight effects. However, this is not true with NASA Land PEATE data.

Disclosure statement

No potential conflict of interest was reported by the authors.

References

Baker, N., 2013, National Aeronautics and Space Administration. *Joint Polar Satellite System (JPSS) VIIRS Radiometric Calibration Algorithm Theoretical Basis Document (ATBD)*. Joint Polar Satellite System (JPSS) Ground Project Code 474 474-00027. Greenbelt, Maryland: Goddard Space Flight Center, Accessed May 2013. https://jointmission.gsfc.nasa.gov/sciencedocs/2015-06/474-00027_ATBD-VIIRS-Radiometric-Calibration_C.pdf

Cao, C., F. J. De Luccia, X. Xiong, R. Wolfe, and F. Weng. 2014. "Early On-Orbit Performance of the Visible Infrared Imaging Radiometer Suite Onboard the Suomi National Polar-Orbiting Partnership (S-NPP) Satellite." *IEEE Transactions on Geoscience and Remote Sensing* 52: 1142–1156. doi:10.1109/TGRS.2013.2247768.

Cao, C., X. Shao, and S. Uprety. 2013. "Detecting Light Outages after Severe Storms Using the S-NPP/VIIRS Day/Night Band Radiances." *IEEE Geoscience and Remote Sensing Letters* 10: 1582–1586. doi:10.1109/LGRS.2013.2262258.

Cao, C., S. Uprety, J. Xiong, A. Wu, P. Jing, D. Smith, G. Chander, N. Fox, and S. Ungar. 2010. "Establishing the Antarctic Dome C Community Reference Standard Site Towards Consistent Measurements from Earth Observation Satellites." *Canadian Journal of Remote Sensing* 36: 498–513. doi:10.5589/m10-075.

Geis, J., C. Florio, D. Moyer, K. Rausch, and F. J. De Luccia. 2012. "VIIRS Day-Night Band Gain and Offset Determination and Performance." *Proceedings of the International Society for Optical Engineering 8510, Earth Observing Systems XVII* 8510: 851012–851012–10. doi:10.1117/12.930078.

Hagelin, S., E. Masciadri, F. Lascaux, and J. Stoesz. 2008. "Comparison of the Atmosphere above the South Pole, Dome C and Dome A: First Attempt." *Monthly Notices of the Royal Astronomical Society* 387: 1499–1510. doi:10.1111/mnr.2008.387.issue-4.

Koç, N., E. Jansen, and H. Haflidason. 1993. "Paleoceanographic Reconstructions of Surface Ocean Conditions in the Greenland, Iceland and Norwegian Seas through the Last 14 Ka Based on Diatoms." *Quaternary Science Reviews* 12: 115–140. doi:10.1016/0277-3791(93)90012-B.

Lawrence, J. S., M. C. B. Ashley, A. Tokovinin, and T. Travouillon. 2004. "Exceptional Astronomical Seeing Conditions above Dome C in Antarctica." *Nature* 431: 278–281. doi:10.1038/nature02929.

Lee, S., K. Chiang, X. Xiong, C. Sun, and S. Anderson. 2014. "The S-NPP VIIRS Day-Night Band On-Orbit Calibration/Characterization and Current State of SDR Products." *Remote Sensing* 6: 12427–12446. doi:10.3390/rs61212427.

Lee, S., J. McIntire, H. Oudrari, T. Schwarting, and X. Xiong. 2015. "A New Method for Suomi-NPP VIIRS Day and Night Band On-Orbit Radiometric Calibration." *IEEE Transactions on Geoscience and Remote Sensing* 53: 324–334. doi:10.1109/TGRS.2014.2321835.

Letu, H., M. Hara, H. Yagi, K. Naoki, G. Tana, F. Nishio, and O. Shuhei. 2010. "Estimating Energy Consumption from Night-Time DMPS/OLS Imagery after Correcting for Saturation Effects." *International Journal of Remote Sensing* 31: 4443–4458. doi:10.1080/01431160903277464.

Li, X., and D. Li. 2014. "Can Night-Time Light Images Play a Role in Evaluating the Syrian Crisis?" *International Journal of Remote Sensing* 35: 6648–6661. doi:10.1080/01431161.2014.971469.

Liao, L. B., S. Weiss, S. Mills, and B. Hauss. 2013. "SUOMI NPP VIIRS Day-Night Band On-Orbit Performance." *Journal of Geophysical Research: Atmospheres* 118: 2013JD020475.

Mills, S., S. Weiss, and C. Liang. 2013. "VIIRS Day/Night Band (DNB) Stray Light Characterization and Correction." *Proceedings of the International Society for Optical Engineering 8866, Earth Observing Systems XVIII* 8866: 88661P–88661P–18. doi:10.1117/12.2023107.

Shao, X., C. Cao, B. Zhang, S. Qiu, C. Elvidge, and M. Von Hendy. 2014. "Radiometric Calibration of DMSP-OLS Sensor Using VIIRS Day/Night Band." *Proceedings of the International Society for Optical Engineering 9264, Earth Observing Missions and Sensors: Development, Implementation, and Characterization III* 9264: 92640A–92640A–8. doi:10.1117/12.2068999.

Thuillier, G., M. Hersé, D. Labs, T. Foujols, W. Peetermans, D. Gillotay, P. C. Simon, and H. Mandel. 2003. "The Solar Spectral Irradiance from 200 to 2400 Nm as Measured by the SOLSPEC Spectrometer from the Atlas and Eureca Missions." *Solar Physics* 214: 1–22. doi:10.1023/A:1024048429145.

Uprety, S., and C. Cao. 2010. "A Comparison of the Antarctic Dome C and Sonoran Desert Sites for the Cal/Val of Visible and near Infrared Radiometers." *Proceedings of the International Society for Optical Engineering 7811, Atmospheric and Environmental Remote Sensing Data Processing and Utilization VI: Readiness for GEOSS IV* 7811: 781106–781106–10. doi:10.1117/12.859148.

Uprety, S., and C. Cao. 2011. "Using the Dome C Site to Characterize AVHRR Near-Infrared Channel for Consistent Radiometric Calibration." *Proceedings of the International Society for Optical Engineering 8153, Earth Observing Systems XVI* 8153: 81531Y–81531Y–9. doi:10.1117/12.892481.

Uprety, S., and C. Cao. 2012. "Radiometric and Spectral Characterization and Comparison of the Antarctic Dome C and Sonoran Desert Sites for the Calibration and Validation of Visible and Near-Infrared Radiometers." *Journal of Applied Remote Sensing* 6: 063541. doi:10.1117/1.JRS.6.063541.

Improving accuracy of economic estimations with VIIRS DNB image products

Naizhuo Zhao ⓘ, Feng Chi Hsu, Guofeng Cao ⓘ and Eric L. Samson

ABSTRACT

A new-generation of night-time lights (NTL) image products, the Visible Infrared Imaging Radiometer Suite (VIIRS) Day/Night Band (DNB) monthly composites, have been produced and released by the National Oceanic and Atmospheric Administration's National Centers for Environmental Information. Compared with the last generation NTL image products, the Defense Meteorological Satellite Program's Operational Linescan System stable light composites, the new NTL image products have finer spatial resolution with compatible radiance values across different month/year images. However, the current defects in VIIRS DNB monthly composites show ephemeral lights, relatively high radiance in winter months, and missing data over the high-latitude regions of the northern hemisphere in summer months. This study presents a method to improve the accuracy of the new NTL image products by statistically modelling the time series VIIRS NTL images and uses the improved imagery to estimate socio-economic factors. In this method, we first estimate radiance for each pixel with 'no data' in May, June, July, and August images and then exponentially smooth the monthly time series images to produce a 2014 annual VIIRS DNB image for the contiguous USA. Sum radiance derived from the smoothed annual image shows stronger correlations with gross domestic product at the state level and smaller standard errors of the estimate at the metropolitan and county levels compared with that extracted from the annual image produced by simply averaging the original monthly DNB composites. Such results infer that exponential smoothing effectively improves the quality of the VIIRS DNB images for annual economic estimation.

1. Introduction

Night-time lights (NTL) imagery is a powerful tool to estimate socio-economic factors (e.g. gross domestic product [GDP], electric power consumption, and fossil fuel carbon dioxide emission) (Chen and Nordhaus 2011; Doll, Muller, and Elvidge 2000; Doll, Muller, and Morley 2006; Ghosh et al. 2009; Ghosh et al. 2010a; 2010b; Letu et al. 2010; Lo 2002; Oda and Maksyutov 2011; Sutton, Elvidge, and Ghosh 2007; Zhao, Currit, and Samson

2011; Zhao, Ghosh, and Samson 2012; Zhao, Samson, and Currit 2015) as well as map urban expansion and population distribution (Liu et al. 2012; Sutton et al. 1997; Sutton et al. 2001). The most widely used NTL image data have been collected by the Defense Meteorological Satellite Program's Operational Linescan System (DMSP-OLS). Raw DMSP-OLS NTL images contain many ephemeral lights (typically fires and lightning). Accurate quantitative studies using NTL images to assess human socio-economic activities were not possible until the National Oceanic and Atmospheric Administration's (NOAA) National Centers for Environmental Information (NCEI) (formerly National Geophysical Data Center) released DMSP-OLS annual NTL image products. These annual image products were produced by all the available cloud-free NTL images for the particular calendar years in the NCEI's digital archive with ephemeral lights such as fire and lightning removed from the products. The NCEI has produced three different types of DMSP-OLS NTL image products named frequency detection, radiance calibrated, and stable light image composites. Of these image products, stable light image composites are most widely used because this type of image product covers a longer period (from 1992 to 2013) compared with the other two types of NTL image products. However, after 2013, the NCEI stopped producing average stable light image composites because the DMSP-OLS has been replaced by a new-generation low-light sensor, the Day/Night Band (DNB) of the Visible Infrared Imaging Radiometer Suite (VIIRS), carried by the Suomi National Polar-orbiting Partnership (S-NPP) satellite (Miller et al. 2012).

The NCEI recently released the first version VIIRS DNB composites. The version 1 VIIRS DNB composite spans the globe from −65° to 75° latitude. Compared to the previous DMSP-OLS NTL image products, the present VIIRS DNB composites have finer spatial resolution (15 *vs.* 30 as). Additionally, the NPP-VIIRS has a larger quantization range than the DMSP-OLS (14 *vs.* 6 bit) with saturated pixels being nonexistent in the VIIRS DNB composites (Elvidge et al. 2013). Moreover, with an on-board calibration system, pixels are given values in radiance (nW cm^{-2} sr^{-1}) and are compatible across different VIIRS DNB composites (Liao et al. 2013). Although the composites have been used to quanti-tatively estimate socio-economic parameters (Levin and Zhang 2017; Ma et al. 2014; Shi et al. 2014), drawbacks of the current monthly VIIRS DNB image products are critical. First, the VIIRS DNB composites for May, June, July, and August contain numerous pixels in high-latitude regions of the northern hemisphere with no data because solar illumi-nation seriously contaminates these regions in the summer. Additionally, ephemeral lights are not removed from the image products reducing accuracy of quantitative estimation for socio-economic parameters. Finally and most importantly, brightness of artificial NTLs that is useful information for quantitatively assessing socio-economic factors is contaminated by seasonal noise, demonstrated by relatively high radiance in winter-month images and relatively low radiance in summer-month images (Levin 2017). Given these problems, elaborate processing is needed before the current VIIRS DNB composites can be used to estimate socio-economic factors with more accuracy.

Exponential smoothing is a widely used technique to smooth and forecast time series data based on a recursive computing algorithm in which each new incoming observa-tion is forecast by a current observation and the last-period smoothed observation (Gelper, Fried, and Croux 2010). Giving exponentially decaying weights to past observa-tions is the most notable feature of exponential smoothing (Holt 2004). This distinctive weighting pattern with a recursive computing scheme makes exponential smoothing act as a low-pass filter to remove high-frequency noise (Casiez, Roussel, and Vogel 2012; LaViola 2003; Holloway 1958). Each pixel has a time series set of radiance values when the present monthly VIIRS DNB composites are stacked together. Relative to persistent anthropogenic NTL observed in cities and other sites, lightning, fire, and other ephemeral lights act as high-frequency noise and can be removed by exponential smoothing. It can be expected that the anthropogenic NTL should not have large changes in brightness across different months. Moreover, the closer the temporal proximity of each 2 month period, the closer the economic and population situations and consequently the closer the brightness of the stable NTL. Thus, the exponential smoothing can also be used to estimate approximate brightness of NTLs for the missing data in the May, June, July, and August VIIRS DNB composites and consequently obtain relatively accurate average brightness of NTLs for a calendar year.

The USA is one of the major regions studied by NTL imagery because of its central role in the global economy (Zhao, Zhou, and Samson 2015). The major objective of the present study is to address a method processing the present VIIRS NTL image products (i.e. the version 1 VIIRS DNB composites) to estimate annual GDP of the USA more accurately at different geographic scales. To fulfil this study objective, we first use complete January, February, March, April, September, October, November, and December (non-summer) VIIRS DNB composites to estimate the missing radiance values in May, June, July, and August (summer) composites. We then combine the original monthly image composites with the estimated images and apply exponential smoothing to remove ephemeral lights and smooth the exceptionally large fluctuations in radiance values. Next, we average the smoothed monthly images to produce a new 2014 VIIRS DNB image product. Finally, we use the newly produced DNB image to estimate GDP at the metropolitan and the county levels and compare the results with those obtained by using the original monthly VIIRS DNB composites.

2. Data

The monthly average night-time version 1 VIIRS DNB composites for the 12 months of 2014 were obtained from NOAA's NCEI (available from: http://ngdc.noaa.gov/eog/viirs/download_monthly.html; accessed 27 April 2017). Each of the composites was produced by all available cloud-free VIIRS DNB images for that particular calendar month in the NCEI's digital archive. The current version 1 VIIRS DNB composites contain two different configurations. The first excludes any data contaminated by stray light (typically solar illumination) while in the second configuration, data impacted by stray light are corrected but not removed. Thus, the second configuration composites contain more data coverage in high-latitude regions but of worse quality (see note for the Version 1 Nighttime VIIRS DNB Composites, available: http://ngdc.noaa.gov/eog/viirs/download_monthly.html, last access: 27 April 2017). In this study, we selected the DNB composites with the first configuration to pursue higher economic estimation accuracy. Lightning and lunar illumination also have been removed from the first configuration composites besides stray light. However, fires, aurora, and other temporal lights were not filtered. The S-NPP VIIRS contains on-board calibration systems and consequently, in monthly VIIRS DNB image composites, pixels' values represent averaged radiances of observed NTL in 1 month. A marked superiority of the current VIIRS DNB composite is that values of pixels across different monthly images are compatible compared with the most widely used last-generation NTL image product, the version 4 DMSP-OLS stable light image composite. Additionally, compared with the stable light image composite, the current VIIRS DNB composite has a finer spatial resolution (15 as) and larger quantization value range. The quantization range of the DMSP-OLS stable light image is from 0 to 63 and 63 is particularly assigned for saturated pixels. In the current VIIRS DNB composites, saturated pixels are nonexistent. The largest pixel value is commonly larger than 5000 in a monthly VIIRS DNB composite for the USA even though such high pixel values are nearly all derived from oil or natural gas flaring (Elvidge et al. 2015).

Data of GDP were taken from the US Department of Commerce's Bureau of Economic Analysis (available from: http://www.bea.gov/regional/index.htm; accessed 27 April 2017). These data were used to establish correlations with sum radiance (i.e. summed pixel value representing accumulated brightness of NTLs in an area) derived from the original and the exponentially smoothed VIIRS DNB images at the state, metropolitan, and county levels. One metropolitan region is the composite of at least one county. The list of counties in metropolitan statistical areas was also obtained from the US Department of Commerce's Bureau of Economic Analysis (available from: http://www. bea.gov/regional/docs/msalist.cfm; accessed 27 April 2017).

3. Methodology

When NPP satellites pass through high-latitude regions of the northern hemisphere in summer, these regions are still affected by solar illumination causing noise and making clean nocturnal images seasonally unavailable for a large area of the northern USA. Figure 1 clearly shows that the problem of missing data exists in the original May, June, July, and August VIIRS DNB composites. The most serious missing data are in the June VIIRS DNB composite in which the regions north to a line generally passing through Phoenix, Arizona and Atlanta, Georgia are completely dark with pixel values of zero (Figure 1(b)). Different from the zero value of pixels south of the line representing no

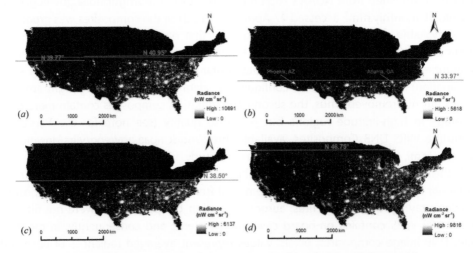

Figure 1. Original VIIRS DNB composites for May (a), June (b), July (c), and August (d).

light observed, the zero value of pixels north of the line is caused by missing data (see note for the Version 1 Nighttime VIIRS DNB Composites: '…it is imperative that users of these data utilize the cloud-free observations file and not assume a value of zero in the average radiance image means that no lights were observed.' Available from: http://ngdc.noaa.gov/eog/viirs/download_monthly.html, last access 27 April 2017).

It is common that brightness of urban NTL exhibits small variations across different months of a year. However, if a country or region with relatively large area does not experience any severe circumstance, such as natural catastrophe, economic collapse, or political unrest, accumulated brightness of NTL for the country or region should not show drastic changes in the same year (Zhao, Zhou, and Samson 2015). Brightness of NTL is a good indicator of economic level and population (Chen and Nordhaus 2011; Doll, Muller, and Elvidge 2000; Doll, Muller, and Morley 2006; Ghosh et al. 2010b; Sutton et al. 1997; Sutton et al. 2001). From 2013 to 2014, GDP and population of the USA increased 3.88% and 0.7%, respectively (The World Bank 2016). However, changes in brightness of NTL in the 2014 monthly VIIRS DNB composites are much larger than those in GDP and population. Figure 2 shows monthly changes of sum radiance of the contiguous USA in the original 2014 VIIRS DNB composites of January to December. Excluding the very small sum radiances in May, June, July, and August due to the lack of uncontaminated data, large variance in sum radiance exists among the other eight months, demonstrated by the exceptionally high sum radiances in January and February contrast with those in April and September. Pixel values in monthly VIIRS DNB image composites represent averaged radiance of NTL. Thus, the large changes in sum radiance among the VIIRS DNB images are not due to the uses of different gain values as is the situation occurring with DMSP-OLS stable light images (Elvidge et al. 2009). Levin (2017) demonstrated that observed radiance in VIIRS DNB images positively correlates with albedo of land surface across different months. Snow significantly increases backscatter of NTL and consequently leads to the increased brightness of

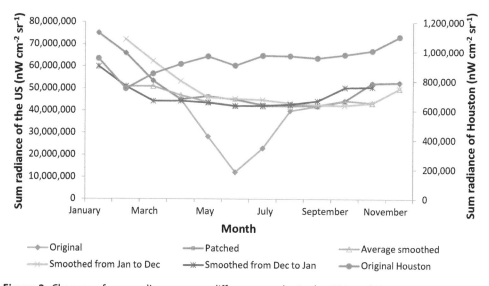

Figure 2. Changes of sum radiance across different months in the USA and Houston, TX.

NTL collected by the S-NPP satellites. Levin's (2017) findings soundly explain the phenomenon shown in Figure 2 that sum radiances of winter-month images are apparently higher than those of spring or fall-month images; yet, the real factors leading to the large variance in radiance are more complicated than what Levin (2017) explored. Besides the monthly changes in sum radiance of the USA, Figure 2 also shows monthly changes in sum radiance of Houston, Texas. Houston is located in a low-latitude region of the USA and thus does not have pixels with missing data and did not have snow in 2014. Sum radiance of Houston for January (952,816 nW cm^{-2} sr^{-1}) is very close to those for April and September (914,748 and 955,792 nW cm^{-2} sr^{-1}, respectively), but large changes in radiance still exist in Houston. For example, from January to February, sum radiance of Houston decreased 21.45%. Hence, besides the changes in land-cover situations, many other uncertain factors (e.g. ephemeral lights and varied atmospheric conditions) are also likely to influence the observed brightness of NTL considerably. The current monthly VIIRS DNB image composites have finer temporal resolution than the version 4 DMSP-OLS stable light image products (monthly vs. annual). The enhancement in temporal resolution leads to a relative reduction in available cloud-free and stray-light-free images and so increases impacts of the uncertain factors on the average radiance values in the VIIRS DNB composites.

Therefore, to improve the current VIIRS DNB image products for the USA, we need to (1) patch the summer images of May, June, July, and August with large areas of 'no data' and (2) smooth the large changes in radiance. The 12 monthly VIIRS DNB images can be deemed as a set of time series data showing changes in brightness of NTL across 2014. Exponential smoothing has two basic functions of forecasting and smoothing time series data and so can be used to accomplish the patching and smoothing jobs of our major objectives. The overall flow chart of producing the smoothed annual VIIRS DNB image product is exhibited in Figure 3.

3.1. Patching the missing data

Original monthly VIIRS DNB composites of January, February, March, and April were stacked sequentially and then, each pixel had four time series DN values applied. The exponential smoothing can be expressed by Equation (1) (Holt 2004):

$$S_t = ax_t + (1 - a)S_{t-1}$$
$$= a\left[x_t + (1 - a)x_{t-1} + (1 - a)^2 x_{t-2} + \cdots + (1 - a)^{t-1}x_1\right] + (1 - a)^t x_0 \quad (0 < a < 1).$$

$$(1)$$

where x denotes an observation value (i.e. the radiance of a pixel in an original monthly VIIRS DNB image), S represents a smoothed (or predicted) value, t denotes a period (i.e. a month in this study), and a is the smoothing factor. Equation (1) demonstrates that in exponential smoothing, a new smoothed (or predicted) value S_t is affected by a current observation value x_t and the last period smoothed value S_{t-1} and that the effect of an old observation value on a new smoothed (or predicted) value declines exponentially with period t continuously updated. The value of a determines the smoothing effect: a relatively large value (i.e. close to 1) of a provides a small smoothing effect while a relatively small value (i.e. close to 0) of a indicates that the estimate at the current time

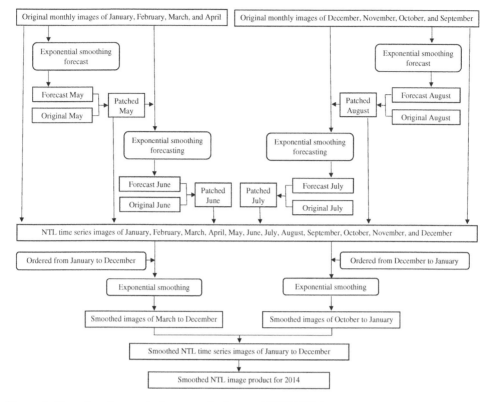

Figure 3. Flow chart for producing smoothed annual VIIRS DNB image.

point is influenced considerably by both current and the previous observations. The value of a is usually (but not exclusively) determined by an optimization method (Hyndman and Athanasopoulos 2016). In this study, the optimal value of a for each exponential smoothing was selected by minimizing the sum of the squared errors (SSE):

$$\text{SSE} = \sum_{t=1}^{T} (S_t - x_t)^2. \tag{2}$$

Usually, a region's economic level and population in 1 month are closer to its adjacent months than in later/earlier months and consequently, brightness of NTL in May should be closer to brightness in April than that in March, February, or January. Based on Equation (1), radiances of May were estimated by radiances of April and smoothed radiances of March. In this study, we defined the starting value for S_1 as

$$S_1 = \frac{(x_1 + x_0)}{2}. \tag{3}$$

where S_1 is the smoothed radiance of February, and x_1 and x_0 denote radiance of February and January, respectively. The estimations (including optimizing the values of a) were accomplished by the 'forecast.HoltWinters' function in the *forecast* package of R (Hyndman et al. 2014). The 'forecast.HoltWinters' function provides confidence intervals for the forecast. We selected the mean values in the prediction intervals as the final estimate values. Since in the original May image, only a portion of pixels suffer from 'no data' (see Figure 1(a)), a patched image for May was produced by combining the original and the estimated May images with the following equation:

$$L_{pat} = \begin{cases} L_{ori}, L_{ori} \neq 0 \\ L_{est}, L_{ori} = 0 \end{cases} \tag{4}$$

where L_{pat}, L_{ori}, and L_{est} represent radiances in the patched, the original, and the estimated May images.

The patched May image was added into the stacked images of January, February, March, and April resulting in each pixel having five time series radiance values. The new time series data set was processed by exponential smoothing to estimate radiances of June. A patched image for June was produced by Equation (4) where L_{ori} and L_{est} denote radiances in the original and estimated June images for this incidence.

Repeating the above steps, the original images of December, November, October, and September were sequentially stacked to produce a patched image of August. The patched image of August was added into the stacked images of December, November, October, and September to produce a patched image of July. At this point in the data manipulation, the 'no data' problem becomes nonexistent in the 12 (eight original and four patched) monthly VIIRS DNB images. Henceforth, the 12 monthly images are referred to as patched time series images to discriminate from the original time series VIIRS DNB images even though actually eight of the 12 images were reviewed but not patched.

3.2. Smoothing the abnormally large changes

The 12 patched time series images were stacked sequentially from January to December resulting in each pixel having 12 time series radiance values. The exponential smoothing was run to smooth the time series radiance values and produce 11 smoothed images for February–December (see Equations (1) and (3)). The first smoothed value (i.e. S_1) is for the second stacked month (i.e. February, as stacking is sequential from January–December) and so, there are no smoothed values for the first stacked month. The image for February was generated by simply averaging the patched images of January and February rather than by exponential smoothing (see Equation (3)). Consequently, we obtained 10 exponentially smoothed images for March–December.

To produce smoothed images for January and February, we reversed the order of the patched time series images and stacked them sequentially from December to January. The exponential smoothing was run again to smooth the radiance values of the con-versely ordered images and generate 11 smoothed images for November–January. (With this process, December is the first stacked month, and so, there are no smoothed values for December.) The image for November was produced by averaging the patched images of December and November; so, 10 exponentially smoothed images for October–January were achieved.

There is only one smoothed image for each month of January, February, November, and December and there are two smoothed images for each month of March, April, May, June, July, August, September, and October. If one month has two smoothed images, then the two images were averaged. The 12 smoothed or smoothed-and-averaged images were combined together to establish smoothed time series VIIRS DNB images while ultimately, the smoothed time series images were averaged to produce a 2014 annual VIIRS DNB image for the USA (Figure 4(b)).

3.3. Evaluating the patched data

The problem of missing data never appeared in Houston, Texas, or Orlando, Florida, in any of the 12 monthly VIIRS DNB composites because Houston and Orlando are located in relatively low-latitude regions of the northern hemisphere. Thus, we can compare estimated radiances with original radiance values in these areas. In this study, the Houston area is composed of Harris, Fort Bend, and Montgomery counties with 51,846 pixels of which approximately 89.46% of the pixels have non-zero radiances in the original monthly VIIRS DNB composites for May, June, July, and August. The Orlando area (i.e. Orange County) contains 14,028 pixels of which approximately 86.24% of the pixels have non-zero radiances in the four original summer VIIRS DNB composites. We used Equation (4) to calculate average difference rate (ADR) between the estimated radiances (L_{est}) and the original radiances (L_{ori}) for each month of May, June, July, and August:

$$ADR = \frac{(AD)}{(ADN)} = \frac{\left(\sum_{k=1}^{n}|L_{est} - L_{ori}|\right)/n}{\left(\sum_{k=1}^{n}L_{ori}\right)/n} \tag{5}$$

where AD denotes average difference between radiances of the estimated and the original images, ADN denotes average radiances in the original image, and n represents the number of pixels included in the calculation.

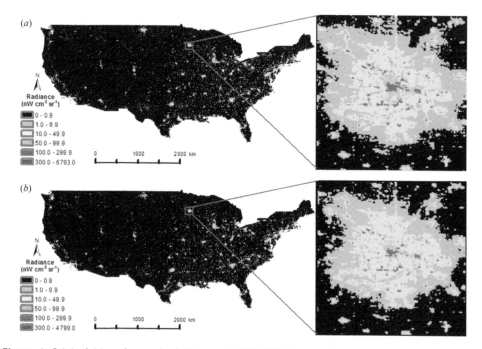

Figure 4. Original (a) and smoothed (b) annual VIIRS DNB images for 2014.

3.4. Estimating GDP

We averaged eight original monthly VIIRS DNB images without the problem of missing data to produce a 2014 annual original VIIRS DNB image for the USA (Figure 4(a)). Strong relationships between GDP and summed brightness of NTL at the national and the state/province levels have been supported by many previous studies using DMSP-OLS NTL image composites (Chen and Nordhaus 2011; Doll, Muller, and Morley 2006; Ghosh et al. 2009, 2010b; Sutton, Elvidge, and Ghosh 2007; Wu et al. 2013; Zhao, Currit, and Samson 2011). In this study, we tested the similar correlation between GDP and sum radiance at the state level. If the correlation is significant, then this examination will provide a substantiation that the new-generation NTL image products still have the capability to estimate economy. More importantly, if the exponential smoothing effectively reduces ephemeral lights, sum radiance derived from the annual smoothed image should have a closer correlation with GDP at the state level than that derived from the original image.

Before testing the relationship between sum radiance and GDP at the state level, we need to consider a common issue in NTL imagery: blooming. Blooming is a phenomenon in NTL imagery where urban peripheries and waterbodies inside cities are brightened by urban lights (Imhoof et al. 1997). An easy and most widely used method to eliminate the effects of blooming is to set thresholds of pixel values (Imhoof et al. 1997; Liu et al. 2016; Small, Pozzi, and Elvidge 2005; Sutton et al. 2010; Zhao, Ghosh, and Samson 2012; Zhou et al. 2015). To make conclusions of the comparison more cogent, in this study, we first set one as the threshold of pixel value to process the annual images. Pixels with values smaller than one were revalued to be 0. Then, Pearson correlation coefficients between GDP and sum radiances derived from the thresholded annual images were compared. We repeated the above steps and another 14 different thresholds (i.e. 2–15) were set and corresponding Pearson correlation coefficients were obtained. We did not set more thresholds because preliminary methodological assessments show that the correlation coefficients reached their maximum when the threshold was set as nine (or eight for the annual original image). The correlation coefficients began to decrease with further increase in the threshold value.

The logarithms of sum radiance extracted from the original and the smoothed DNB images were regressed on the logarithm of GDP at the metropolitan and the county levels (see Figures 5 and 6 for the specific regression functions). In this study, we selected linear functions for the regressions based on experience of previous NTL imagery studies and, more importantly, preliminary methodological assessments that linear functions have higher regression accuracy (i.e. larger R^2) than any other function formats (e.g. exponential, power, and polynomial) to estimate GDP at the metropolitan and the county levels. To further improve accuracy of the linear regression functions, we conducted logarithmic transformations to make distributions of the data (i.e. sum radiance and GDP) less skewed (Doll, Muller, and Elvidge 2000; Lo 2002; Sutton, Elvidge, and Ghosh 2007; Tian et al. 2014). We selected 10 as the base of the logarithms to ensure that transformed sum radiance and GDP have nearly the same value ranges. Preliminary assessments showed that the logarithmic transformations effectively improved accuracy of the regressions. For example, after the logarithmic transformations, the R^2 values of the regression functions for the original and the smoothed images were increased from 0.71 to 0.85 and from 0.75 to 0.86,

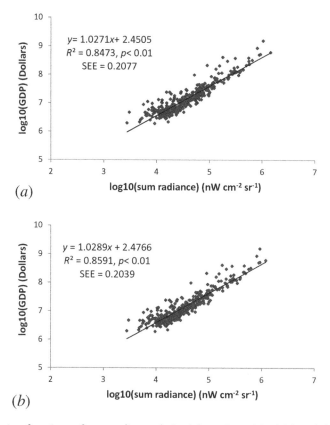

Figure 5. Regression functions of sum radiance derived from the original (a) and the smoothed (b) annual images to GDP at the metropolitan level.

respectively, at the metropolitan level. At the county level, the R^2 values of the regression functions for the original and the smoothed images were increased from 0.69 to 0.81 and from 0.71 to 0.83, respectively. The accuracy of the regression functions was evaluated by the R^2 value and the standard error of the estimate (SEE). The SEE (σ_{est}) was calculated as follows:

$$\sigma_{est} = \sqrt{\frac{\sum_{i=1}^{n} (Y'_i - Y_i)^2}{n}}. \tag{6}$$

where Y' represents the estimated logarithm of GDP, Y denotes the logarithm of GDP reported by the US Department of Commerce's Bureau of Economic Analysis, i represents a metropolis or county, and n is the number of metropolizes (382) or counties (3058) involved in the regression. We also produce a map (Figure 7) showing a change in absolute error of prediction (AEP) in each county. The change in absolute error of prediction (CAEP) was computed as follows:

$$CAEP = |EP_S| - |EP_O|. \tag{7}$$

in which EP_S and EP_O are errors of prediction of GDP estimated by the smoothed and the original DNB images. The error of prediction (EP) was calculated as follows:

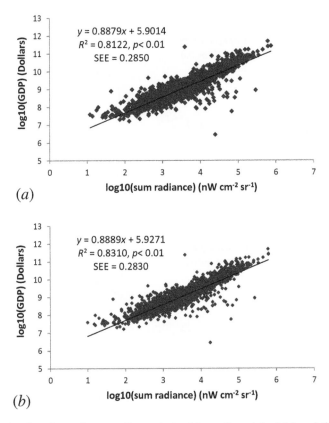

Figure 6. Regression functions of sum radiance derived from the original (a) and the smoothed (b) annual images to GDP at the county level.

$$EP = Y' - Y \qquad (8)$$

Consequently, a county with a negative change in AEP value indicates that sum radiance derived from the smoothed DNB image can more accurately estimate GDP of the county than that derived from the original DNB image, whereas a positive change in AEP value suggests that the exponential smoothing did not improve but reduced estimation accuracy in that county. The counts of counties with available official GDP data (i.e. 3058 in this study) and with negative (in this case, or positive) changes in AEP were input into a sign test to judge whether the exponential smoothing significantly improved the ability of the DNB image to estimate GDP at the county level. The null hypothesis was that the exponential smoothing generated equal influences on improving and reducing estimation accuracy at the county level. It needs to be particularly explained that although at the state level, the original (smoothed) DNB image thresholded by eight (perhaps nine) should be the most suitable for estimating GDP, at the metropolitan and the county levels, only one but no larger values were set to threshold the images. The reason for this thresholding is that preliminary methodological assessments show that if the threshold value is set larger than one, there will be countries with sum radiance equal to zero.

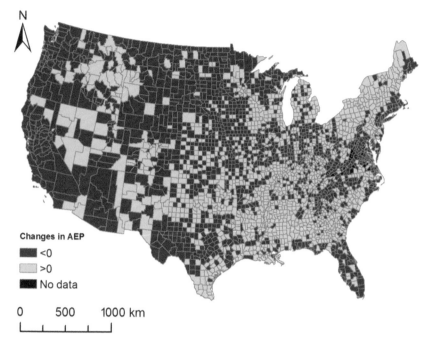

Figure 7. The map showing changes in absolute error of prediction (AEP).

4. Results and discussion

4.1. *Evaluating the patched results*

Tables 1 and 2 exhibit differences between estimated radiances and radiances in the original monthly image composites in Houston and Orlando, respectively. In the original images of May, June, July, and August, average radiances of all pixels located in the Houston area are around 18 nW cm^{-2} sr^{-1}. Average differences (i.e. AD in Equation (4)) for the 4 months are between 2.02 and 2.95 nW cm^{-2} sr^{-1} and ADRs (i.e. ADR in Equation (4)) vary from 0.11 to 0.16. By considering pixels with non-zero radiance values in the original images and their corresponding pixels in the estimated images, average differences are about 2.21–3.21 nW cm^{-2} sr^{-1} but ADRs still fall in the 0.11–0.16 range (Table 1). Compared with Houston, Orlando's ADRs are relatively small. Regardless of the consideration of all pixels or only non-zero pixels, in Orlando, the ADRs of the four summer months vary from 0.11 to 0.12 (Table 2). These comparisons demonstrate that

Table 1. Differences between estimated radiances and radiances in the original DNB composites in the Houston area.

	All pixels			Non-zero pixels		
Month	Average radiance (nW cm^{-2} sr^{-1})	Average difference (nW cm^{-2} sr^{-1})	Average difference rate	Average radiance (nW cm^{-2} sr^{-1})	Average difference (nW cm^{-2} sr^{-1})	Average difference rate
May	18.66	2.95	0.16	20.35	3.21	0.16
June	17.46	2.54	0.14	20.57	2.93	0.14
July	18.76	2.05	0.11	20.93	2.28	0.11
August	18.72	2.02	0.11	20.42	2.21	0.11

Table 2. Differences between estimated radiances and radiances in the original DNB composites in the Orlando area.

	All pixels			Non-zero pixels		
Month	Average radiance (nW cm^{-2} sr^{-1})	Average difference (nW cm^{-2} sr^{-1})	Average difference rate	Average radiance (nW cm^{-2} sr^{-1})	Average difference (nW cm^{-2} sr^{-1})	Average difference rate
May	14.88	1.77	0.12	17.07	2.03	0.12
June	14.30	1.79	0.12	16.89	2.10	0.12
July	14.26	1.60	0.11	16.62	1.86	0.11
August	14.18	1.68	0.12	16.24	1.92	0.12

the exponential-smoothing estimated radiances are very close to radiances in the original VIIRS DNB composites. Thus, it can be inferred that using the estimated radiances to patch the previously described missing summer values in the northern USA is also dependable.

4.2. Evaluating the smoothed results

The most notable feature of changes in sum radiance across original VIIRS DNB time series images is that sum radiances of January and February are considerably high due to the large snow coverage in northern USA (Figure 2) (Levin 2017). Thus, if we only exponentially smooth the time series images from January to December, aside from the problem of the unavailable smoothed image for January, sum radiances in smoothed images for February and March are still exceptionally large in comparison (Figure 2). That is because the smoothed image for February is only a simple average of the original images of January and February and radiance values in the smoothed image of March are greatly affected by those in the original images of January and February. Thus, it is necessary to exclude the first-time (i.e. sequentially stacked images from January to December) smoothed image for February from the final smoothed time series images and to reverse the order of the patched time series images to perform another exponential smoothing to insert into the time series.

Sum radiances of the reversely (i.e. sequentially stacking images from December to January) smoothed images for March, February, and January are considerably reduced compared to those of the first-time smoothed images. That is because in the second-time exponential smoothing, radiances in the smoothed images of March, February, and January are considerably influenced by those in original (or patched) images of the later months (i.e. from April to December). Sum radiance in these later-months images does not demonstrate large fluctuations except a small increase in the last two months. During Thanksgiving to New Year, many American families use lights to decorate their houses. Consequently, the relatively high sum radiances in November and December are predictable. However, in the reversely smoothed time series images, the image for December is unavailable while the image for November is a simple average of the original November and December images and radiances in the image of October are greatly influenced by those in the original images of November and December images. It can be inferred that if relatively large errors exist in the original November and December images, smoothed images for October cannot be very accurate. Thus, the first-time exponential smoothing is also necessary. Combining and averaging the two sets of smoothed images can further enhance accuracy of the final time series images.

After patching, the sum radiances of the summer months are very close to those of April, September, and October, but the sum radiances of January and February are markedly higher than those of the other months (Figure 2). After smoothing (and averaging), the sum radiance of February has been decreased to nearly equal with those of March and December. We admit that after averaging the two sets of exponentially smoothed images, the sum radiance of January is still high. However, the differences in sum radiance between January and the other months have been apparently reduced (Figure 2). For example, the sum radiance of the original January image is 1.79 times higher than that of the original September image. After smoothing, the sum radiance of January decreased from 75,161,945 to 59,970,159 nW cm^{-2} sr^{-1} while the sum radiance of September slightly increased from 42,054,538 to 42,287,976 nW cm^{-2} sr^{-1} making the sum radiance of January 1.42 times higher than that of September after exponential smoothing. Meanwhile, exponential smoothing did not destroy normal changes in sum radiance across different months. Notably, the sum radiances of November and December are still slightly higher than those of summer or fall months (Figure 2). It has been explained that these increases in sum radiance correctly reflect actual changes in brightness of NTLs because from Thanksgiving to New Year, many American families use lights to decorate their houses as well as public light displays.

4.3. *Estimating/Correlating to GDP at different geographic levels*

Differences in radiance between the original and the smoothed annual images are difficult to discern by visual observation except an apparent decrease in maximum radiance in the images (Figure 4). This decrease suggests that ephemeral lights (e.g. fires) have been effectively reduced. Admittedly, several pixels with very high radiances (e.g. >1000 nW cm^{-2} sr^{-1}) still exist in the smoothed images. (See the legend in Figure 4 (b). The highest radiance is 4799 nW cm^{-2} sr^{-1}.) The extremely bright pixels nearly all correspond to gas flaring (e.g. in west Texas) rather than stable urban NTL. The gas flaring associated with year-round oil production occurs in every month and consequently cannot be removed as high-frequency noise. Oil production generates considerably large amounts of GDP and as such should be considered in this study.

The differences in radiance and consequently in representing brightness of actual NTL between the two annual (i.e. original and smoothed) images can be distinguished by investigating correlations with GDP data. Table 3 shows that when the threshold increases to nine (eight for the original annual image), the Pearson correlation coefficients reach maximums while increasing the threshold value to 15, the coefficients of the two annual images both decrease. Thus, at the state level, the best threshold value for the contiguous USA should be approximately nine. Whatever the best threshold value selected is to remove lit rural areas, sum radiance evaluated with the closer relationship to GDP is always from the smoothed but not the original annual image (Table 3). These closer relationships suggest that the exponential-smoothing manipulation effectively improves the quality of the current VIIRS DNB images to represent actual situations of stable NTL more accurately.

Table 3. Pearson correlation coefficients between GDP and sum radiance at the state level (all significant at the $p = 0.01$ level).

Threshold	Pearson correlation coefficient, r	
	Original	Smoothed
1	0.8194	0.8328
2	0.8266	0.8426
3	0.8297	0.8457
4	0.8317	0.8476
5	0.8329	0.8488
6	0.8335	0.8496
7	0.8338	0.8500
8	0.8339	0.8504
9	0.8338	0.8505
10	0.8334	0.8504
11	0.8330	0.8501
12	0.8323	0.8497
13	0.8315	0.8493
14	0.8305	0.8485
15	0.8293	0.8478

At the metropolitan and the county levels, sum radiance derived from the original DNB image was regressed with GDP and the amounts of GDP generated in each metropolitan and county area were estimated by the regression functions. The relatively high regression accuracy (see R^2 values in Figures 5(a) and 6(a)) and the strong correlations (see Pearson correlation coefficients in Table 3) illustrate that the new-generation NTL image products still have a strong capacity to measure economy at different geographic scales. However, compared to the sum radiance derived from the original DNB image, sum radiance extracted from the exponentially smoothed DNB image has higher regression accuracy (i.e. the larger R^2 and the smaller SEE values) (see Figures 5 and 6). Such comparison demonstrates that the current version 1 VIIRS DNB image products are defective and the processing of exponential smoothing is effective to further improve quality of the DNB images for annual economic estimation.

Figure 7 shows that after performing the exponential smoothing, estimation accuracy of 1609 counties was improved. The null hypothesis of the sign test was rejected ($p = 0.004$) with the counts of 1609 and 3058 (i.e. the number of total counties calculated for the estimation) input into the test, indicating that the exponential smoothing did significantly improve estimation accuracy of GDP at the county level. Although a joint consideration of the result of the sign test and the comparison of the R^2 value and SEE of the regression functions makes us more solidly believe that the exponential smoothing can improve the ability of the VIIRS DNB images estimating GDP at the county level, there are still a considerable number of counties with decreased or unchanged estimation accuracy after the exponential-smoothing process. The reasons that so many counties do not have improved estimation accuracy probably come from two aspects. First, as previously stated, many uncertain factors such as varied atmospheric conditions and different numbers of cloud-free images result in the different qualities of the VIIRS DNB composite across different areas of the USA. Thus, the exponential smoothing cannot improve every county's estimation accuracy of GDP. More importantly, although brightness of NTLs strongly correlates to the amount of GDP, sum radiances of all regions do not correspond to their GDPs in the same proportion. Many social and natural factors such as mainstay industry, population density, and natural environment are all likely to affect the quantitative relationship between sum radiance and the amount of GDP (Liu et al. 2016; Zhao, Currit, and Samson 2011; Zhao, Ghosh, and Samson 2012). Moreover, such influences of the uncertain factors on the quantitative correlation between GDP and sum radiance may become more intense considering scale and consequently, sum radiance becomes progressively lower. For example, if the pillar industry of a county is agriculture, most GDP of the county cannot be reflected by its NTLs and thus, the county's GDP is likely to be underestimated. If such a county coincidentally has a relatively large number of pixels with overvalued radiances, then after the exponential-smoothing processing, estimation accuracy of the county is more likely to be decreased rather than increased.

Despite the uncertain factors, efficiency of the proposed method on correcting radiance values can be substantiated by more accurate estimations of the counties located in North Dakota and west Texas. A number of very bright pixels (radiance higher than 1000 nW cm^{-2} sr^{-1}) exist in North Dakota and west Texas. The extremely bright pixels nearly all correspond to gas flaring associated with fossil fuel production. Although the fossil fuel production generates considerable amounts of GDP, these amounts are smaller than the values estimated by the sum radiance given that the gas flaring is much brighter than the usual NTL from urban areas. Exponential smoothing can effectively reduce these exceptionally high radiance values and consequently bring about more accurate estimations for the counties containing such extremely bright pixels. For example, after exponential smoothing, sum radiances of William, North Dakota, and Midland, Texas, are reduced from 21,166 and 41,532 nW cm^{-2} sr^{-1} to 18,131 and 38,666 nW cm^{-2} sr^{-1} respectively. However, the exponential smoothing does not necessarily lead to a decrease in sum radiance in every county. For example, sum radiance of Milam that is located in central Texas is increased from 2245 to 2307 nW cm^{-2} sr^{-1}.

5. Conclusions

Compared to the old-generation NTL images, the new-generation NTLVIIRS DNB images have a finer spatial resolution. Additionally, with an on-board calibration system, pixel values across different VIIRS DNB images are compatible. However, we find that it is not as accurate as possible to directly use the current VIIRS DNB monthly image composites produced by NOAA's NCEI for studying socio-economic systems due to the existence of ephemeral lights, large variance in radiance, and missing data spreading over the high-latitude regions of the northern hemisphere in summer months. In this study, we stack 2014 monthly VIIRS DNB images as time series data, use exponential smoothing as an estimating process and adjusting tool to patch the pixels with 'no data,' and finally produce a smoothed annual VIIRS DNB image for the contiguous USA to estimate GDP. Sum radiance derived from the smoothed annual image has higher correlations with GDP at the state level and can be used to more accurately estimate GDP at the metropolitan and the county levels compared with that derived from the original annual VIIRS DNB image product.

The smoothed annual VIIRS DNB images referenced in this study can be obtained from GitHub (https://github.com/thestarlab/data). We did not release any patched/smoothed monthly images generated in this study because a certain number of pixels, especially in the months of March–October, may be over-smoothed. These patched/smoothed monthly images are only exemplary data and should not be used individually for any specific studies. In the spring of 2017, NOAA's NCEI began to produce annual VIIRS DNB image composites. Yet, the release of the annual VIIRS DNB image composites seriously lags behind the release of the monthly VIIRS DNB image composites for the same year. Thus, the exponential smoothing method can be treated as an expediency to improve the monthly VIIRS DNB image products for achieving timely estimations of annual GDP (or other socio-economic factors) with greater accuracy before the temporally corresponding annual VIIRS DNB composites are released. More importantly, each of the annual image composites is for a fixed time period (i.e. from January of a year to December of the same year) while the exponential smoothing method addressed in this study can be used to process monthly VIIRS DNB time-series images to obtain average brightness of NTLs for more flexible periods (e.g. from July of a year to June of the next year).

Acknowledgements

We thank the Nighttime Lights Special Issue and technical editors for their comments and suggestions to improve this manuscript.

Disclosure statement

No potential conflict of interest was reported by the authors.

ORCID

Naizhuo Zhao ⓘ http://orcid.org/0000-0002-1778-2112
Guofeng Cao ⓘ http://orcid.org/0000-0003-4827-1558

References

Casiez, G., N. Roussel, and D. Vogel. 2012. "1€ Filter: A Simple Speed-Based Low-Pass Filter for Noisy Input in Interactive Systems." CHI'12, Austin, TX, May 5–10, 2527–2530.
Chen, X., and W. D. Nordhaus. 2011. "Using Luminosity Data as a Proxy for Economic Statistics." *Proceedings of the National Academy of Sciences* 108 (21): 8589–8594. doi:10.1073/pnas.1017031108.
Doll, C. H., J.-P. Muller, and C. D. Elvidge. 2000. "Night-Time Imagery as a Tool for Global Mapping of Socio-Economic Parameters and Greenhouse Gas Emissions." *AMBIO: A Journal of the Human Environment* 29 (3): 157–162. doi:10.1579/0044-7447-29.3.157.
Doll, C. N. H., J.-P. Muller, and J. G. Morley. 2006. "Mapping Regional Economic Activity from Night-Time Light Satellite Imagery." *Ecological Economics* 57 (1): 75–92. doi:10.1016/j.ecolecon.2005.03.007.
Elvidge, C. D., K. Baugh, M. Zhizhin, and F.-C. Hsu. 2013. "Why VIIRS Data are Superior to DMSP for Mapping Nighttime Lights." *Proceedings of the Asia-Pacific Advanced Network 2013* 35: 62–69. doi:10.7125/APAN.35.7.
Elvidge, C. D., M. Zhizhin, K. Baugh, F.-C. Hsu, and T. Ghosh. 2015. "Methods for Global Survey of Natural Gas Flaring from Visible Infrared Imaging Radiometer Suite Data." *Energies* 9 (1): 14. doi:10.3390/en9010014.

Elvidge, C. D., D. Ziskin, K. E. Baugh, B. T. Tuttle, T. Ghosh, D. W. Pack, E. H. Erwin, and M. Zhizhin. 2009. "A Fifteen Year Record of Global Natural Gas Flaring Derived from Satellite Data." *Energies* 2 (3): 595–622. doi:10.3390/en20300595.

Gelper, S., R. Fried, and C. Croux. 2010. "Robust Forecasting with Exponential and Holt-Winters Smoothing." *Journal of Forecasting* 29 (3): 285–300. doi:10.1002/for.1125.

Ghosh, T., S. Anderson, R. L. Powell, P. C. Sutton, and C. D. Elvidge. 2009. "Estimation of Mexico's Informal Economy and Remittances Using Nighttime Imagery." *Remote Sensing* 1 (3): 418–444. doi:10.3390/rs1030418.

Ghosh, T., C. D. Elvidge, P. C. Sutton, K. E. Baugh, D. Ziskin, and B. T. Tuttle. 2010a. "Creating a Global Grid of Distributed Fossil Fuel CO2 Emissions from Nighttime Satellite Imagery." *Energies* 3 (12): 1895–1913. doi:10.3390/en3121895.

Ghosh, T., R. L. Powell, C. D. Elvidge., K. E. Baugh, P. C. Sutton, and S. Anderson. 2010b. "Shedding Light on the Global Distribution of Economic Activity." *The Open Geography Journal* 3: 147–160. doi:10.2174/1874923201003010147.

Holloway, J. L. 1958. "Smoothing and Filtering of Time Series and Space Fields." *Advances in Geophysics* 4: 351–389. doi:10.1016/S0065-2687(08)60487-2.

Holt, C. C. 2004. "Forecasting Seasonals and Trends by Exponentially Weighted Moving Averages." *International Journal of Forecasting* 20 (1): 5–10. doi:10.1016/j.ijforecast.2003.09.015.

Hyndman, R. J., and G. Athanasopoulos. 2016. "Forecasting: Principles and Practice." Accessed April 26 2017. http://otexts.com/fpp/

Hyndman, R. J., G. Athanasopoulos, S. Razbash, D. Schmidt, Z. Zhou, Y. Khan, and E. Wang. 2014. "R Package: Forecast." http://cran.r-project.org/web/packages/forecast/index.html

Imhoof, M. L., W. T. Lawrence, D. C. Stutzer, and C. D. Elvidge. 1997. "A Technique for Using Composite DMSP/OLS 'City Lights' Satellite Data to Accurately Map Urban Areas." *Remote Sensing of Environment* 61 (3): 361–370. doi:10.1016/S0034-4257(97)00046-1.

LaViola, J. J. 2003. "Double Exponential Smoothing: An Alternative to Kalman Filter-Based Predictive Tracking." In *Proceedings of EGVE' 03*, edited by J. Deisinger, and A. Kunz, 199–206. New York, NY: ACM. doi:10.1145/769953.769976

Letu, H., M. Hara, H. Yagi, K. Naoki, G. Tana, F. Nishio, and O. Shuhei. 2010. "Estimating Energy Consumption from Night-Time DMPS/OLS Imagery after Correcting for Saturation Effects." *International Journal of Remote Sensing* 31 (16): 4443–4458. doi:10.1080/01431160903277464.

Levin, N. 2017. "The Impact of Seasonal Changes on Observed Nighttime Brightness from 2014 to 2015 Monthly VIIRS DNB Composites." *Remote Sensing of Environment* 193: 150–164. doi:10.1016/j.rse.2017.03.003.

Levin, N., and Q. Zhang. 2017. "A Global Analysis of Factors Controlling VIIRS Nighttime Light Levels from Densely Populated Areas." *Remote Sensing of Environment* 190: 366–382. doi:10.1016/j.rse.2017.01.006.

Liao, L. B., S. Weiss, S. Mills, and B. Hauss. 2013. "Suomi NPP VIIRS Day-Night Band On-Orbit Performance." *Journal of Geophysical Research: Atmospheres* 118 (22): 12705–12718. doi:10.1002/2013JD020475.

Liu, Y., T. Delahunty, N. Zhao, and G. Cao. 2016. "These Lit Areas are Undeveloped: Delimiting China's Urban Extents from Thresholded Nighttime Light Imagery." *International Journal of Applied Earth Observation and Geoinformation* 50: 39–50. doi:10.1016/j.jag.2016.02.011.

Liu, Z., C. He, Q. Zhang, Q. Huang, and Y. Yang. 2012. "Extracting the Dynamics of Urban Expansion in China Using DMSP-OLS Nighttime Light Data from 1992 to 2008." *Landscape and Urban Planning* 106 (1): 62–72. doi:10.1016/j.landurbplan.2012.02.013.

Lo, C. P. 2002. "Urban Indicators of China from Radiance-Calibrated Digital DMSP-OLS Nighttime Images." *Annals of the Association of American Geographers* 92 (2): 225–240. doi:10.1111/1467-8306.00288.

Ma, T., Y. Zhou, Y. Wang, C. Zhou, S. Haynie, and T. Xu. 2014. "Diverse Relationships between Suomi-NPP VIIRS Night-Time Light and Multi-Scale Socioeconomic Activity." *Remote Sensing Letters* 5 (7): 652–661. doi:10.1080/2150704X.2014.953263.

Miller, S. D., S. P. Mills, C. D. Elvidge, D. T. Lindsey, T. F. Lee, and J. D. Hawkins. 2012. "Suomi Satellite Brings to Light a Unique Frontier of Nighttime Environmental Sensing Capabilities."

Proceedings of the National Academy of Sciences 109 (39): 15706–15711. doi:10.1073/pnas.1207034109.

Oda, T., and S. Maksyutov. 2011. "A Very High-Resolution (1 km×1 km) Global Fossil Fuel CO_2 Emission Inventory Derived Using a Point Source Database and Satellite Observations of Nighttime Lights." *Atmospheric Chemistry and Physics* 11: 543–556. doi:10.5194/acp-11-543-2011.

Shi, K., B. Yu, Y. Huang, Y. Hu, B. Yin, Z. Chen, L. Chen, and J. Wu. 2014. "Evaluating the Ability of NPP-VIIRS Nighttime Light Data to Estimate the Gross Domestic Product and the Electric Power Consumption of China at Multiple Scales: A Comparison with DMSPOLS Data." *Remote Sensing* 6: 1705–1724. doi:10.3390/rs6021705.

Small, C., F. Pozzi, and C. D. Elvidge. 2005. "Spatial Analysis of Global Urban Extent from DMSP-OLS Night Lights." *Remote Sensing of Environment* 96 (3–4): 277–291. doi:10.1016/j.rse.2005.02.002.

Sutton, P., C. Roberts, C. Elvidge, and H. Meij. 1997. "A Comparison of Nighttime Satellite Imagery and Population Density for the Continental United States." *Photogrammetric Engineering and Remote Sensing* 63 (11): 1303–1313. doi:0099-1112/97/6311-1303$3.00/0.

Sutton, P. C., C. D. Elvidge, and T. Ghosh. 2007. "Estimation of Gross Domestic Product at Sub-National Scales Using Nighttime Satellite Imagery." *International Journal of Ecological Economics and Statistics* 8: 5–21.

Sutton, P. C., A. R. Goetz, S. Fildes, C. Forster, and T. Ghosh. 2010. "Darkness on the Edge of Town: Mapping Urban and Peri-Urban Australia Using Nighttime Satellite Imagery." *The Professional Geographer* 62 (1): 119–133. doi:10.1080/00330120903405006.

Sutton, P. C., D. Roberts, C. D. Elvidge, and K. Baugh. 2001. "Census from Heaven: An Estimate of the Global Human Population Using Night-Time Satellite Imagery." *International Journal of Remote Sensing* 22 (16): 3061–3076. doi:10.1080/01431160010007015.

Tian, J., N. Zhao, E. L. Samson, and S. Wang. 2014. "Brightness of Nighttime Lights as a Proxy for Freight Traffic: A Case Study of China." *IEEE Journal of Selected Topics in Applied Earth Observations and Remote Sensing* 7 (1): 206–212. doi:10.1109/JSTARS.2013.2258892.

The World Bank. 2016. "World Bank Open Data." Accessed April 26 2017. http://data.worldbank.org/

Wu, J., Z. Wang, W. Li, and J. Peng. 2013. "Exploring Factors Affecting the Relationship between Light Consumption and GDP Based on DMSP/OLS Nighttime Satellite Imagery." *Remote Sensing of Environment* 134: 111–119. doi:10.1016/j.rse.2013.03.001.

Zhao, N., N. Currit, and E. Samson. 2011. "Net Primary Production and Gross Domestic Product in China Derived from Satellite Imagery." *Ecological Economics* 70 (5): 921–928. doi:10.1016/j.ecolecon.2010.12.023.

Zhao, N., T. Ghosh, and E. L. Samson. 2012. "Mapping Spatio-Temporal Changes of Chinese Electric Power Consumption Using Night-Time Imagery." *International Journal of Remote Sensing* 33 (20): 6304–6320. doi:10.1080/01431161.2012.684076.

Zhao, N., E. L. Samson, and N. A. Currit. 2015. "Nighttime-Lights-Derived Fossil Fuel Carbon Dioxide Emission Maps and Their Limitations." *Photogrammetric Engineering & Remote Sensing* 81 (12): 935–943. doi:10.14358/PERS.81.12.935.

Zhao, N., Y. Zhou, and E. L. Samson. 2015. "Correcting Incompatible DN Values and Geometric Errors in Nighttime Lights Time-Series Images." *IEEE Transactions on Geoscience and Remote Sensing* 53 (4): 2039–2049. doi:10.1109/TGRS.2014.2352598.

Zhou, Y., S. J. Smith, K. Zhao, M. Imhoff, A. Thomson, B. Bond-Lamberty, G. R. Asrar, X. Zhang, C. He, and C. D. Elvidge. 2015. "A Global Map of Urban Extent from Nightlights." *Environmental Research Letters* 10: 054011. doi:10.1088/1748-9326/10/5/054011.

A new multichannel threshold algorithm based on radiative transfer characteristics for detecting fog/low stratus using night-time NPP/VIIRS data

Shensen Hu, Shuo Ma, Wei Yan, Jun Jiang and Yunxian Huang

ABSTRACT

The Visible Infrared Imager Radiometer Suite (VIIRS) with 22 imagery and radiometric spectral bands on board the Suomi National Polar-orbiting Partnership (NPP) provides imagery products with accurate radiometric calibration and high resolution in space and time. In this article, a multichannel threshold algorithm based on radiative transfer characteristics (MRTC) using VIIRS products is proposed to monitor fog and low stratus at night. Sensitivity analysis of the top-of-atmosphere (TOA) reflectance of fog/low stratus on various influential factors is performed; thus, a reflectance threshold lookup table from various observation geometries is established. Pixels with suitable reflectance are detected as fog/low stratus preliminarily. The final results are confirmed by removing snow pixels and high/medium cloud pixels, and testing the surface homogeneity. Validation experiments are then performed on five typical cases of heavy fog/low stratus with more than a half moon in China using three detection algorithms. This indicates that in some examples the MRTC algorithm achieves some improvements over the existing two algorithms with about a 0.1323 average false alarm ratio (FAR), a 0.8587 average probability of detection (POD), and a 0.7595 average critical success index (CSI).

1. Introduction

Fog and low stratus are common natural obstacles for human traffic because of the poor visibility and air quality (Bendix et al. 2006). Therefore, fog and low stratus must be described in time and space to avoid or reduce the potential loss (Pagowski, Gultepe, and King 2004). Compared with the detection results from ground meteorological stations, the remote-sensing results of fog and low stratus from meteorological satellites are obviously more continuous in special coverage and more frequent in time coverage, and show great promise in detecting fog and low stratus with the development of radiation detectors on board the satellites (Cermak and Bendix 2005).

Research on fog and low-stratus detection based on satellite remote-sensing data began in the 1970s and has focused on the daytime monitoring and forecasting of fog and low stratus for the limitations of early infrared radiation detectors (Gurka 1974; Olivier 1995). Bendix computed the maximum and minimum albedo thresholds based on radiative transfer calculations and proposed a novel approach using the threshold technique for the daytime detection of fog/low stratus (Bendix et al. 2006). With the development of infrared radiation detectors, the detection of fog/low stratus at night or twilight gradually came to people's attention. It was theoretically proved that small droplets of fog/low stratus are obviously less emissive at a 3.9 μm channel than at a 10.8 μm channel (Hunt 1973). The dual-channel difference (DCD) algorithm for detecting night-time fog/low stratus is proposed, applied, and promoted over years, using satellite products from different radiation detectors on board several meteorological satellites (Eyre et al. 1984; Dybbroe 1993; Ellrod 1995; Elvidge et al. 1997; Bendix 2002; Cermak and Bendix 2007; Chaurasia et al. 2011; Mosher 2013).

The Visible Infrared Imager Radiometer Suite (VIIRS) with 22 imagery and radiometric spectral bands (including the Day Night Band (DNB) covering almost seven orders of magnitude in dynamic range and other bands) on board the Suomi National Polar-Orbiting Partnership (NPP) provides imagery products with accurate radiometric calibration and high resolution in space and time (Miller et al. 2013; Cao et al. 2013; Ma et al. 2015; Qiu et al. 2016). Based on the advantages of VIIRS products, Jiang proposed a multichannel threshold algorithm based on image cutting (MIC) and achieved better accuracy compared with the existing DCD algorithm (Jiang et al. 2015).

In our study, we combine DNB with other VIIRS channels and propose a multichannel threshold algorithm based on radiative transfer characteristics (MRTC) to monitor fog and low stratus at night. This article is organized as follows. Section 2 introduces the data used in the MRTC algorithm. Section 3 explains the procedures of detecting fog and low stratus using VIIRS products. Section 4 presents the validation experiments on five typical cases using the MRTC algorithm and analyses the detection results of fog/low stratus compared with the existing DCD and MIC algorithms. Section 5 summarizes the conclusions of the study.

2. Data

The VIIRS data product comprises the raw data record (RDR), the sensor data record (SDR), and the environment data record (EDR). In this article, four types of VIIRS data products are used to identify fog/low-stratus pixels: VIIRS/DNB SDR (SVDNB), VIIRS/DNB SDR geolocation content summary (GDNBO), VIIRS/Moderate Resolution Band SDR (SVM5, SVM7, SVM10, and SVM16), and VIIRS/Moderate Resolution Band SDR geolocation content summary (GMODO). The selected channels are presented and summarized in Table 1. The SVDNB product provides the calibrated top-of-atmosphere (TOA) radiance for each pixel and quality flag information. The GDNBO product provides the corresponding latitude, longitude, solar zenith angle, lunar zenith angle (LZA), lunar azimuth angle, satellite zenith angle, satellite azimuth angle, moon phase angle, and moon illumination fraction. The Moderate Resolution Band SDR products contain the brightness temperatures, radiances, and quality flag information. The GMODO product contains the corresponding latitude and longitude. Note that the charge-coupled device of DNB and the M-band are mounted in different focal planes (Cao et al. 2014), so the two SDR geolocation summaries are used to match the spatial pixels.

Table 1. Channels used in the algorithm.

VIIRS channel	Wavelength range (μm)	Central wavelength (μm)	Spatial resolution at nadir (m)	Application
DNB	0.500–0.900	0.700	750	Detection preliminary
M5	0.662–0.682	0.672	750	Removal of snow
M7	0.846–0.885	0.865	750	Removal of snow
M10	1.580–1.640	1.61	750	Removal of snow
M16	11.538–12.488	12.01	750	Removal of high/medium clouds

3. Algorithm description

The overall structure of the MRTC algorithm is outlined in Figure 1. First, the sensitivity of the TOA reflectance to the variation of different factors is studied, such as moon parameters, satellite observation geometry, optical properties of fog/low stratus, surface reflectance, atmospheric profiles, aerosol optical thickness, and so on. Thus the TOA reflectance threshold lookup table from various observation geometries is established. Then we preprocess the SDR products of several VIIRS channels and compute the TOA reflectance of candidate pixels. According to the restriction of the maximum and minimum reflectance thresholds in the aforementioned lookup table, eligible fog/low-stratus pixels are identified preliminarily. To achieve a better detection result, we further remove the pixels of snow and high/medium clouds, and test the surface homogeneity

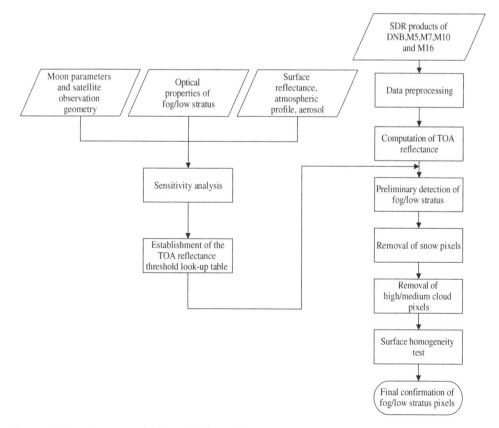

Figure 1. Overall structure of the MRTC algorithm.

to eliminate noise. Finally, a pixel that meets all the prescribed criteria of each step is ultimately identified as a fog/low-stratus pixel.

3.1. Sensitivity analysis

Owing to the relatively stable optical properties of fog/low stratus, the uncertainties of TOA reflectance of fog/low stratus with variable influential factors can be meaningfully studied. In the sensitivity analysis, SCIATRAN (Rozanov et al. 2014), a comprehensive software package applicable for modelling radiative transfer processes, is used to model the TOA reflectance of fog/low stratus. However, the software package is designed for use only in the daytime. To realize the detection of night-time fog/low stratus, Miller's lunar irradiance model (Miller and Turner 2009), which can produce the irradiance spectra of 1 nm resolution between 0.3 μm and 1.2 μm for a given time, is introduced into SCIATRAN to yield the lunar irradiance spectra. Owing to the high relative humidity (RH) of fog/low stratus, a water cloud model calculated by another radiative transfer software package, libRadtran (Mayer and Kylling 2005), is used to describe the scattering properties of fog/low-stratus droplets. In the radiative transfer process, thousands of Legendre polynomials describe the bulk-phase function, indicating a strong forward scattering (Nakajima and Tanaka 1988). In addition, the delta-M method (Wiscombe 1977), which can approximate the forward scattering peak as a delta function and allow for the remaining energy to be parameterized with fewer expansion terms, is used in the simulation to truncate the phase function and improve the computational efficiency.

According to previous statistical results (Bendix et al. 2006), the reference values of effective radius (R_e), cloud optical thickness (COT), and cloud top height (Z_T) are shown in Table 2, along with other influential parameters such as surface reflectance, atmospheric profiles, and aerosol. The atmospheric profiles are taken from a climatological data base from SCIATRAN. The contents of water vapour and various gases in these monthly average profiles are different for various latitudes. The aerosol concentration data is obtained from LOWTRAN 7 data. In addition, the aerosols in Table 2 are lower tropospheric aerosols specifically.

The results of the uncertainty ranges in the simulation of the TOA reflectance of fog/low stratus caused by several factors are shown in Table 2. When the simulation focuses on the uncertainty range for one parameter, the other values of influential factors are input with reference values. Thus, we can calculate the uncertainty range caused by one

Table 2. The maximum uncertainty ranges of simulated TOA reflectance for various LZAs, caused by changes in the reference values of surface reflectance of 0.15 (0–0.3), atmospheric profile type of 5°S (35°S-35°N), no aerosol type (Urban-Marine), R_e of 7 μm (4-12 μm), COT of 10 (1–30), and Z_T of 200 m (100–400 m).

LZA	Surface reflectance (%)	Atmospheric profile type (%)	Aerosol type (%)	R_e (%)	COT (%)	Z_T (%)
0°	6.59	1.52	2.79	6.43	77.63	0.19
10°	6.62	1.33	2.45	6.87	75.41	0.23
20°	6.78	1.46	2.36	6.72	79.37	0.21
30°	6.32	1.21	2.77	7.23	85.59	0.17
40°	6.58	1.53	2.65	6.34	83.43	0.15
50°	6.92	1.26	2.37	6.91	81.22	0.21
60°	6.20	1.43	2.76	6.12	78.45	0.16

certain influential factor. We use six LZAs to ensure sufficient moonlight conditions in the simulation of various lunar illuminations. Regarding satellite observation geometry, the viewing zenith angle (VZA) defines the value of the line-of-sight zenith angle, and the relative azimuth angle (RAA) indicates the value of the relative azimuth angle of line-of-sight with respect to the moon. In the step of parameter setting, six VZAs ranging from 0° to 50° with a 10° interval and 19 RAAs ranging from 0° to 180° with a 10° interval are used. Therefore, for each input parameter of LZA and other influential factors, 114 reflectances are computed, respectively.

Uncertainties influenced by various parameters in the simulation are all examined, and the uncertainty ranges in Table 2 all indicate the largest simulation difference from the reference values within the input range. Table 2 shows that, within the input range of atmospheric profile type and aerosol type, the maximum uncertainty is up to only 1.53% and 2.79%, respectively. Such minor changes, which can be negligible, are reasonable because contributions from the atmospheric molecular scattering are far smaller than those from the fog/low-stratus reflection in radiative transfer. Similarly, within the input range of 0–0.3 surface reflectance, the maximum uncertainty is up to 6.92%. The fact that the reflection from fog/low stratus is much larger than the surface reflection can account for this. Regarding the properties of fog/low stratus, within the input range of Z_T and R_e, the maximum uncertainty can be up to 0.23% and 7.23%, respectively. Thus, the effects of Z_T and R_e within the input ranges seem negligible, compared with that of COT in which the maximum uncertainty is up to a very poor 85.59% within the input range.

Based on the above analysis, the simulated uncertainties for the variation of surface reflectance, atmospheric profiles, aerosol, R_e, and Z_T are relatively small. For R_e and surface reflectance, only in the worst case can the uncertainty be up to about 7%, which is quite small and can be negligible, whereas the other parameters have even smaller uncertainties. Therefore, COT is obviously the main factor affecting the TOA reflectance of fog/low stratus compared with others in the sensitivity analysis.

To further understand the impacts of COT as well as R_e and LZA on TOA reflectance of fog/low stratus in the simulation, the simulated bidirectional reflectance factors (BRFs) with various observation geometries (0° ≤ VZA ≤ 60°, 0° ≤ RAA ≤ 360°) are displayed in Figure 2, with VZA and RAA presented in the radial and tangential directions, respectively. Assuming all other parameters are set as the reference values, Figures 2(a) and (b) describe that BRFs are calculated with two R_e values, which are equal to 4 μm and 12 μm. Similarly, Figures 2(c) and (d) present BRFs with COT values equal to 1 and 30, respectively. On the bottom panel, Figure 2(e) is BRF calculated with LZA = 60°. Finally, Figure 2(f) is the basic reference BRF to the other five BRFs, calculated with all the influential parameters set as reference values.

As seen from Figure 2, BRF obviously varies much with various VZAs and RAAs owing to the variable colours in the different areas of a certain picture. The larger BRF is generally found in the area where VZA = 30° and RAA = 0° on the condition of LZA = 30°, indicating that fog/low stratus has a strong backward scattering. Meanwhile, attention needs to be paid to the absolute values of BRFs instead of the visual colours due to the quite different colour codes on the medium panel. For example, in the same area of the two pictures on the top panel, BRFs vary little between 0.48 and 0.60 regardless of the R_e difference. In comparison, in the same

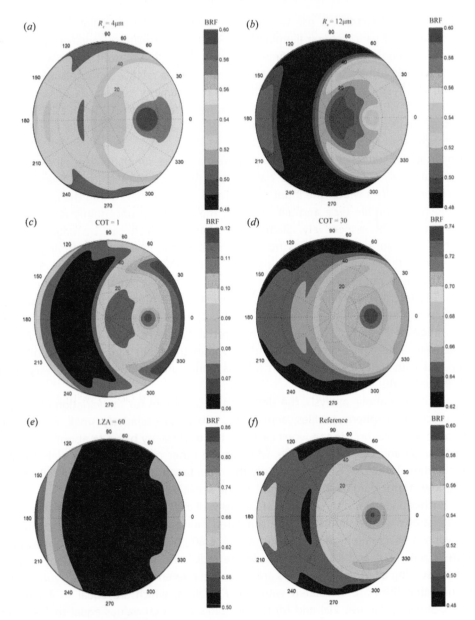

Figure 2. Simulated TOA BRFs of fog/low stratus for various R_e, COT, and LZA values.

area of the two pictures on the middle panel, the BRFs calculated with various COTs differ so much that the maximum BRF with COT = 1 is equal to about 0.12, and the maximum BRF with COT = 30 is equal to about 0.74, even not in an order of magnitude in some areas. The calculation result confirms the above assumption that COT is the main factor affecting the TOA reflectance of fog/low stratus. In addition, by comparing the BRFs calculated with LZA = 60° and 30° on the bottom panel, we can find that the two colourful shapes are totally different, indicating that various LZAs also greatly affect the TOA reflectance of fog/low stratus.

Overall, surface reflectance, atmospheric profiles, aerosol, R_e, and Z_T have relatively minor impacts on the simulation of the TOA reflectance of fog/low stratus whereas COT and observation geometry are the most sensitive factors. Thus, a reflectance threshold lookup table must be established, with the following input to the radiative transfer model to obtain the TOA reflectance of fog/low stratus: LZA taken every 5° from 0° to 80°, RAA taken every 10° from 0° to 180°, VZA taken every 5° from 0° to 50°, and maximum and minimum COT values, with other influential factors set as reference values. Thus, for a certain satellite observation geometry, the maximum and minimum reflectance thresholds of fog/low stratus are obtained.

3.2. Reflectance calculation

SVDNB products provide the at-aperture spectra radiance received in the DNB channel, but we need the TOA reflectance of pixels in our algorithm. Therefore, DNB spectra radiance must be converted into TOA reflectance ρ_{DNB}.

$$\rho_{DNB} = L_{DNB}/L_{LTOA}, \tag{1}$$

where L_{DNB} is the at-aperture spectra radiance received by VIIRS/DNB, which has been preprocessed to eliminate the N/A fill values and match the special pixels (Jiang et al. 2015), and L_{LTOA} is the downwelling TOA lunar radiance computed by

$$L_{LTOA} = \frac{E_L}{\pi} \cos(\theta_l), \tag{2}$$

where θ_l is the LZA, and E_L is the downwelling TOA lunar irradiance expressed by

$$E_L = \frac{\int_{\lambda_1}^{\lambda_2} I_L(\lambda) r(\lambda) d\lambda}{\int_{\lambda_1}^{\lambda_2} r(\lambda) d\lambda}, \tag{3}$$

where $I_L(\lambda)$ is the lunar irradiance spectra calculated by the Miller lunar irradiance model, which varies with the moon phase angle, LZA, and several other distance factors (Miller and Turner 2009). Meanwhile, $r(\lambda)$ is the relative spectra response of VIIRS/DNB released by the VIIRS science team. Note that λ is the wavelength of radiation.

3.3. Post-processing

With the reflectance threshold lookup table for various satellite observation geometries established, the DNB pixels with reflectance located in the interval of the maximum and minimum thresholds from the lookup table are identified, and thus fog/low-stratus pixels are detected preliminarily. However, the preliminary results of the detected pixels are often mixed with those pixels in which TOA reflectance appears to be similar. The post-processing procedure consisting of the removal of snow pixels and high/medium cloud pixels, and the surface homogeneity test, is obviously useful to further improve the accuracy of the detection method.

3.3.1. Removal of snow pixels

In 0.4–1.3 μm, both clouds and snow have large reflectance, whereas in 1.3–3 μm clouds obviously have a larger reflectance than snow (Crane and Anderson 1984). Based on the reflectance difference of two targets in various spectra bands, the normalized difference snow index (NDSI) and normalized difference vegetation index (NDVI) defined by Equation (4) are used to identify the snow pixels, referring to the snow classification techniques adopted by the Moderate Resolution Imaging Spectroradiometer (MODIS) (Hall, Riggs, and Salomonson 1995).

$$NDSI = \frac{R_{M5} - R_{M10}}{R_{M5} + R_{M10}}, NDVI = \frac{R_{M7} - R_{M5}}{R_{M7} + R_{M5}}, \tag{4}$$

where R_{M5}, R_{M7}, and R_{M10} are the reflectance in M5 (0.672 μm), M7 (0.865 μm), and M10 (1.61 μm) of VIIRS channels, respectively. Based on the thresholds of the NDSI and NDVI values (Baker 2011), snow pixels are removed from the preliminary result of fog/low stratus. For more details, readers can refer to Jiang et al. (2015)

3.3.2. Removal of high/medium cloud pixels

The fact that high/medium cloud pixels have a much higher Z_T than fog/low-stratus pixels is well recognized. Based on the difference of Z_T, we first obtain reanalysed data at 0000 UTC of the corresponding date from the European Centre for Medium-Range Weather Forecasts (ECMWF) and then interpolated the temperature profiles into latitude–longitude grids. When latitude and longitude are known for a certain pixel, combining the pixel's temperature profile and the brightness temperature from an SVM16 product produces the pixel's Z_T and thus removes high/medium cloud pixels with a certain Z_T threshold.

3.3.3. Surface homogeneity test

In general, fog/low-stratus areas appear to be continuous and homogeneous in space. However, numerous isolated pixels appear in the detection result after the above-mentioned removal procedures. To further improve the accuracy of the detection of fog/low stratus, the surface homogeneity (Cermak and Bendix 2005) is tested by

$$SH = \frac{\mu}{3\sigma}, \tag{5}$$

where SH is the surface homogeneity level, whereas μ and σ are, respectively, the mean value and standard deviation of grey values of a 3 × 3 pixel area with the candidate pixel centred, assuming 1 for fog/low-stratus pixels and 0 for others. As a result, if SH<0.22, the candidate central pixel is regarded as noise from experience; otherwise, it is finally identified as a fog/low-stratus pixel.

4. Validation experiments

To validate the feasibility of the MRTC algorithm proposed in this study, experiments have been performed on five cases of heavy fog/low stratus with more than a half moon in China using the three various algorithms. Additionally, the corresponding ground-based meteorological data collected on the hour is used to validate the accuracy. Using ground-measured data, fog and low stratus are identified traditionally by three indices: horizontal visibility (HV)≤10000 m, RH≥95%, and cloud base height (CBH)≤2500 m.

Figures 3 and 4 describe two typical cases among the five, occurring at 1839 UTC 27 October 2012 and 1811 UTC 27 January 2013, respectively. The lines in the figures are used as provincial boundaries to better compare the locations of fog/low-stratus area. For case 1, Figures 3(a) and (b) are, respectively, an original low-level-light image of DNB and an infrared image of M16. Meanwhile, Figures 3(c)–(e) describe the monitoring results of fog/low stratus using the DCD, MIC, and MRTC algorithms, respectively. In addition, Figures 3(f) and (g) indicate the monitoring results of ground-measured meteorological stations, showing fog in Area C for HV<10000 m and low stratus in Areas A and B for CBH<2500 m. Figure 4 describes the conditions of case 2, but the structure of the seven images is similar.

Comparing the remote-sensing results to the ground-measured results, three remote-sensing monitoring results generally correspond to the ground-measured results. However, several false fog/low-stratus pixels exist using the DCD algorithm so that the whole image appears white in most areas of Figure 3(c). In addition, there are some undetected fog/low-stratus pixels using the MIC algorithm, especially in Area A. In comparison, fog/low-stratus pixels using the MRTC algorithm visually agree better with the ground-measured results in most areas, except for some white and relatively isolated pixels that are in the lower-left area in Figure 3(e). However, from further observation, the lack of ground-measured data of CBH in the corresponding areas of Figure 3(f) makes it difficult to confirm whether fog/low stratus is in those areas and whether the detection result of this area is accurate.

For case 2, the ground-based products in Figures 4(f) and (e) show that fog is in all the three areas of A, B, and C for HV<10000 m, and low stratus is only in Area A for CBH<2500 m. Comparing the remote-sensing results shown in Figures 4(c)–(e) with the ground-measured results, we find the difference of the monitoring results. For example, using the MIC algorithm some undetected fog/low-stratus pixels are seen in Area C in Figure 4(d), whereas using the DCD algorithm, nearly all pixels in Area C are undetected and appear dark in Figure 4(c). The monitoring results are similar in other areas of Figures 4(c) and (d). However, the monitoring result is somewhat opposite using the MRTC algorithm shown in Figure 4(e). Note that many pixels in Area C are regarded as fog/low stratus, but they are actually medium clouds pixels for 2500 m<CBH<5000 m and HV>10000 m. Moreover, in the lower-right part of Area B, the MRTC algorithm is the only one in which those pixels are identified as fog/low stratus corresponding to the ground-based results.

It is difficult, however, to conclude visually which algorithm among the three achieves the most accurate monitoring results of fog/low stratus based on the above-mentioned analysis only. To further quantifiably study the accuracy of the three algorithms for the detection of fog/low stratus in night-time, three indices are used (Bendix, Cermak, and Thies 2004): false alarm ratio (FAR), probability of detection (POD), and critical success index (CSI), which are defined by

$$FAR = \frac{N_F}{N_H + N_F}$$

$$POD = \frac{N_H}{N_H + N_M} \, , \tag{6}$$

$$CSI = \frac{N_H}{N_H + N_M + N_F}$$

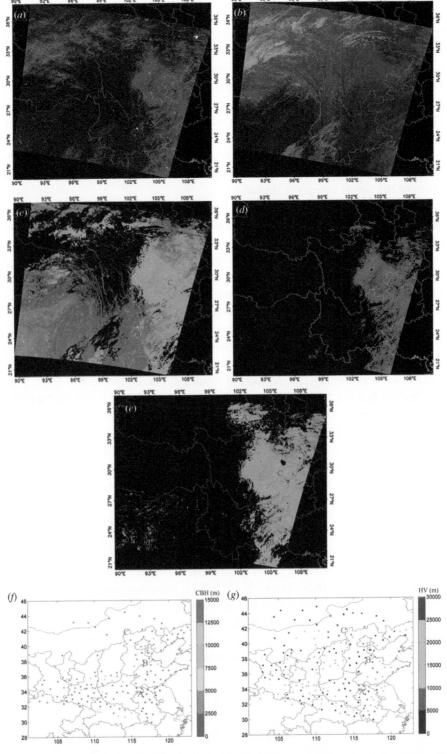

Figure 3. Experimental results of case 1. (*a*) Low-light-level image of DNB. (*b*) Infrared image of M16. (*c*) Monitoring result of the DCD algorithm. (*d*) Monitoring result of the MIC algorithm. (*e*) Monitoring result of the MRTC algorithm. Ground-measured results of (*f*) CBH and (*g*) HV.

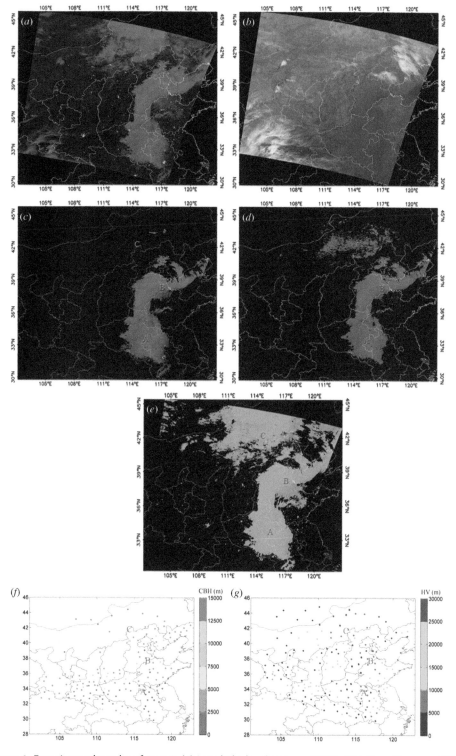

Figure 4. Experimental results of case 2. (*a*) Low-light-level image of DNB. (*b*) Infrared image of M16. (*c*) Monitoring result of the DCD algorithm. (*d*) Monitoring result of the MIC algorithm. (*e*) Monitoring result of the MRTC algorithm. Ground-measured results of (*f*) CBH and (*g*) HV.

Table 3. Comparison of the accuracy of three algorithms.

Algorithm	Case 1 accuracy (1839 UTC 27 Oct 2012)			Case 2 accuracy (1811 UTC 27 Jan 2013)			Case 3 accuracy (1900 UTC 27 Sep 2013)		
	FAR	POD	CSI	FAR	POD	CSI	FAR	POD	CSI
DCD	0.3330	0.7143	0.5263	0.1524	0.7721	0.6938	0.1374	0.7194	0.6457
MIC	0.1216	0.7654	0.6781	0.1053	0.8342	0.7465	0.1513	0.8217	0.7167
MRTC	0.1326	0.7842	0.7035	0.1178	0.8514	0.7643	0.1203	0.8667	0.7748

Algorithm	Case 4 accuracy (1904 UTC 02 Dec 2012)			Case 5 accuracy (1824 UTC 30 Aug 2012)			Mean accuracy		
	FAR	POD	CSI	FAR	POD	CSI	FAR	POD	CSI
DCD	0.1326	0.8093	0.7202	0.2128	0.7255	0.6066	0.1936	0.7481	0.6385
MIC	0.1150	0.8762	0.7867	0.1827	0.8095	0.6855	0.1352	0.8214	0.7227
MRTC	0.1071	0.9021	0.814	0.1837	0.8889	0.7407	0.1323	0.8587	0.7595

where N_F is the numbers of false pixels classified as fog/low stratus, but free in the ground-measured results; N_H is the numbers of hit pixels classified as fog/low stratus in accordance with the ground-measured results; N_M is the numbers of missed pixels classified as fog/low stratus-free, but reported by the ground-measured results.

For the five cases studied, FAR, POD, and CSI using three different algorithms are shown in Table 3. In general, the MIC and the MRTC algorithms achieve much smaller FARs than does the DCD algorithm, and the MRTC algorithm's FAR is slightly smaller than that of the MIC algorithm. For POD and CSI, the MRTC algorithm achieves the largest value, with a minor advantage over the MIC algorithm, whereas POD and CSI of the DCD algorithm are much smaller, generally corresponding to the above-mentioned monitoring results of fog/low-stratus pixels in Figures 3 and 4. Therefore, the three indices show that the accuracy of the MRTC algorithm is slightly better than the existing MIC algorithm and much better than the existing DCD algorithm for detecting night-time fog/low stratus.

5. Conclusions

A new multichannel threshold algorithm based on MRTC for detecting fog/low stratus using night-time NPP/VIIRS data is presented in this study. First, we study the uncertainty ranges of TOA reflectance of fog/low stratus with a variety of factors, such as moon parameters, satellite observation geometry, optical properties of fog/low stratus, surface reflectance, atmospheric profile, and aerosol type. According to the reflectance threshold lookup table from various observation geometries we established, fog/low-stratus pixels with eligible reflectance are detected preliminarily. After the post-processing steps consisting of the removal of snow pixels and high/medium clouds pixels, and the surface homogeneity test, the pixel that meets all the prescribed criteria is ultimately identified as a fog/low-stratus pixel.

To validate the accuracy of the MRTC algorithm we presented, experiments have been performed on five cases of heavy fog/low stratus with more than a half moon in China. The experimental result indicates that the MRTC algorithm achieves about a 0.1323 average FAR, a 0.8587 average POD, and a 0.7595 average CSI. It is concluded that in some example cases, the MRTC algorithm is more accurate in detecting night-time fog/low stratus. In general, the MRTC algorithm has achieved some improvements over the existing two algorithms.

There are, however, still limitations in the MRTC algorithm and several problems to address in future study. For one, similar to the MIC algorithm, the MRTC algorithm does not apply under low lunar illumination. The condition of night-time lunar illumination of more than a half moon restricts the potential operational application of the algorithm. Meanwhile, the reference values of the optical properties of fog/low stratus, such as R_e, COT, and Z_T, are input straight as Bendix's statistical results. It is difficult to determine whether those statistical parameters of fog/low stratus are suitable when applied in the analysis of typical fog/low-stratus cases in China due to the variation of time and space. In addition, uncertainties of TOA reflectance caused by the uncertainty of the simulation model as well as the above-mentioned fog/low-stratus statistics parameters remain in the sensitivity analysis. If we can use a more stable and accurate fog/low-stratus model to establish the reflectance threshold lookup table in future study, the detection results will undoubtedly be more satisfying.

Acknowledgements

The authors thank Dr S. Miller for making his lunar irradiance model publicly available. We also thank Dr V. V. Rozanov and the SCIATRAN working group for making SCIATRAN publicly available. Thanks are extended to the National Oceanic and Atmospheric Administration for making VIIRS products available for download. We appreciate the China Meteorological Administration for validation data from ground-measured meteorological stations. This work was supported by the National Natural Science Foundation of China (Grants 41375029 and 41575028).

Disclosure statement

No potential conflict of interest was reported by the authors.

Funding

This work was supported by the National Natural Science Foundation of China [Grant Numbers: 41375029 and 41575028].

References

Baker, N. 2011. "Joint Polar Satellite System (JPSS) VIIRS Snow Cover Algorithm Theoretical Basis Document (ATBD)." *Northrop Grumman Aerospace System*, Redondo Beach, CA, 1–52. https://jointmission.gsfc.nasa.gov/sciencedocs/2015-06/474-00038_VIIRS_Snow_Cover_ATBD_Rev-_20110422.pdf

Bendix, J. 2002. "A Satellite-Based Climatology of Fog and Low-Level Stratus in Germany and Adjacent Areas." *Atmospheric Research* 64: 3–18. doi:10.1016/S0169-8095(02)00075-3.

Bendix, J., J. Cermak, and B. Thies. 2004. "New Perspectives in Remote Sensing of Fog and Low Stratus-Terra/Aqua-Modis and MSG." In Proceedings of Third International Conference on Fog, Fog Collection and Dew, edited by H. Rautenbach, G2.1–G2.4. CapeTown, South Africa: University of Pretoria.

Bendix, J., B. Thies, T. Nauß, and J. Cermak. 2006. "A Feasibility Study of Daytime Fog and Low Stratus Detection with TERRA/AQUA-MODIS over Land." *Meteorological Applications* 13 (2): 111–125. doi:10.1017/S1350482706002180.

Cao, C., F. J. De Luccia, X. Xiong, R. Wolfe, and F. Weng. 2014. "Early On-Orbit Performance of the Visible Infrared Imaging Radiometer Suite Onboard the Suomi National Polar-Orbiting

Partnership (S-NPP) Satellite." *IEEE Transactions on Geoscience & Remote Sensing* 52 (2): 1142–1156. doi:10.1109/TGRS.2013.2247768.

Cao, C., J. Xiong, S. Blonski, Q. Liu, S. Uprety, X. Shao, Y. Bai, and F. Weng. 2013. "Suomi NPP VIIRS Sensor Data Record Verification, Validation, and Long-Term Performance Monitoring." *Journal of Geophysical Research: Atmospheres* 118 (20): 11,664–11,678. doi:10.1002/2013JD020418.

Cermak, J., and J. Bendix, 2005. "Fog/low Stratus Detection and Discrimination Using Satellite Data." Proceeding of COST722: Midterm Workshop on Short-Range Forecasting Methods of Fog, Visibility and Low Clouds, Langen, Germany, COST, 50.

Cermak, J., and J. Bendix. 2007. "Dynamical Nighttime Fog/Low Stratus Detection Based on Meteosat SEVIRI Data: A Feasibility Study." *Pure Applications Geophysics* 164: 1179–1192. doi:10.1007/s00024-007-0213-8.

Chaurasia, S., V. Sathiyamoorthy, B. Shukla, B. Simon, P. C. Joshi, and P. K. Pal. 2011. "Night Time Fog Detection Using MODIS Data over Northern India." *Meteorological Applications* 18: 483–494. doi:10.1002/met.v18.4.

Crane R. G., and M. R. Anderson. 1984. "Satellite Discrimination of Snow/Cloud Surfaces." *International Journal of Remote Sensing* 5 (1): 213–223. doi:10.1080/01431168408948799.

Dybbroe, A. 1993. "Automatic Detection of Fog at Night Using AVHRR Data." Proceedings of Sixth AVHRR Data Users' Meeting, Belgirate, Italy, 245–252. European Organisation for the Exploitation of Meteorological Satellites.

Ellrod, G. P. 1995. "Advances in the Detection and Analysis of Fog at Night Using GOES Multispectral Infrared Imagery." *Weather and Forecasting* 10: 606–619. doi:10.1175/1520-0434 (1995)010<0606:AITDAA>2.0.CO;2.

Elvidge C D, Baugh K E, Kihn E A, Kroehl H. W., and Davis, E. R. 1997. "Mapping of City Lights Using DMSP Operational Linescan System Data." *Photogramm Engineering Remote Sensing* 63 (6): 727–734.

Eyre, J. R., J. L., Brownscombe and R. J., Allam. 1984. "Detection of Fog at Night Using Advanced Very High Resolution Radiometer (AVHRR) Imagery." *Meteorological Magazine* 113: 266–271.

Gurka, J. J. 1974. "Using Satellite Data for Forecasting Fog and Stratus Dissipation." Preprints, Fifth Conference on Weather Forecasting and Analysis, St. Louis, MO, 54–57. American Meteorological Society.

Hall, D. K., G A Riggs, and V. V. Salomonson. 1995. "Development of Methods for Mapping Global Snow Cover Using Moderate Resolution Imaging Spectroradiometer Data." *Remote Sensing of Environment* 54 (2): 127–140. doi:10.1016/0034-4257(95)00137-P.

Hunt, G. E. 1973. "Radiative Properties of Terrestrial Clouds at Visible and Infra-Red Thermal Window Wavelengths." *Quart Journal Royal Meteorological Society* 99: 346–369. doi:10.1002/qj.49709942013.

Jiang, J., W. Yan, S. Ma, Y. Jie, X. Zhang, S. Hu, L. Fan, and L. Xia. 2015. "Three Cases of a New Multichannel Threshold Technique to Detect Fog/Low Stratus during Nighttime Using SNPP Data." *Weather and Forecasting* 30: 1763–1780. doi:10.1175/WAF-D-15-0050.1.

Ma, S., W. Yan, Y.-X. Huang, W.-H. Ai, and X. Zhao. 2015. "Vicarious Calibration of S-NPP/VIIRS Day–Night Band Using Deep Convective Clouds." *Remote Sensing of Environment* 158: 42–55. doi:10.1016/j.rse.2014.11.006.

Mayer, B., and A. Kylling. 2005. "Technical Note: The Libradtran Software Package for Radiative Transfer Calculations - Description and Examples of Use." *Atmospheric Chemistry & Physics* 5 (7): 1855–1877. doi:10.5194/acp-5-1855-2005.

Miller, S., W. Straka, S. Mills, C. Elvidge, T. Lee, J. Solbrig, A. Walther, A. Heidinger, and S. Weiss. 2013. "Illuminating the Capabilities of the Suomi National Polar-Orbiting Partnership (NPP) Visible Infrared Imaging Radiometer Suite (VIIRS) Day/Night Band." *Remote Sensing* 5 (12): 6717–6766. doi:10.3390/rs5126717.

Miller, S. D., R. E. Turner. 2009. "A Dynamic Lunar Spectral Irradiance Data Set for NPOESS/VIIRS Day/Night Band Nighttime Environmental Applications." *IEEE Transactions on Geoscience & Remote Sensing* 47 (7): 2316–2329. doi:10.1109/TGRS.2009.2012696.

Mosher, F. R. 2013. "Attempting To Turn Night Into Day; Development Of Visible Like Nighttime Images." Proceedings 16th Conference On Aviation, Range, And Aerospace Meteorology, Austin,

TX. American Meteorological Society 5.6. http://commons.erau.edu/cgi/viewcontent.cgi?article= 1023&context=db-applied-aviation

Nakajima, T., and M. Tanaka. 1988. "Algorithms for Radiative Intensity Calculations in Moderately Thick Atmospheres Using a Truncation Approximation." *Journal of Quantitative Spectroscopy & Radiative Transfer* 40 (1): 51–69. doi:10.1016/0022-4073(88)90031-3.

Olivier, J. 1995. "Spatial Distribution of Fog in the Namib." *Journal of Arid Environments* 29: 129–138. doi:10.1016/S0140-1963(05)80084-9.

Pagowski, M., I., Gultepe, and P., King. 2004. "Analysis and Modeling of an Extremely Dense Fog Event in Southern Ontario." *Journal Applications Meteor* 43: 3–16. doi:10.1175/1520-0450(2004) 043<0003:AAMOAE>2.0.CO;2.

Qiu, S., X. Shao, C. Cao, and S. Uprety. 2016. "Feasibility Demonstration for Calibrating Suomi-National Polar-Orbiting Partnership Visible Infrared Imaging Radiometer Suite Day/Night Band Using Dome C and Greenland under Moon Light." *Journal of Applied Remote Sensing* 10 (1): 016024–016024. doi:10.1117/1.JRS.10.016024.

Rozanov, V. V., A. V., Rozanov, A. A., Kokhanovsky, and J. P., Burrows. 2014. "Radiative Transfer through Terrestrial Atmosphere and Ocean: Software Package SCIATRAN." *Journal of Quantitative Spectroscopy & Radiative Transfer* 133 (2): 13–71. doi:10.1016/j.jqsrt.2013.07.004.

Wiscombe, W. J. 1977. "The Delta-M Method: Rapid yet Accurate Radiative Flux Calculations for Strongly Asymmetric Phase Functions." *Journal of the Atmospheric Sciences* 34 (9): 1408–1422. doi:10.1175/1520-0469(1977)034<1408:TDMRYA>2.0.CO;2.

Intercalibration between DMSP/OLS and VIIRS night-time light images to evaluate city light dynamics of Syria's major human settlement during Syrian Civil War

Xi Li, Deren Li, Huimin Xu and Chuanqing Wu

ABSTRACT

Monthly composites of night-time light acquired from the Meteorological Satellite Program's Operational Linescan System (DMSP/OLS) had been used to evaluate socio-economic dynamics and human rights during the Syrian Civil War, which started in March 2011. However, DMSP/OLS monthly composites are not available subsequent to February 2014, and the only available night-time light composites for that period were acquired from the Suomi National Polar-orbiting Partnership satellite's Visible Infrared Imaging Radiometer Suite (Suomi NPP/VIIRS). This article proposes an intercalibration model to simulate DMSP/OLS composites from the VIIRS day-and-night band (DNB) composites, by using a power function for radiometric degradation and a Gaussian low pass filter for spatial degradation. The DMSP/OLS data and the simulated DMSP/OLS data were combined to estimate the city light dynamics in Syria's major human settlement between March 2011 and January 2017. Our analysis shows that Syria's major human settlement lost about 79% of its city light by January 2017, with Aleppo, Daraa, Deir ez-Zor, and Idlib provinces losing 89%, 90%, 96%, and 99% of their light, respectively, indicating that these four provinces were most affected by the war. We also found that the city light in Syria and 12 provinces rebounded from early 2016 to January 2017, possibly as a result of the peace negotiation signed in Geneva.

1. Introduction

The Syrian Civil War, which broke out in March 2011, has been one of the most severe humanitarian disasters since World War II. Human rights groups have estimated that by September 2016, the war had caused about 430,000 deaths and more than 11,000,000 displaced persons (Syrian Observatory For Human Rights 2016). The Syrian Civil War is very complex that there are multiple sides, including the Assad Regime, the Free Syrian Army, Jihad groups (e.g. the Islamic State [IS] and al-Nusra Front), and ethnic minority (e.g. Kurdish forces), with international military intervention from both US-led coalition and Russia. Syria is extremely dangerous for journalists (BBC 2014) and aid workers (Young and Cunningham 2016); therefore, it is difficult to monitor humanitarian conditions in Syria during the war. Remotely sensed imagery, which can record Earth's surface information from outer space, has been proved to be an efficient and objective data source for evaluating armed conflicts and human rights (AAAS 2013; Sulik and Edwards 2010). In particular, night-time light images have played an important role in evaluating the Syrian Civil War (Li and Li 2014; Corbane et al. 2016), because these images can directly capture a measure of human activities with large spatial cover and low cost.

Night-time light images record visible light at night and have been widely used in research fields of economics (Chen and Nordhaus 2011; Li, Xu et al. 2013), human geography (Liu et al. 2012; Pandey, Joshi, and Seto 2013; Yu et al. 2014), energy (Elvidge et al. 2009, 2016), light pollution (Jiang et al. 2017), and medical sciences (Bauer et al. 2013). As night-time light is strongly correlated to the economy in both spatial (Elvidge et al. 1997; Li, Xu et al. 2013) and temporal dimensions (Ma et al. 2012), a decline of night-time light can be used to measure the economic effect of the armed conflict and natural disasters (Witmer and O'Loughlin 2011; Li et al. 2015; Li and Li 2014; Kohiyama et al. 2004; Li, Chen, and Chen 2013).

The study by Li et al. (2015) used night-time light monthly composites from Meteorological Satellite Program's Operational Linescan System (DMSP/OLS) between January 2008 and February 2014 to evaluate the humanitarian crisis in Syria, showing that the spatiotemporal dynamics of human rights conditions in Syria can be retrieved by these images. Following this study, Corbane et al. (2016) evaluated recent dynamics of Syrian Civil War by Suomi National Polar-orbiting Partnership satellite's Visible Infrared Imaging Radiometer Suite (Suomi NPP/VIIRS).

Unfortunately, the National Geophysical Data Center (NGDC) of the USA stopped producing monthly composites of DMSP/OLS after February 2014. The successor product is the Suomi NPP/VIIRS day-and-night band (DNB) monthly composite, which has been systematically produced for the period started at April 2012. Therefore, each data set only covers a part of the Syrian Civil War, but intercalibration between DMSP/OLS and S-NPP/VIIRS data sets for change detection has not previously published. This article aims to evaluate the Syrian Civil War from March 2011 to January 2017 by combining DMSP/OLS and Suomi NPP/VIIRS DNB monthly composites with an intercalibration model.

2. Study area and data

Syria is located in the Middle East, with Turkey, Jordan, Israel, Iraq, and Lebanon as its neighbouring countries. Syria is made up of 14 provincial regions with Damascus city as its capital. This study makes use of three types of data, administrative borders, night-time light images, and land-cover maps. Syria's international and provincial border data, shown in Figure 1, were retrieved from the Global Administrative Areas (http://www.gadm.org/). The night-time light images include DMSP/OLS and Suomi NPP/VIIRS DNB monthly composites. In rest of this article, we use 'DMSP/OLS images' and 'VIIRS images' to represent these two data sets briefly. The DMSP/OLS images with spatial resolution of 0.008333° were ordered from NGDC and have been intercalibrated (Li and Li 2014), and the VIIRS images with spatial resolution of 0.004167° were downloaded from NGDC. Both data sets were retrieved from the panchromatic band, with a wavelength range of 0.5–0.9 μm. The land-cover map of

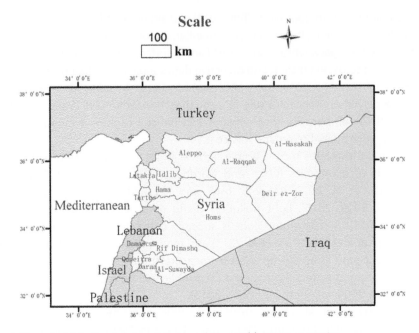

Figure 1. The administrative border of Syria and its neighbouring countries.

Syria for 2010 was retrieved from a global land-cover map in 30 m resolution (GLC30) produced by National Geomatics Centre of China (Chen et al. 2015). An urban proportion map in resolution of 0.008333° for 2010 was derived from the GLC30 using spatial aggregation. From these data, an urban mask was derived by defining an urban pixel as one where the urban proportion is no less than 10%. In this study, we only analyse the city light dynamics in this urban mask, by ignoring human settlement of which the urban proportion is less than 10%; thus, only major human settlement in Syria is analysed in this study.

As designed by NGDC, the VIIRS images have two types of data sets, VIIRS Cloud Mask Configuration (VCMCFG) and VIIRS Cloud Mask Stray-light Corrected Configuration

Table 1. The availability of night-time light images in this study.

	January	February	March	April	May	June	July	August	September	October	November	December
2011	DMSP	DMSP	DMSP	–	–	–	–	–	DMSP	DMSP	DMSP	DMSP
	–	–	–	–	–	–	––	–	–	–	–	–
2012	DMSP	DMSP	DMSP	–	–	–	–	–	DMSP	DMSP	DMSP	DMSP
	–	–	–	NPP	–	–	NPP	NPP	NPP	NPP	NPP	NPP
2013	DMSP	DMSP	DMSP	–	–	–	–	–	DMSP	DMSP	DMSP	DMSP
	NPP	NPP	NPP	NPP	NPP	–	–	NPP	NPP	NPP	NPP	NPP
2014	DMSP	DMSP	–	–	–	–	–	–	–	–	–	–
	NPP	NPP	NPP	NPP	NPP	–	–	NPP	NPP	NPP	NPP	NPP
2015	–	–	–	–	–	–	–	–	–	–	–	–
	NPP	NPP	NPP	NPP	–	–	–	NPP	NPP	NPP	NPP	NPP
2016	–	–	–	–	–	–	–	–	–	–	–	–
	NPP	NPP	NPP	NPP	NPP	–	NPP	NPP	NPP	NPP	NPP	NPP
2017	–											
	NPP											

DMSP denotes the DMSP/OLS images, and NPP represents the Suomi NPP/VIIRS images. For each month, there are two grids showing the availability of the two satellite images. '–' Denotes that the image is not available. We have discarded VIIRS images which do not covered the whole Syria, so VIIRS images were not temporally continuous.

(VCMSLCFG). In this study, we choose VCMCFG which has much longer temporal coverage than VCMSLCFG. For some summer months, the VCMCFG data are unavailable for a part of Syria, and we have discarded the images for 3 months. After the selection, the availability of the DMSP/OLS and VIIRS images is shown in Table 1, and there are 13 months covered by both of the two types of images.

3. Intercalibration between the two data sets

3.1. *Background*

DMSP/OLS has provided the main source for remote sensing of night-time light because NGDC has produced annual composites of DMSP/OLS night-time light for 1992–2013. Radiometric intercalibration of DMSP/OLS composites is an important issue since there is no on-board radiometric calibration on DMSP satellites. The radiometric intercalibration is based on the invariant region method, which assumes there are invariant pixels in multi-temporal night-time light images. These invariant pixels are used as training samples to generate an intercalibration function (Elvidge et al. 2009; Wu et al. 2013; Li, Chen, et al. 2013; Zhang, Pandey, and Seto 2016).

The launch of Suomi NPP brought a new era for night-time light remote sensing, as the VIIRS DNB images have higher spatial and radiometric quality than the DMSP/OLS images (Elvidge et al. 2013). The Suomi NPP was launched at the end of 2011, and NGDC has produced VIIRS DNB monthly composites from April 2012 to present. The NGDC had stopped producing DMSP/OLS monthly composites later than February 2014. The Syrian Civil War, which broke out in March 2011, shows no sign of stopping, and therefore, DMSP/OLS and VIIRS images both cover a part of the war. It is therefore necessary to combine DMSP/OLS and VIIRS images to evaluate the night-time light dynamics of the war. However, because of the differences in the data, the two data sets must be intercalibrated to a consistent data set.

The intercalibration between DMSP/OLS and VIIRS images has received limited prior attention. Shao et al. (2014) calibrated single-day DMSP/OLS images with VIIRS images using Dome C in Antarctic as the calibration site. However, their method was not designed for DMSP/OLS and VIIRS composites which are mixture of daily images.

Differences between DMSP/OLS and VIIRS images are potentially derived from the following differences between the two sensors: (1) the spatial resolutions of original DMSP/OLS and VIIRS images are 742 m and 2.7 km respectively, thus VIIRS images provide more spatial detail; (2) spectral response of the two data are different; (3) the point of spread function of the two sensors are different; (4) overpass time at night for the two satellites is different, with DMSP around 9:30 pm and Suomi NPP around 1:30 am; (5) VIIRS DNB has a wider radiance range than DMSP/OLS so that VIIRS has stronger low light detection ability and therefore, unlike DMSP/OLS, does not suffer from frequent saturation; and (6) there is no on-board calibration for DMSP/OLS resulting in products based on uncalibrated digital numbers, while VIIRS has on-board calibration, and thus VIIRS images are radiance products.

These factors make the two data sets very different in both spatial and radiometric properties. Figure 2(a,b) illustrate the DMSP/OLS and VIIRS composites of Syria for November 2014. Figure 2(c) shows the urban extent map of major human settlement

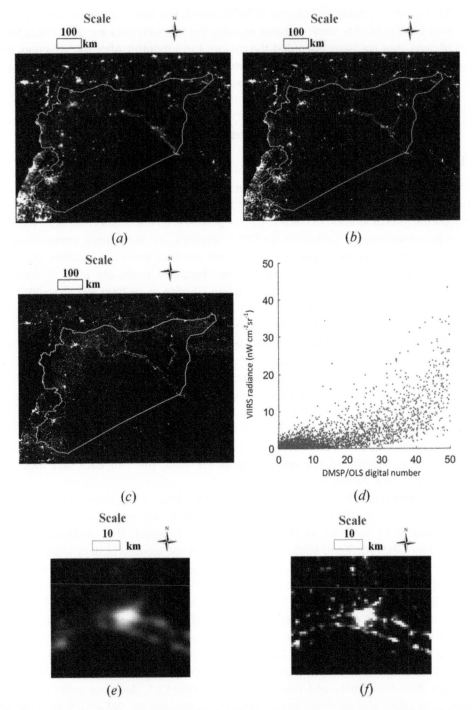

Figure 2. Night-time light images, urban extent map, and related analysis in Syria: (a) DMSP/OLS image of Syria for November 2014, (b) VIIRS image of Syria for November 2014, (c) the urban extent map of Syria for 2010, (d) scatter diagram for the DMSP/OLS and VIIRS images in the urban extent (in fact, the largest values of DMSP/OLS and VIIRS composites are exceeding the bound in the scatter diagram), (e) DMSP/OLS image of Raqqa city for November 2014, and (f) VIIRS image of Raqqa city for November 2014.

with a resolution of 0.008333°. Figure 2(d) shows a scatter diagram for the urban pixels in these two images, where the VIIRS composite in original resolution of 0.004167° is aggregated to the resolution of DMSP/OLS. As we only analyse the city light in this study, we used the above strategy to generate a scatter diagram. Figure 2(d) shows that the DMSP/OLS-VIIRS points for the same month are widely scattered, and their data ranges are very different so that the original DMSP/OLS and VIIRS night-time light images cannot be compared quantitatively. Raqqa city, currently the *de facto* capital of the IS, was used as an example to show the spatial details of the simulated DMSP/OLS image in Figure 2(e,f). Comparing the DMSP/OLS and VIIRS image, the VIIRS image provides more details than the DMSP/OLS image so that they are quite different. In summary, the VIIRS images are quite different from the DMSP/OLS images in both spatial and radiometric dimension.

3.2. Preprocessing the data sets

The spatial resolution of VIIRS composites is 0.004167°, while that of the DMSP/OLS is 0.008333°. Therefore, the VIIRS composites were aggregated to a spatial resolution of 0.008333°. VIIRS, which can detect moonlit reflected by natural land cover such as vegetation, water, and desert, is able to detect lower level of light than DMSP/OLS. We therefore remove VIIRS low light signal that cannot be detected by DMSP/OLS, by simply subtracting a threshold from the VIIRS image. The threshold is set to 0.3 nW cm^{-2} sr^{-1} based on our experience. For the pixels with negative values in the subtracted image, the values are set to 0.

3.3. Nonlinear relationship of DMSP/OLS and VIIRS images

The intercalibration model is based on the temporal overlap of DMSP/OLS and VIIRS images. Three factors should be considered when building the model: (f_1) the nonlinear relationship between the VIIRS radiance and DMSP/OLS digital number, (f_2) spatial degradation of VIIRS images to match DMSP/OLS images, and (f_3) oversaturation occurring in DMSP/OLS composites but not in VIIRS composites.

To investigate (f_1), we should alleviate the effect of (f_2) since the two factors are mixed. The following steps are designed to explore the relationship: (1) for each DN value of a DMSP/OLS image, all pixels in this value are extracted, and the corresponding pixels in the same locations in the VIIRS image for the same month are also extracted; (2) the mean value of the extracted VIIRS pixels was calculated; and (3) for each DN value in the DMSP/OLS image, a corresponding VIIRS value is calculated so that a curve is generated, with DMSP/OLS value as the independent variable and the VIIRS value as the dependent variable. The curve for November 2012 is drawn in Figure 3. This curve is similar to a ridgeline (Zhang, Pandey, and Seto 2016), which can be used to show the general relationship between two data sets.

Figure 3 shows that the two data sets can potentially be fitted by a power function:

$$y = ax^b \tag{1}$$

where *x* denotes the DMSP/OLS value, *y* denotes the VIIRS value, *a* and *b* are coefficients. The form of this function still exists when the variables of DMSP/OLS

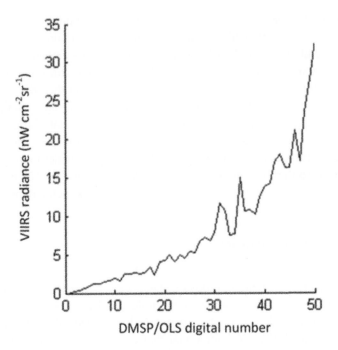

Figure 3. The relationship between DMSP/OLS value and corresponding averaged VIIRS radiance. The DMSP/OLS and S-NPP/VIIRS data are for November 2012.

and VIIRS values are exchanged. We find optimal combinations of the coefficients of the power function to best fit the two data sets for each month by generating a value for coefficient of determination, R^2, and the averaged R^2 value for all the 13 months is 0.6393. In comparison, the linear regression model without a constant, $y = ax$, is not appropriate for the two data sets (we use MATLAB function for modelling and a negative R^2 value comes out, indicating that the model is not appropriate for the data). This analysis suggests that the power function can potentially improve the comparability between the two data sets. Therefore, we use the power function to describe the nonlinear relationship between the VIIRS radiance and DMSP/OLS digital number.

3.4. The intercalibration model

Firstly, we define G as a function for matrix transformation as

$$G(\mathbf{X}, b) = \begin{pmatrix} x_{11}^b & \cdots & x_{1n}^b \\ \cdots & \cdots & \cdots \\ x_{m1}^b & \cdots & x_{mn}^b \end{pmatrix} \quad (2)$$

where b denotes the coefficient in this function, and $\mathbf{X}$ denotes a matrix as the variable:

$$\mathbf{X} = \begin{pmatrix} x_{11} & \cdots & x_{1n} \\ \cdots & \cdots & \cdots \\ x_{m1} & \cdots & x_{mn} \end{pmatrix} \quad (3)$$

Therefore, we define G as a power function for a matrix. Considering (f_1) and (f_2), we simulate a DMSP/OLS image $\mathbf{Y}$ from VIIRS image $\mathbf{X}$:

$$Y = aG(\mathbf{X}, b) * \mathbf{M} \tag{4}$$

$\mathbf{M}$ denotes a normalized matrix which stands for a low pass filter, '*' denotes spatial convolution, a and b are coefficients. Considering (f_3), the simulated DMSP/OLS image should be corrected using the following equation:

$$y'_{i,j} = \begin{cases} y_{i,j} & \text{if } y_{i,j} \leq c \\ c & \text{if } y_{i,j} > c \end{cases} \tag{5}$$

where $y_{i,j}$ denotes value of pixel in location of ith row and jth column in image $\mathbf{Y}$, $y'_{i,j}$ denotes the saturation-corrected pixel value in image $\mathbf{Y}'$, and C denotes the threshold for the saturation. Using this equation, the simulated DMSP/OLS image $\mathbf{Y}$ is corrected to $\mathbf{Y}'$. In this study, we set c to 50 based on our experiences. Equations (3)–(5) are used to simulate a DMSP/OLS image $\mathbf{Y}'$ from VIIRS composite $\mathbf{X}$. If we can get the coefficients a, b and matrix $\mathbf{M}$, the model can be determined.

3.5. The method to estimate the model coefficients

The DMSP/OLS and VIIRS images for the overlapped months are used to construct training data sets to estimate the coefficients in the intercalibration. First, we derive a non-saturated mask for DMSP/OLS image, where the digital numbers are equal or less than c ($c = 50$ as described early). Second, only urban pixels are selected, because the night-time light in nonurban areas tends to be unstable across different time. We multiply the non-saturated mask with the urban extent map, getting a new mask $\mathbf{K}$. $\{\mathbf{D}^{(1)}, \dots \mathbf{D}^{(k)}\}$ denote the DMSP/OLS images in k months, and $\{\mathbf{V}^{(1)}, \dots \mathbf{V}^{(k)}\}$ denote the VIIRS images in the same period. Therefore, our purpose is to estimate coefficients a, b and matrix $\mathbf{M}$ using training data sets $\{\mathbf{D}^{(1)}, \dots \mathbf{D}^{(k)}\}$, $\{\mathbf{V}^{(1)}, \dots \mathbf{V}^{(k)}\}$, and $\mathbf{K}$, with the following steps:

(1) We choose a Gaussian low pass filter for $\mathbf{M}$, as this filter has been widely used for smoothing an image. In this filter, there are two parameters, window size and σ (Fisher et al. 2003). We set window size to 13 pixels. Thus, σ is the only variable for $\mathbf{M}$ so that $\mathbf{M}$ can be rewritten as $\mathbf{M}_\sigma$.
(2) Select an original value for (σ, b) randomly.
(3) We generate a set of matrices $\{\mathbf{Z}_{\sigma,b}^{(1)}, \dots, \mathbf{Z}_{\sigma,b}^{(k)}\}$, where $.\mathbf{Z}_{\sigma,b}^{(i)} = G(\mathbf{V}^{(i)}, b) * \mathbf{M}_\sigma$
(4) Based on Equation (4), we use linear regression analysis to estimate coefficient a from $\{\mathbf{Z}_{\sigma,b}^{(1)}, \dots, \mathbf{Z}_{\sigma,b}^{(k)}\}$, $\{\mathbf{D}_{\sigma,b}^{(1)}, \dots, \mathbf{D}_{\sigma,b}^{(k)}\}$, and mask $\mathbf{K}$: (A) For each pixel which falls into mask $\mathbf{K}$, we select the pixel value pair from $\mathbf{Z}_{\sigma,b}^{(i)}$ and $\mathbf{D}_{\sigma,b}^{(i)}$ ($i = 1, \dots, k$), and then we use all the selected pixel value pairs from k image pairs to construct a training set, which is made up of two vectors $\mathbf{Z}_{\sigma,b}$ and $\mathbf{D}_{\sigma,b}$; (B) we use a linear regression model $d = az$ to fit the set $\mathbf{Z}_{\sigma,b}$ and $\mathbf{D}_{\sigma,b}$, and the root-mean-square deviation (RMSE) of this regression is recorded as $RMSE_{\sigma,b}$.

(5) For coefficients (σ, b), we finally find the optimal solution $(\hat{\sigma}, \hat{b})$ with the lowest RMSE, by repeating step 3–4 and using a fast optimization algorithm. And the coefficient a is estimated as $\hat{a}$ based on $(\hat{\sigma}, \hat{b})$.

3.6. Evaluation of the model

The proposed model makes use of a number of DMSP/OLS and VIIRS image pairs to estimate the coefficients in the intercalibration model, and a simulated DMSP/OLS image is generated from a VIIRS image by using the model. There are totally 13 months when the DMSP/OLS and VIIRS images coexist. A cross-validation process is used to evaluate the model:

(1) In a scenario, image pairs of DMSP/OLS and VIIRS for 12 months are selected to construct a training set, and the remaining image pair is used as the validation set.
(2) For each scenario, we estimate the model coefficients from the training set and generate a simulated DMSP/OLS image from the VIIRS image in the validation set. There are totally 13 scenarios so that there are 13 pairs of DMSP/OLS and simulated DMSP/OLS images.
(3) The two indices, Pearson correlation coefficient, denoted as r, and RMSE, were calculated in the defined urban area for each pair of images. Image pairs before and after intercalibration are compared by the two indices, shown in Table 2.

The RMSE is defined as

$$RMSE = \sqrt{\frac{\sum_{i=1}^{n}(x_i - \hat{x}_i)^2}{n}} \qquad (6)$$

where x_i is the DMSP/OLS image value in the ith urban pixel, $\hat{x}_i$ is the simulated DMSP/OLS image value in the ith urban pixel, and n is the number of urban pixels. After applying the proposed model, the simulated DMSP/OLS image value is calculated from the VIIRS image value. For the original DMSP/OLS and VIIRS images, the simulated DMSP/OLS image value is equal to the VIIRS image value as there was no transformation on the original VIIRS image.

Table 2. Comparison between original and intercalibrated data sets by r and RMSE.

Month	Original data sets		Intercalibrated data sets	
	r	RMSE	r	RMSE
September 2012	0.1979	40.5282	0.9221	4.3740
October 2012	0.2687	26.8691	0.9349	4.5427
November 2012	0.6276	12.5101	0.9180	4.8481
December 2012	0.2649	20.8784	0.8872	4.7852
January 2013	0.1279	40.8895	0.8931	4.5722
February 2013	0.4849	11.7144	0.9169	3.8384
March 2013	0.4659	12.1966	0.9187	4.1357
September 2013	0.3312	20.0306	0.9310	5.5019
October 2013	0.5952	9.8622	0.9261	5.6274
November 2013	0.4592	12.4806	0.9312	6.3140
December 2013	0.3877	12.0908	0.9351	5.4357
January 2014	0.3474	13.3966	0.9071	5.7447
February 2014	0.2602	20.3114	0.9192	5.2365
Average	0.3707	19.5199	0.9158	4.9967

From Table 2, the *r* value for the data set in all 13 months has been improved, from 0.3707 to 0.9158 for the average value. In addition, the RMSE for all the months has also been improved from 19.5199 to 4.9967. These findings indicate that the proposed intercalibration model helps to improve the consistency between the DMSP/OLS and VIIRS images.

To show some details of the intercalibration process, we select the scenario in which November 2014 was selected as the validation set and select the rest of the month as the training set. The DMSP/OLS, VIIRS, and simulated DMSP/OLS image for November 2014 were shown in Figures 2(a,b) and 4(a), respectively, and the scatter diagram showing the relationship between the DMSP/OLS and simulated DMSP/OLS images for the urban area is plotted in Figure 4(b). Comparing Figures 2(d) and 4(b), the simulated DMSP/OLS generated from the VIIRS image is well correlated to the DMSP/OLS image, while the correlation between the original VIIRS and DMSP/OLS image is much weaker. In summary, the intercalibration model has improved the comparability between the DMSP/OLS and VIIRS images.

3.7. Generating a time series night-time light data set for Syria

We use image pairs in all the overlapped 13 months to estimate the coefficients in the intercalibration model and get a solution of $\hat{\sigma} = 1.7142$, $\hat{b} = 0.4436$, $\hat{a} = 11.7319$. We use the intercalibration model and the estimated coefficients to generate simulated DMSP/OLS images after February 2014. Therefore, a time series night-time light data set between January 2011 and January 2017 is generated and denoised with a temporal median filter in size of 3 for further analysis.

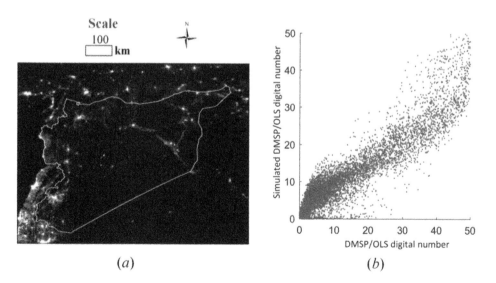

Figure 4. Comparison between simulated DMSP/OLS and DMSP/OLS images. (a) The simulated DMSP/OLS image for November 2014; (b) scatter diagram for DMSP/OLS and simulated DMSP/OLS images in the urban area for November 2014.

4. City light dynamics of major human settlement during the Syrian Civil War

4.1. Analysis at provincial and national level

Based on the time series night-time light data set, we analyse the city light dynamics in major human settlement for Syria (for convenience, 'major human settlement' will be omitted in following content). For each province in Syria, we calculate its total city light (TCL) for each month, which is defined as follows:

$$t = \sum_i x_i u_i \tag{7}$$

where t denotes the TCL, x_i denotes the pixel value for ith pixel, u_i denotes the urban/nonurban value (if ith pixel falls into urban area, $u_i = 1$, otherwise, $u_i = 0$). To analyse the relative change of TCL, March 2011, when the Syrian Civil War broke out, is used as the baseline. For a certain area, the relative TCL in a month is calculated as the ratio of the TCL in the month to TCL in March 2011. The relative TCL for the whole Syria and its 14 provinces were calculated and illustrated in Figure 5.

Figure 5 shows a general decline of city light for Syria and all the provinces between March 2011 and January 2017. The city light has lost by 65–99% for all the provinces in Syria. Among these provinces, Aleppo, Daraa, Deir ez-Zor, and Idlib are the most affected provinces, losing 89%, 90%, 96%, and 99% of their city light during the period, respectively. These four provinces were reported to have major battles and massacres during the civil war. It is very striking to see Idlib was reduced to less than 1% of its city light during April 2015 to January 2017 compared to the pre-war level, suggesting that this province suffered even more than Aleppo.

Although the city light generally declined for each province, the city light in most of the provinces, such as Al-Hasakah, Al-Raqqah, Al-Suwayda, Daraa, Hama, and Homs, rebounded in early 2014. The rebound may be explained by peace talks in January 2014 that resulted in a ceasefire in some regions (CBS 2014). However, the rebound in city light also occurred in Al-Raqqah, which has been controlled by the IS since January 2014. It has been reported that the IS is not only a military group but also an effective ruler which attempts to provide social services including food and electricity, so we can infer that the IS repaired the damaged power infrastructure after it occupied Al Raqqah (Caris and Reynolds 2014). Nevertheless, the city light in Al Raqqah has seen a general decline, corroborating the claim that the IS-controlled area is suffering from an electricity shortage (Syrian Observatory For Human Rights 2014). Although there was a sharp rise of city light of Al Raqqah in October 2015, it fell back within several months, most likely due to Russia and US-led airstrike in Raqqah city and its surrounding areas.

To demonstrate how the city light changes from year to year, we extract the relative TCL for every March from 2011 to 2016, and also for January 2017. From Table 3, we can find that Syria has lost about 60% of its city lights in the first two years of the war, and this pattern also exists for seven provinces, Al-Hasakah, Aleppo, Al-Raqqah, Daraa, Deir ez-Zor. Idlib, Quneitra. Combining Figure 5 and Table 3, the city light for Syria as a whole and 12 provinces rebounded from the early 2016 to July 2016. This recovery of city light may be a result of the reduced violence associated with the peace negotiation, signed in Geneva and started in February 2016, between the Syrian government and a number of rebel groups.

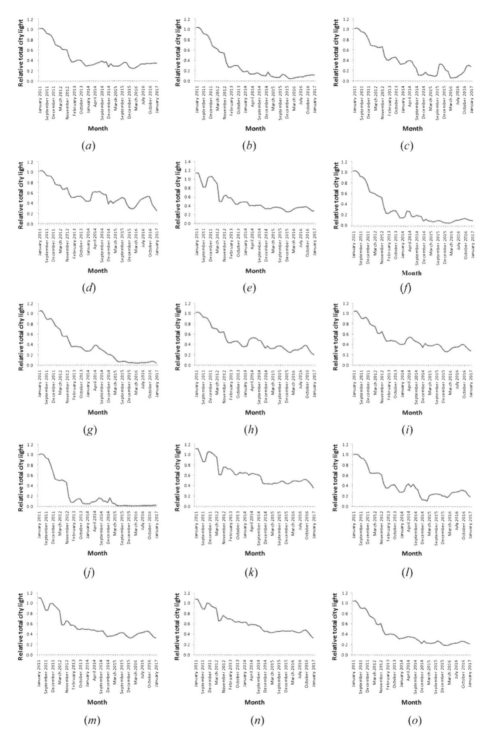

Figure 5. Relative total city light of 14 provinces in Syria between January 2011 and January 2017: (a) Al-Hasakah, (b) Aleppo, (c) Al-Raqqah, (d) Al-Suwayda, (e) Damascus, (f) Daraa, (g) Deir ez-Zor, (h) Hama, (i) Homs, (j) Idlib, (k) Latakia, (l) Quneitra, (m) Rif Dimashq, (n) Tartus, and (o) Syria.

Table 3. The relative total city light for Syria and its provinces.

Province	Relative total city light (%)						
	March 2011	March 2012	March 2013	March 2014	March 2015	March 2016	January 2017
Al-Hasakah	100	66	39	30	28	25	33
Aleppo	100	62	29	13	8	5	11
Al-Raqqah	100	67	43	37	10	5	27
Al-Suwayda	100	74	52	60	42	32	24
Damascus	100	90	51	40	35	31	27
Daraa	100	60	27	27	9	6	10
Deir ez-Zor	100	68	36	30	13	3	4
Hama	100	70	45	49	31	27	19
Homs	100	74	47	50	37	28	27
Idlib	100	49	11	8	5	0	1
Latakia	100	95	67	60	42	48	35
Quneitra	100	64	39	39	19	20	18
Rif Dimashq	100	85	53	47	37	36	32
Tartus	100	90	67	57	43	46	32
Syria	100	70	40	33	22	18	21

4.2. Analysis at city level

Six Syrian cities, Kobani, Aleppo, Homs, Raqqah, Deir ez-Zor, and Hama which suffered from intensive battles, are selected to analyse more spatial details of city light dynamics. For each city, we first find the location of its centre by Google Earth and generate a circle to include the urban extent. Radii of these circles are 3, 20, 8, 8, 8, and 8 km for Kobani, Aleppo, Homs, Raqqah, Deir ez-Zor, and Hama, respectively. In each city, the TCL for each month was calculated and recorded.

The city light dynamics was shown in Figure 6 and Table 4. Among these cities, only Kobani city, located in the Syria–Turkey border, is not the provincial capital, but this city was world-famous for 'Battle of Kobani.' The IS launched the 'Siege of Kobani' on September 2014, resulting in a severe humanitarian crisis. The Kurdish forces, along with Free Syrian Army and international allies, finally win the battle in April 2015. We can see that the city light was steadily declined during March 2011 to April 2015, showing that the long-lasted Syrian civil war and Battle of Kobani had destroyed the city's power supply system, but it rebounded after April 2015, showing that the city was under reconstruction after the IS retreated. Although there were some fluctuations of the city light during the reconstruction, the city light reached to 36% of the pre-war level in January 2017.

For the five provincial capitals, the city light in Aleppo city and Deir ez-Zor city was declined continuously, only remaining 6% and 1%, respectively, comparing to the pre-war level, showing that these two cities suffered from seemingly endless conflicts. For cities of Homs, Raqqah, and Hama, the city light have a general decline, losing 69%, 70%, and 73%, respectively. In addition, there were obvious fluctuations of the city light in these cities such as the three rebounds in Homs, suggesting that the temporary peace occurred and stopped for several time in these cities.

4.3. Comparison on variation of the three types of images

Although the VIIRS, DMSP/OLS, and simulated DMSP/OLS images in time series can track the dynamics of city light in Syria, their variation of dynamics may be different.

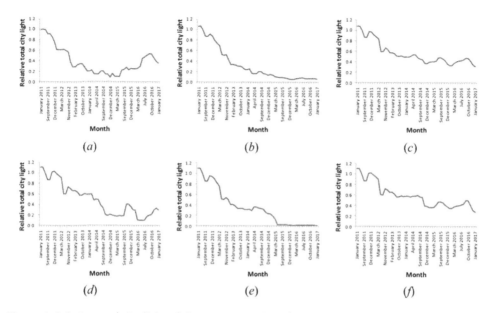

Figure 6. Relative total city light of the six cities in Syria between January 2011 and January 2017: (a) Kobani, (b) Aleppo, (c) Homs, (d) Raqqah, (e) Deir ez-Zor, and (f) Hama.

Table 4. The relative total city light for six Syrian cities.

City	Relative total city light (%)						
	March 2011	March 2012	March 2013	March 2014	March 2015	March 2016	January 2017
Kobani	100	62	33	14	10	25	36
Aleppo	100	70	33	16	10	6	6
Homs	100	84	55	50	40	37	31
Raqqah	100	91	63	48	17	10	30
Deir ez-Zor	100	76	38	29	13	1	1
Hama	100	90	61	56	36	38	27

As there are 13 months when the DMSP/OLS and VIIRS images coexist, the images of these months are used for analysis. We use the intercalibration model in Section 3.7 to generate simulated DMSP/OLS images in the 13 months, and then there are three final data sets, DMSP/OLS, VIIRS, and simulated DMSP/OLS. In addition, we are interested in the simulated DMSP/OLS data set without operation by Equation (5) so that we add a simulated DMSP/OLS data set without saturation for analysis. For a certain region and a data set, we used coefficient of variation (CV) to measure its dynamics:

$$V_t = \frac{\sigma_t}{\mu_t} \qquad (8)$$

where t denotes time series TCL of n months, which can be expressed as $\{t_1, \ldots, t_n\}$, σ_t denotes the standard deviation of t, μ_t denotes the mean of t, and v_t denotes the CV. Therefore, for each region and each data set, a value for CV is derived and shown in Figure 7.

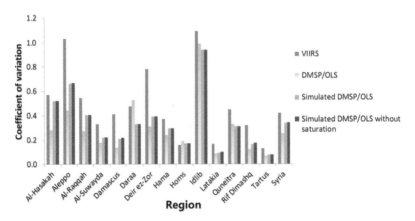

Figure 7. Coefficient of variation for three types of time series night-time light data.

For 14 provinces, there are 12 provinces where the variation of VIIRS data set is larger than that of the DMSP/OLS data set, and Daraa and Homs are two exceptions. In addition, the CV values of VIIRS and DMSP/OLS for Syria are 0.4221 and 0.2535, respectively, indicating that the VIIIRS data set has much higher dynamics than DMSP/OLS data set. For the simulated DMSP/OLS data set, the CV value is between those of DMSP/OLS and VIIRS data sets, with Daraa, Idlib, and Quneitra as exceptions, and the CV value of this data set for Syria is 0.3407, indicating that the dynamics of simulated DMSP/OLS is larger than DMSP/OLS data set and less than the VIIRS data set. This analysis shows that VIIRS data set has larger variation than the DMSP/OLS data set to track the dynamics of city light, which is likely to be explained by the reason that VIIRS has more information than DMSP/OLS. We also found that the simulated DMSP/OLS images without saturation have very similar CV value to the simulated DMSP/OLS images (saturation-corrected images) for all the provinces, but CV value of the former one for Syria is 0.3433, a little larger than that of the latter one (e.g. 0.3407), suggesting that the cutting high radiance values of VIIRS images will result in reduced dynamic information.

5. Conclusion

Our previous study has proved that the night-time light images from DMSP/OLS can play an important role in evaluating the undergoing Syrian Crisis (Li and Li 2014). VIIRS monthly composites have taken the place of DMSP/OLS composites since early 2014, so it is a great challenge to continuously monitor the Syrian Civil War using night-time light images. DMSP/OLS and VIIRS composites are quite different in both spatial and radio-metric dimensions, and how to estimate the city light loss from March 2011 to the present is challenging.

In this study, we made use of a power function and a Gaussian low pass filter to simulate DMSP/OLS images from VIIRS images so that the DMSP/OLS and VIIRS images can be combined to evaluate city light dynamics. However, we have not proved whether or not the power function and Gaussian filter are optimal choices for the intercalibration, and therefore finding an optimal intercalibration model will be an important work in future studies. In addition, the estimated coefficients in the model depend on the training data sets, and therefore the coefficients will change with the change of training data set. As VIIRS images have higher spatial resolution than DMSP/OLS images, they are more effective to detect small objects that emit light at night. However, in this study, we only analyse the TCL of major human settlement in administrative regions or cities, the size of which is not small, so that the advantage of VIIRS is the higher radiometric sensitivity, which is helpful to track the temporal variation of city light.

This study employed an urban mask map to extract city light from the night-time light images. This is a newly developed strategy that was recently used in the Iraq conflict study (Li et al. 2015) but not yet used in previous Syria conflict study (Li and Li 2014). The urban mask can help to exclude ephemeral night-time light (e.g. wildfire), flaring gas light, reflected moonlight, and blooming light in suburban areas, which are not directly related to electricity supply and human rights, so excluding them is useful for monitoring the human activities in Syria. Due to this strategy, the estimated city light, named city light in major human settlement, between March 2011 and February 2014, differs somewhat from that of the previous study (Li and Li 2014), but the temporal trends are similar. This study is the first attempt to intercalibrate DMSP/OLS between VIIRS composites to analyse the night-time light dynamics and the regression analysis shows that the simulated DMSP/OLS and real DMSP/OLS images were consistent. Furthermore, the overpass times for the two satellites are different, likely contributing some errors in the intercalibration method. Therefore, more efforts should be paid to build a more effective and accurate intercalibration model to evaluate night-time light changes from DMSP/OLS and VIIRS images in future studies. However, it is important to mention that some information is missing when VIIRS composites are used to simulate DMSP/OLS composites.

The Syrian Civil War covers a long period from 2011 to present, so that we should use two data sets to evaluate this conflict. More information on the Syrian Civil War will be retrieved if the proposed intercalibration process is combined with the population data (Corbane et al. 2016), and this will be a valuable work. For future conflict, which will be covered only by VIIRS images, the intercalibration between the two data sets is not necessary. On the other side, the annual DMSP/OLS composites are only available for 1992–2013 and VIIRS images will play a unique role for the years after that period; therefore, intercalibration between the annual DMSP/OLS and VIIRS images is particularly valuable to evaluate regional development and urbanization from 1992 to present.

Acknowledgements

The DMSP/OLS and VIIRS images were acquired from National Geophysical Data Center (NGDC) of the USA, and the GLC30 product was acquired from National Geomatics Centre of China.

Disclosure statement

No potential conflict of interest was reported by the authors.

Funding

This research was supported by Key Laboratory of Spatial Data Mining & Information Sharing of Ministry of Education, Fuzhou University: [Grant Number 2016LSDMIS03], the Fundamental Research Funds for the Central Universities: [Grant Number 2042016kf0162], Natural Science Foundation of Hubei Province: [Grant Number 2014CFB726].

References

AAAS. 2013. "Conflict in Aleppo, Syria: A Retrospective Analysis." Accessed 4 May 2017. http://www.aaas.org/aleppo_retrospective

Bauer, S. E., S. E. Wagner, J. Burch, R. Bayakly, and J. E. Vena. 2013. "A Case-Referent Study: Light at Night and Breast Cancer Risk in Georgia." *International Journal of Health Geographics* 12 (1): 1–10. doi:10.1186/1476-072X-12-23.

BBC. 2014. "Syria Most Dangerous Place in the World for Journalists." Accessed 4 May 2017. http://www.bbc.com/news/world-middle-east-28865514

Caris, C. C., and S. Reynolds. 2014. "ISIS Governance in Syria." https://www.understandingwar.org/sites/default/files/ISIS_Governance.pdf

CBS. 2014. "Syria Peace Talks Score Small First Success." Accessed 4 May 2017. http://www.cbsnews.com/news/syria-peace-talks-score-small-first-success/

Chen, J., J. Chen, A. Liao, X. Cao, L. Chen, X. Chen, C. He, et al. 2015. "Global Land Cover Mapping at 30 M Resolution: A Pok-Based Operational Approach". *ISPRS Journal of Photogrammetry and Remote Sensing* 103: 7–27. doi:10.1016/j.isprsjprs.2014.09.002.

Chen, X., and W. D. Nordhaus. 2011. "Using Luminosity Data as a Proxy for Economic Statistics." *Proceedings of the National Academy of Sciences* 108 (21): 8589–8594. doi:10.1073/pnas.1017031108.

Corbane, C., T. Kemper, S. Freire, C. Louvrier, and M. Pesaresi. 2016. *Monitoring the Syrian Humanitarian Crisis with the JRC's Global Human Settlement Layer and Night-Time Satellite Data.* Luxembourg: Publications Office of the European Union.

Elvidge, C. D., D. Ziskin, K. E. Baugh, B. T. Tuttle, T. Ghosh, D. W. Pack, E. H. Erwin, and M. Zhizhin. 2009. "A Fifteen Year Record of Global Natural Gas Flaring Derived from Satellite Data." *Energies* 2 (3): 595–622. doi:10.3390/en20300595.

Elvidge, C. D., K. E. Baugh, E. A. Kihn, H. W. Kroehl, E. R. Davis, and C. W. Davis. 1997. "Relation between Satellite Observed Visible-Near Infrared Emissions, Population, Economic Activity and Electric Power Consumption." *International Journal of Remote Sensing* 18 (6): 1373–1379. doi:10.1080/014311697218485.

Elvidge, C. D., K. E. Baugh, M. Zhizhin, and F.-C. Hsu. 2013. "Why VIIRS Data are Superior to DMSP for Mapping Nighttime Lights." *Proceedings of the Asia-Pacific Advanced Network* 35: 62–69. doi:10.7125/APAN.35.7.

Elvidge, C. D., M. Zhizhin, K. Baugh, F. C. Hsu, and T. Ghosh. 2016. "Methods for Global Survey of Natural Gas Flaring from Visible Infrared Imaging Radiometer Suite Data." *Energies* 9 (1): 14.

Fisher, R., S. Perkins, A. Walker, and E. Wolfart. 2003. "Gaussian Smoothing." Accessed 4 May 2017. http://homepages.inf.ed.ac.uk/rbf/HIPR2/gsmooth.htm

Jiang, W., G. He, T. Long, C. Wang, Y. Ni, and R. Ma. 2017. "Assessing Light Pollution in China Based on Nighttime Light Imagery." *Remote Sensing* 9 (2): 135. doi:10.3390/rs9020135.

Kohiyama, M., H. Hayashi, N. Maki, and M. Higashida. 2004. "Early Damaged Area Estimation System Using DMSP-OLS Night-Time Imagery." *International Journal of Remote Sensing* 25 (11): 2015–2036. doi:10.1080/01431160310001595033.

Li, X., and D. Li. 2014. "Can Night-Time Light Images Play a Role in Evaluating the Syrian Crisis?" *International Journal of Remote Sensing* 35 (18): 6648–6661. doi:10.1080/01431161.2014.971469.

Li, X., F. Chen, and X. Chen. 2013. "Satellite-Observed Nighttime Light Variation as Evidence for Global Armed Conflicts." *IEEE Journal of Selected Topics in Applied Earth Observations and Remote Sensing* 6 (5): 2302–2315. doi:10.1109/JSTARS.2013.2241021.

Li, X., H. Xu, X. Chen, and C. Li. 2013. "Potential of NPP-VIIRS Nighttime Light Imagery for Modeling the Regional Economy of China." *Remote Sensing* 5 (6): 3057–3081. doi:10.3390/rs5063057.

Li, X., R. Zhang, C. Huang, and D. Li. 2015. "Detecting 2014 Northern Iraq Insurgency Using Night-Time Light Imagery." *International Journal of Remote Sensing* 36 (13): 3446–3458. doi:10.1080/01431161.2015.1059968.

Li, X., X. Chen, Y. Zhao, J. Xu, F. Chen, and H. Li. 2013. "Automatic Intercalibration of Night-Time Light Imagery Using Robust Regression." *Remote Sensing Letters* 4 (1): 45–54. doi:10.1080/2150704X.2012.687471.

Liu, Z., C. He, Q. Zhang, Q. Huang, and Y. Yang. 2012. "Extracting the Dynamics of Urban Expansion in China Using DMSP-OLS Nighttime Light Data from 1992 to 2008." *Landscape and Urban Planning* 106 (1): 62–72. doi:10.1016/j.landurbplan.2012.02.013.

Ma, T., C. Zhou, T. Pei, S. Haynie, and J. Fan. 2012. "Quantitative Estimation of Urbanization Dynamics Using Time Series of DMSP/OLS Nighttime Light Data: A Comparative Case Study from China's Cities." *Remote Sensing of Environment* 124: 99–107. doi:10.1016/j.rse.2012.04.018.

Pandey, B., P. K. Joshi, and K. C. Seto. 2013. "Monitoring Urbanization Dynamics in India Using DMSP/OLS Night Time Lights and SPOT-VGT Data." *International Journal of Applied Earth Observation and Geoinformation* 23: 49–61. doi:10.1016/j.jag.2012.11.005.

Shao, X., C. Cao, B. Zhang, S. Qiu, C. Elvidge, and M. Von Hendy. 2014. "Radiometric Calibration Of Dmsp-ols Sensor Using Viirs Day/night Band." *Proceedings of SPIE 9264, Earth Observing Missions and Sensors: Development, Implementation, and Characterization III, 92640A,* Beijing, November 19 . doi: 10.1117/12.2068999.

Sulik, J. J., and S. Edwards. 2010. "Feature Extraction for Darfur: Geospatial Applications in the Documentation of Human Rights Abuses." *International Journal of Remote Sensing* 31 (10): 2521–2533. doi:10.1080/01431161003698369.

Syrian Observatory For Human Rights. 2014. "Darkness Descends over Raqqa." Accessed 4 May 2017. http://syrianobserver.com/EN/News/27921/Darkness+Descends+Over+Raqqa

Syrian Observatory For Human Rights. 2016. "About 430 Thousands Were Killed since the Beginning of the Syrian Revolution." Accessed 4 May 2017. http://www.syriahr.com/en/?p=50612

Witmer, F. D. W., and J. O'Loughlin. 2011. "Detecting the Effects of Wars in the Caucasus Regions of Russia and Georgia Using Radiometrically Normalized DMSP-OLS Nighttime Lights Imagery." *Giscience & Remote Sensing* 48 (4): 478–500. doi:10.2747/1548-1603.48.4.478.

Wu, J., S. He, J. Peng, W. Li, and X. Zhong. 2013. "Intercalibration of DMSP-OLS Night-Time Light Data by the Invariant Region Method." *International Journal of Remote Sensing* 34 (20): 7356–7368. doi:10.1080/01431161.2013.820365.

Young, K. D., and E. Cunningham. 2016. "At Least 12 Aid Workers Killed in Syria Airstrike." *Washington Post.* Accessed 4 May 2017. https://www.washingtonpost.com/world/syrias-7-day-grace-period-ends-with-no-aid-and-cease-fire-in-tatters/2016/09/19/633ecd72-7dea-11e6-ad0e-ab0d12c779b1_story.html

Yu, B., S. Shu, H. Liu, W. Song, J. Wu, L. Wang, and Z. Chen. 2014. "Object-Based Spatial Cluster Analysis of Urban Landscape Pattern Using Nighttime Light Satellite Images: A Case Study of China." *International Journal of Geographical Information Science* 28 (11): 2328–2355. doi:10.1080/13658816.2014.922186.

Zhang, Q., B. Pandey, and K. Seto. 2016. "A Robust Method to Generate a Consistent Time Series from DMSP/OLS Nighttime Light Data." *IEEE Transactions on Geoscience and Remote Sensing* 54 (10): 1–11. doi:10.1109/LGRS.2016.2519241.

Outdoor light and breast cancer incidence: a comparative analysis of DMSP and VIIRS-DNB satellite data

Nataliya A. Rybnikova and Boris A. Portnov

ABSTRACT

Several population-level studies explored the association between breast cancer (BC) incidence and artificial light-at-night (ALAN), and found higher BC rates in more lit areas. Most of these studies used ALAN satellite data, available from the United States Defence Meteorological Satellite Program (US-DMSP), while, in recent years, higher-resolution ALAN data sources, such as Visible Infrared Imaging Radiometer Suite Day-Night Band (VIIRS-DNB), have become available. The present study aims to determine whether the use of different ALAN data sources may affect the BC–ALAN association. As the test case, we use data on BC incidence rates in women residing in the Greater Haifa Metropolitan Area (GHMA; Israel), matching them with US-DMSP and VIIRS-DNB data on ALAN intensities, and controlling for several potential confounders, including age, fertility, and socio-economic status (SES). Both ordinary least squares (OLS) and spatial dependency models were used in the analysis. ALAN emerged as a stronger predictor of BC rates in models based on better-resolution VIIRS-DNB estimates ($t > 6.035$; $p < 0.01$) than in models based on coarser US-DMSP data ($t < 4.196$; $p < 0.01$).

1. Introduction

Recent empirical studies established significant links between artificial light-at-night (ALAN) exposure and several adverse health effects, including decreasing sleep quality and insomnia (Martin et al. 2012; Obayashi et al. 2014), increased alertness (Daurat et al. 2000; Chang et al. 2013), depression (Wallace-Guy et al. 2002; Obayashi et al. 2013), Parkinson disease (Romeo et al. 2013), overweight/obesity (Obayashi et al. 2012; McFadden et al. 2014; Rybnikova, Haim, and Portnov 2016), and elevated risk of hormone-dependent cancers, both in the general population (Haim and Portnov 2013; Hurley et al. 2014; Rybnikova, Haim, and Portnov 2015, 2016a) and in night-time shift workers (Bauer et al. 2013; Schernhammer et al. 2003). In the epidemiological literature, these links are attributed to several impact mechanisms, including the suppression of nocturnal melatonin (MLT) production, circadian disruption, attributed to night-time activities enabled by ALAN, and ALAN acting as a general stressor (Stevens and Rea 2001; Haim and Portnov 2013).

In most population-level studies, satellite data, produced by the United States Defence Meteorological Satellite Program (US-DMSP), were used (Kloog et al. 2009, 2010; Hurley et al. 2014; Rybnikova, Haim, and Portnov 2015). These satellite images cover the period of 1992–2013 and have a relatively coarse, 30 arc-second grid resolution, spanning from −180 to 180 degrees longitude and from −65 to 75 degrees latitude (NOAA 2016b).

In recent years, additional ALAN sources have become available. The Visible Infrared Imaging Radiometer Suite Day-Night Band (VIIRS-DNB) is one of such ALAN data sources, which provides worldwide coverage. VIIRS-DNB images are of a 15 arc-second grid resolution (NOAA 2016a), and, compared to the US-DMSP composites, VIIRS-DNB images have a relatively low light detection limit of near 2×10^{-11} W cm^{-2} sr^{-1}, as opposed to around 5×10^{-10} W cm^{-2} sr^{-1} light detection limit, provided by the US-DMSP composites; unlike US-DMSP, VIIRS-DNB images have no saturation (Elvidge et al. 2013).

The emergence of better-quality satellite images raises a question whether the association between ALAN exposure and its health outcomes, detected by previous studies (cf. *inter alia* Kloog et al. 2009, 2010; Hurley et al. 2014; Rybnikova, Haim, and Portnov 2015), alters if higher-resolution ALAN images are used in the analysis. In this study, we attempt to answer this question empirically by using detailed data on breast cancer (BC) incidence rates obtained for the Greater Haifa Metropolitan Area (GHMA) in Israel.

2. Research variables and data sources

2.1. *Dependent variable*

Data of the areal density of BC cases per km^2 recorded in the year 2012 in the Haifa Bay Area were obtained from the Israel National Cancer Registry and mapped, using the kernel interpolation technique in ArcGIS10.x^{TM} software (see Figure 1).

Owing to the fact that the BC rates deviated from normality (Kolmogorov–Smirnov test value KS>2.709, $p < 0.01$), the original BC rates were transformed using the Box–Cox transformation procedure (Li 2005), to comply with the normality assumption stipulated by multivariate regression analysis (IBM SPSS 2015).

2.2. *ALAN data*

ALAN data for the analysis were downloaded from the NOAA and NASA websites (NOAA 2015 and NASA 2015, respectively). Besides being well established, BC has about a 10 year latency period (Aschengrau, Paulu, and Ozonoff 1998; Hoover et al. 1976). However, first VIIRS-DNB images were released in 2012. To ensure comparability, the year 2012 US-DMSP and VIIRS-DNB data were used in the analysis. We also analysed year 2002 US-DMSP data as a predictor for year 2012 BC rates and found the results being essentially similar to those obtained using year 2012 US-DMSP data. For brevity's sake, the model incorporating year 2002 US-DMSP data are not reported in the following discussion and can be obtained from the authors upon request. Both US-DMSP and VIIRS-DNB ALAN

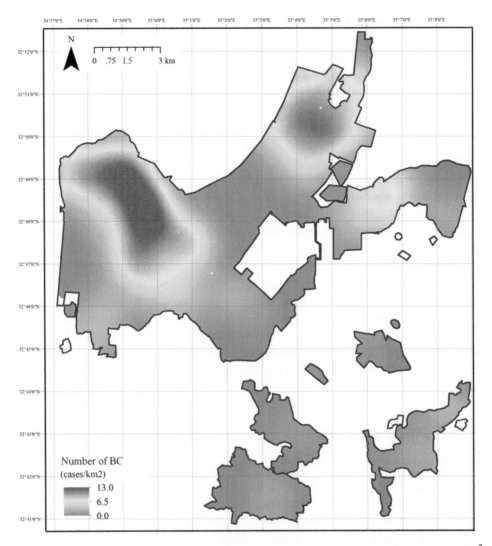

Figure 1. BC incidence rates observed in the study area in 2012 (the number of BC cases per km^2). Bold blue lines define boundaries of localities forming the study area.

levels were normalized, using the standard score technique (Angoff 1984), to ensure comparability.

2.3. Additional explanatory variables

To control for potential confounders, we used several development indicators, such as population density in the neighbourhood, percentage of women over 65 years old, average size of the household, socio-economic status (SES), percentage of Jewish population, as well as distances to the closest main road and to the local industrial zone.

The use of these factors as controls was due to several considerations. In particular, the percentage of women over 65 years old and the percentage of Jewish population in the neighbourhood were used in the analysis as proxies for the number of women in postmenopausal age and for the effect of ethnicity, factors known to influence BC risk (ACS 2016). Average size of the household was considered as an indicator of fertility, a factor also known to be associated with BC incidence (Cowan et al. 1981; Kelsey and Gammon 1990; Parkin 1989; Rosero-Bixby, Oberle, and Lee 1987). SES is a commonly used proxy for population wealth, which reflects differences in lifestyle in general, and differences in diet and healthcare quality, in particular (Hulshof et al. 1991, 2003). High population densities are concomitant with domestic stress (Hall et al. 2005; Han and Naeher 2006), whereas proximities to roads and industries are positively associated with high exposures to air pollutants, which may weaken the immune system and lead to greater disease susceptibility (ACS 2016).

Data on the above control variables were either obtained from the Israel Central Bureau of Statistics or calculated using the ArcGIS10.x^{TM} software (ArcGIS 2015).

The procedure of data linking we used was as follows. First, we converted the satellite images into point data sets, using the *extract multi-values* technique, available in ArcGIS10. x^{TM} software. The extraction provided us with 342 and 215 reference points for the VIIRS-DNB and US-DMSP data sets, respectively (see Figure 2). Next we linked these reference points to the continuous surface of BC rates (see Figure 1) and, then, to the neighbourhood layers of socio-economic data into which individual reference points fall. Distances to roads and industrial areas were calculated using the *join and relate* tool in the ArcGIS10.x^{TM} software. Since some reference points were located outside of the existing residential neighbour-hoods, we obtained 282 and 176 valid observations for VIIRS-DNB and US-DMSP data sets, respectively. Descriptive statistics of the research variables are reported in Table 1.

3. Statistical analysis

The multivariate regressions, we used in the analysis, were given to the following generic equation:

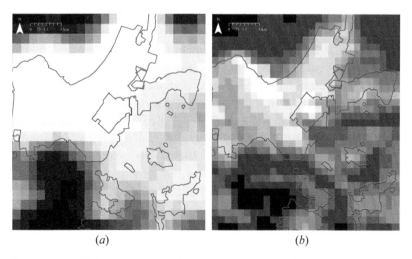

(a)   (b)

Figure 2. Year 2012 ALAN intensities according to US-DMSP (*a*) and VIIRS-DNB (*b*) satellite images. Solid blue lines define the boundaries of localities forming the study area. White pixels mark highly lit areas, dark ones mark low-lit regions; note a high degree of saturation in the left image

Table 1. Descriptive statistics of research variables.

Variable	Minimum	Maximum	Mean	Standard Deviation	Skewness	Kurtosis
Sample used with the US-DMSP ALAN data (Number of observations = 176)						
BC incidence rate in 2012 (cases/km^2)	0.000	13.600	3.523	3.435	0.980	0.011
BC incidence (Box–Cox transformed values, $\lambda = -0.243$)	0.000	1.971	0.996	0.589	−0.133	−1.235
ALAN intensities (dimensionless units)	32.000	63.000	60.157	4.341	−3.091	12.989
ALAN intensities (normalized values)	−7.141	0.594	−0.116	1.083	−3.091	12.989
Population density (persons/km^2)	0.271	20.110	4.173	4.286	1.427	1.447
Women over 65 years old (% of total population)	0.000	36.500	15.541	8.484	0.166	−0.493
Average household size (persons)	1.740	4.440	3.163	0.678	0.468	−0.267
Jewish population (% of total population)	0.000	99.900	79.316	35.610	−1.663	0.950
SES (normalized values)	−1.619	2.883	0.251	1.014	0.563	−0.252
Distance to the closest main road (m)	1.037	1390.854	211.346	228.867	2.265	6.586
Distance to the industrial zone (m)	13.728	6725.761	1986.603	1512.685	1.050	0.532
Sample used with VIIRS-DNB ALAN data (number of observations = 282)						
BC incidence rate in 2012 (cases/km^2)	0.000	14.100	3.377	3.389	1.044	0.157
BC incidence (Box–Cox transformed values, $\lambda = -0.273$)	0.000	1.917	0.946	0.572	−0.108	−1.251
ALAN intensities (dimensionless units)	413.000	763.000	701.126	46.958	−1.805	5.898
ALAN intensities (normalized values)	−6.431	1.229	−0.125	1.028	−1.805	5.898
Population density (persons/km^2)	0.271	22.163	4.206	4.206	1.424	1.683
Women over 65 years old (% of total population)	0.000	41.400	15.606	8.768	0.156	−0.424
Average household size (persons)	1.740	4.840	3.177	0.681	0.485	−0.292
Jewish population (% of total population)	0.000	99.900	78.310	36.323	−1.563	0.627
SES (normalized values)	−1.619	2.883	0.227	1.024	0.585	−0.213
Distance to the closest main road (m)	0.077	1431.757	229.518	268.239	2.040	4.471
Distance to the industrial zone (m)	14.707	6951.688	2074.207	1585.091	1.022	0.344

$$(BC)_i = a_0 + a \times S_i + b \times G_i + c \times (ALAN)_i + \varepsilon, \tag{1}$$

where $(BC)_i$ is breast cancer incidence rates per km^2 in the ith location (Box–Cox transformed values); a_0 is a constant; a, b, and c are regression coefficients; S is variables that account for controls, composed of socio-economic attributes, such as population density (persons/km^2), women over 65 years old (% of population), average size of the household (persons), SES of the neighbourhood population (the values of this measure were obtained from the Israel Central Bureau of Statistics, where they were estimated on a continuous scale, ranging from about −3 to 3, being negative for low-income and positive for high-income areas), neighbourhood share of Jewish population (estimated as % of the total population of the neighbourhood, into which a given pixel falls); G is variables that account for geographical attributes, measuring areal proximity from a pixel to the nearest main road and the main industrial zone (m); (ALAN) is the ALAN levels derived from either US-DMSP or VIIRS-DNB images (dimensionless units); and ε is a random error term.

4. Results

Table 2 reports the results of the analysis, performed separately for VIIRS-DNB and for US-DMSP data, and controlled for potential confounders. Model 1 in Table 2 reports the association between BC and VIIRS-DNB data. In this model, ALAN emerged as a positive and statistically significant predictor of BC rates (regression coefficient $B = 0.286$, $t = 11.887$; $p < 0.01$), which, together with other variables, used in the analysis, helps explain about

Table 2. Factors affecting BC rates in the Great Haifa Metropolitan Area (dependent variable: number of BC cases per km^2, year 2012 data, Box–Cox transformed; method: OLS regression).

	Model 1[g]			Model 2[g]		
Variable	B[a]	t[b]	VIF[c]	B[a]	t[b]	VIF[c]
a_0 (constant)	0.787	(10.051)***	–	0.889	(8.192)***	–
Population density (persons/km^2)	0.045	(8.943)***	1.425	0.068	(9.126)***	1.371
Women over 65 years old (% of total population)	0.010	(4.012)***	1.412	0.002	(0.412)	1.527
Average household size (persons)	−0.036	(−3.261)***	1.124	−0.033	(−2.198)**	1.247
Jewish population (% of total population)	−0.001	(−2.200)**	1.436	−0.001	(−0.657)	1.681
SES (normalized values)	0.061	(3.104)***	1.259	0.099	(3.177)***	1.276
Distance to the closest main road (m)	-3.03×10^{-4}	(−3.870)***	1.438	−0.001	(−3.721)***	1.329
Distance to the industrial zone (m)	9.99×10^{-5}	(2.547)**	12.312	1.27×10^{-5}	(0.197)	12.412
Square of the distance to the industrial zone (m^2)	-1.22×10^{-8}	(−1.830)*	12.783	4.41×10^{-9}	(0.375)	13.823
ALAN intensity[d]	0.286	(11.887)***	1.937	0.158	(4.196)***	2.122
Number of observations[e]	282			176		
R^2	0.738			0.628		
Adjusted R^2	0.729			0.607		
F	(85.015)***			(31.084)***		
SEE[f]	0.297			0.366		
Moran's I	(12.514)***			(10.026)***		

*Indicates a 0.1 two-tailed significance level; **Indicates a 0.05 significance level; ***Indicates a 0.01 significance level. [a] Unstandardized regression coefficient; [b] t-statistic; [c] Variance of inflation; [d] to ensure comparability, ALAN data were normalized using standard score technique; [e] Points within the industrial zone are excluded from the analysis; [f] Standard error of the estimate; [g] Data transformed using the Box–Cox transformation procedure, with λ estimated as −0.243 in Model 1 and −0.273 in Model 2.
Model 1: The year 2012 VIIRS-DNB ALAN data are used as predictor.
Model 2: The year 2012 US-DMSP average visible stable lights levels are used as predictor.

73.8% of the observed BC variance. In Model 2, BC is mutually compared with the US-DMSP ALAN data. As Table 2 shows, the association between BC and US-DMSP is somewhat weaker compared to the VIIRS-DNB–BC association, and the predictive power of the model is lower (coefficient of determination $R^2 = 0.628$), but this association is also positive and statistically significant ($B = 0.158$, $t = 4.196$; $p < 0.01$).

High levels of spatial autocorrelation, reported in Table 2 for ordinary least squares (OLS) models (Moran's $I > 10.0$; $p < 0.01$), naturally occur due to the fact that the observation points are drawn from smooth BC surfaces. The presence of such spatial autocorrelation necessitates the use of spatial dependency models, which take into account the spatial dependency of neighbouring observations (Fotheringham and Rogerson 2009). Table 3 reports the spatial error (SE) models estimated for both VIIRS-DNB (Model 3) and US-DMSP (Model 4) data using the 'Queen' neighbourhood matrix applied to individual pixels of these images. As Table 3 shows, the ALAN–BC association remains statistically significant in the VIIRS-DNB-based model ($t = 6.035$; $p < 0.01$; see Model 3), whereas it is insignificant in the model based on the US-DMSP data ($t = -0.929$; $p > 0.1$; see Model 4). This indicates that, even after controlling for spatial collinearity, the higher-resolution VIIRS-DNB image helps identify a stronger ALAN–BC connection than that detected by the coarser US-DMSP image.

5. Discussion

The present study aimed at assessing the association between BC and night-time light exposure estimates, obtained from two different remote-sensing sources – US-DMSP and

Table 3. Factors affecting BC rates in the Great Haifa Metropolitan Area (dependent variable: number of BC cases per km^2, year 2012 data, Box–Cox transformed; method: spatial error (SE) regression).

Variable	Model 3[g] B^a	Model 3[g] Z^b	Model 4[g] B^a	Model 4[g] Z^b
a_0 (constant)	0.296	(1.003)	0.918	(3.120)***
Population density (persons/km^2)	0.012	(4.104)***	0.018	(3.431)***
Women over 65 years old (% of total population)	0.002	(0.978)	0.003	(1.098)
Average household size (persons)	0.043	(1.601)	−0.022	(−0.480)
Jewish population (% of total population)	0.001	(2.046)**	−0.001	(−0.849)
SES (normalized values)	0.005	(0.365)	0.042	(1.849)*
Distance to the closest main road (m)	-1.39×10^{-4}	(−3.035)***	-2.97×10^{-4}	(−3.463)***
Distance to the industrial zone (m)	1.66×10^{-4}	(3.048)***	1.71×10^{-4}	(2.324)**
Square of the distance to the industrial zone (m^2)	-2.12×10^{-8}	(−2.614)***	-3.66×10^{-8}	(−2.925)***
ALAN intensity[d]	0.121	(6.035)***	−0.034	(−0.929)
Λ	0.965	(69.412)***	0.929	(34.125)***
Number of observations[e]	282		176	
R^2	0.928		0.879	
SEE[f]	0.153		0.204	

*Indicates a 0.1 two-tailed significance level; **Indicates a 0.05 significance level; ***Indicates a 0.01 significance level. [a] Unstandardized regression coefficient; [b] Z-statistic; [d] to ensure comparability, ANAN data were normalized using standard score technique; [e] Points within the industrial zone are excluded from the analysis; [f] Standard error of the estimate; [g] Data transformed using the Box–Cox transformation procedure, with λ estimated as −0.243 in Model 3 and −0.273 in Model 4.
Model 3: The year 2012 VIIRS-DNB ALAN data are used as predictor.
Model 4: The year 2012 US-DMSP average visible stable lights levels are used as predictor.

VIIRS-DNB satellite images. The study goal was to determine whether the association between BC incidence and ALAN, revealed by previous studies, remained coherent if different ALAN data sources are used in the analysis.

According to empirical evidence accumulated to date, exposure to ALAN may increase BC risk due to different impact mechanisms. According to the first and most thoroughly explored causality pathway, ALAN, reaching non-image forming photoreceptors in the human retina, is transferred as neural information to the pineal gland, where it suppresses MLT production, resulting, among other effects, in changing oestrogen receptor affinity and increasing susceptibility to hormone-dependent cancers, such as BC (Schernhammer and Schulmeister 2004; Blask et al. 2011; Haim and Portnov 2013). According to the second potential impact mechanism, humans, being involved in night-time activities, enabled by ALAN, may suffer from daily rhythms disruption, leading to increased susceptibility to various diseases, including BC (Stevens and Rea 2001; Navara and Nelson 2007; Haim and Portnov 2013). According to the third potential impact mechanism, ALAN may work as a general stressor and endocrine disruptor, especially when changes in night-time light intensity are sudden and unexpected (Zubidat, Ben-Shlomo, and Haim 2007; Ashkenazi and Haim 2012; Haim and Portnov 2013).

Several empirical small cohort studies revealed a positive association between BC incidence and ALAN exposure (see *inter alia* Bauer et al. 2013; Kloog et al. 2010; Hurley et al. 2014; Rybnikova, Haim, and Portnov 2015). Being consistent with the results of these studies, the present analysis reconfirms this positive association, showing that this association appears to be largely unaffected by the type of ALAN data used in the analysis.

To the best of our knowledge, this study is the first one testing the BC–ALAN association using ALAN satellite data obtained from different sources and characterized by different quality and spatial resolution. As our study shows, the BC–ALAN association emerges stronger in models based on higher-resolution VIIRS-DNB images than in models based on coarser-resolution US-DMSP data.

Since this study is an epidemiological analysis, it cannot attribute causation to the associations revealed. It is important, however, that the analysis shows consistency in the ALAN–BC association, by demonstrating that the strength of this association increases in line with improving the ALAN data quality. This implies that ALAN does in fact contribute to BC risk and not only correlates with other unrecorded urban parameters, as some researchers (e.g. Kyba and Aronson 2015; Kyba 2016) asserted. However, the ALAN–BC association needs to be investigated further, using specific wavelength intervals of multispectral images, which may become available in the future from the International Space Station (JSC 2016) or from other sources.

Disclosure statement

No potential conflict of interest was reported by the authors.

Funding

This work was supported by the Israel Ministry of Science, Technology and Space.

References

ACS (American Cancer Society). 2016. "What are the Risk Factors for Breast Cancer?" Accessed May 2016. http://www.cancer.org/cancer/breastcancer/detailedguide/breast-cancer-risk-factors

Angoff, W. H. 1984. *Scales, Norms, and Equivalent Scores*. Princeton, NJ: ETS.

ArcGIS (ArcGIS resources). 2015. "About Joining the Attributes of Features by Their Location." Accessed November 2015. http://resources.arcgis.com/EN/HELP/MAIN/10.1/index.html#/About_ joining_the_attributes_of_features_by_their_location/005s00000031000000/

Aschengrau, A., C. Paulu, and D. Ozonoff. 1998. "Tetrachloroethylene Contaminated Drinking Water and Risk of Breast Cancer." *Environmental Health Perspectives* 106 (4): 947–953. doi:10.1289/ehp.98106s4947.

Ashkenazi, L., and A. Haim. 2012. "Light Interference as a Possible Stressor Altering HSP70 and Its Gene Expression Levels in Brain and Hepatic Tissues of Golden Spiny Mice." *Journal of Experimental Biology* 215: 4034–4040. doi:10.1242/jeb.073429.

Bauer, S. E., S. E. Wagner, J. Burch, R. Bayakly, and J. E. Vena. 2013. "A Case-Referent Study: Light at Night and Breast Cancer Risk in Georgia." *International Journal of Health Geographics* 12 (1): 23. doi:10.1186/1476-072X-12-23.

Blask, D. E., S. M. Hill, R. T. Dauchy, S. Xiang, L. Yuan, T. Duplessis, L. Mao, E. Dauchy, and L. A. Sauer. 2011. "Circadian Regulation of Molecular, Dietary, and Metabolic Signaling Mechanisms of Human Breast Cancer Growth by the Nocturnal Melatonin Signal and the Consequences of Its Disruption by Light at Night." *Journal of Pineal Research* 51 (3): 259–269. doi:10.1111/jpi.2011.51.issue-3.

Chang, A. M., F. A. Scheer, C. A. Czeisler, and D. Aeschbach. 2013. "Direct Effects of Light on Alertness, Vigilance, and the Waking Electroencephalogram in Humans Depend on Prior Light History." *Sleep* 36 (8): 1239–1246. doi:10.5665/sleep.289.

Cowan, L. D., L. Gordis, J. A. Tonascia, and G. S. Jones. 1981. "Breast Cancer Incidence in Women with a History of Progesterone Activity." *American Journal of Epidemiology* 114 (2): 209–217.

Daurat, A., J. Foret, O. Benoit, and G. Mauco. 2000. "Bright Light During Nighttime: Effects on the Circadian Regulation of Alertness and Performance." *Neurosignals* 9 (6): 309–318. doi:10.1159/000014654.

Elvidge, C. D., K. E. Baugh, M. Zhizhin, and F. C. Hsu. 2013. "Why VIIRS Data are Superior to DMSP for Mapping Nighttime Lights." *Proceedings of the Asia-Pacific Advanced Network* 35: 62–69. doi:10.7125/APAN.35.7.

Fotheringham, A. S., and P. A. Rogerson. 2009. *The SAGE Handbook of Spatial Analysis*, 528 p. New York: SAGE Publications. ISBN: 978-1-4129-1082-8.

Haim, A., and B. A. Portnov. 2013. *Light Pollution as a New Risk Factor for Human Breast and Prostate Cancers*. Dordrecht: Springer.

Hall, S. A., J. S. Kaufman, R. C. Millikan, T. C. Rickets, D. Herman, and D. A. Sawitz. 2005. "Urbanization and Breast Cancer Incidence in North Carolina, 1995–1999." *Annals of Epidemiology* 15 (10): 796–803. doi:10.1016/j.annepidem.2005.02.006.

Han, X., and L. P. Naeher. 2006. "A Review of Traffic-Related Air Pollution Exposure Assessment Studies in the Developing World." *Environment International* 32: 106–120. doi:10.1016/j.envint.2005.05.020.

Hoover, R., L. A. Gray, P. Cole, and B. MacMahon. 1976. "Menopausal Estrogens and Breast Cancer." *New England Journal of Medicine* 295: 401–405. doi:10.1056/NEJM197608192950801.

Hulshof, K. F., J. H. Brussaard, A. G. Kruizinga, J. Telman, and M. R. Lowik. 2003. "Socio-Economic Status, Dietary Intake and 10 Y Trends: The Dutch National Food Consumption Survey." *European Journal of Clinical Nutrition* 57: 128–137. doi:10.1038/sj.ejcn.1601503.

Hulshof, K. F., M. R. Lowik, F. J. Kok, M. Wedel, H. A. Brants, R. J. Hermus, and F. Ten Hoor. 1991. "Diet and Other Life-Style Factors in High and Low Socio-Economic Groups (Dutch Nutrition Surveillance System)." *European Journal of Clinical Nutrition* 45: 441–450.

Hurley, S., D. Goldberg, D. Nelson, A. Hertz, P. L. Horn-Ross, L. Bernshtein, and P. Reynolds. 2014. "Light at Night and Breast Cancer Risk among California Teachers." *Epidemiology* 25 (5): 697–706. doi:10.1080/07420520802694020.

IBM SPSS Library. 2015. "General linear Modeling in SPSS for Windows." Accessed November 2015. http://www.ats.ucla.edu/stat/spss/library/sp_glm.htm

JSC (Johnson Space Center). 2016. "Gateway to Astronaut Photography of Earth." Accessed May 2016. http://eol.jsc.nasa.gov/

Kelsey, J. L., and M. D. Gammon. 1990. "The Epidemiology of Breast Cancer." *CA: A Cancer Journal for Clinicians* 41: 146–165. doi:10.3322/canjclin.41.3.146.

Kloog, I., A. Haim, R. Stevens, and B. Portnov. 2009. "Global Co-Distribution of Light at Night (LAN) and Cancers of Prostate, Colon, and Lung in Men." *Chronobiology International* 26: 108–125. doi:10.1080/07420520802694020.

Kloog, I., R. Stevens, A. Haim, and B. Portnov. 2010. "Nighttime Light Level Co-Distributes with Breast Cancer Incidence Worldwide." *Cancer Causes & Control* 21: 2059–2068. doi:10.1007/s10552-010-9624-4.

Kyba, C. C. 2016. "Defense Meteorological Satellite Program Data Should No Longer Be Used for Epidemiological Studies." *Chronobiology International* 4: 1–3.

Kyba, C. C. M., and K. J. Aronson. 2015. "Assessing Exposure to Outdoor Lighting and Health Risks." *Epidemiology* 26 (4): e50. doi:10.1097/EDE.0000000000000307.

Li, P. 2005. "Box-Cox Transformations: An Overview." Accessed October 2014. http://www.ime.usp.br/~abe/lista/pdfm9cJKUmFZp.pdf

Martin, J. S., M. Hébert, E. Ledoux, M. Gaudreault, and L. Laberge. 2012. "Relationship of Chronotype to Sleep, Light Exposure, and Work-Related Fatigue in Student Workers." *Chronobiol International* 29 (3): 295–304. doi:10.3109/07420528.2011.653656.

McFadden, E., M. E. Jones, M. J. Schoemaker, A. Ashworth, and A. J. Swerdlow. 2014. "The Relationship Between Obesity and Exposure to Light at Night: Cross-Sectional Analyses of over 100,000 Women in the Breakthrough Generations Study." *American Journal of Epidemiology* 180 (3): 245–250. doi:10.1093/aje/kwu117.

NASA (The National Aeronautics and Space Administration). 2015. "Visible Earth: A Catalog of NASA Images and Animations of Our Home Planet." Accessed November 2015. http://visible earth.nasa.gov/view.php?id=79765

Navara, K. J., and R. J. Nelson. 2007. "The Dark Side of Light at Night: Physiological, Epidemiological, and Ecological Consequences." *Journal of Pineal Research* 43 (3): 215–224. doi:10.1111/jpi.2007.43.issue-3.

NOAA (National Oceanic and Atmospheric Administration). 2015. "Global Radiance Calibrated Night-time Lights." Accessed November 2015. http://ngdc.noaa.gov/eog/dmsp/download_rad cal.html

NOAA (National Oceanic and Atmospheric Administration). 2016a. "Version 1 Nighttime VIIRS Day/ Night Band Composites." Accessed April 2016. http://ngdc.noaa.gov/eog/viirs/download_ monthly.html

NOAA (National Oceanic and Atmospheric Administration). 2016b. "Version 4 DMSP-OLS Nighttime Lights Time Series." Accessed April 2016.http://ngdc.noaa.gov/eog/gcv4_readme.txt

Obayashi, K., K. Saeki, J. Iwamoto, Y. Ikadab, and N. Kurumatania. 2013. "Exposure to Light at Night and Risk of Depression in the Elderly." *Journal of Affective Disorders* 151 (1): 331–336. doi:10.1016/j.jad.2013.06.018.

Obayashi, K., K. Saeki, J. Iwamoto, N. Okamoto, K. Tomioka, S. Nezu, Y. Ikada, and N. Kurumatani. 2012. "Exposure to Light at Night, Nocturnal Urinary Melatonin Excretion, and Obesity/ Dyslipidemia in the Elderly: A Cross-Sectional Analysis of the HEIJO-KYO Study." *The Journal of Clinical Endocrinology & Metabolism* 98 (1): 337–344. doi:10.1210/jc.2012-2874.

Obayashi, K., K. Saeki, J. Iwamoto, N. Okamoto, K. Tomioka, S. Nezu, Y. Ikada, and N. Kurumatani. 2014. "Effect of Exposure to Evening Light on Sleep Initiation in the Elderly: A Longitudinal Analysis for Repeated Measurements in Home Settings." *Chronobiology International* 31 (4): 461–467. doi:10.3109/07420528.2013.840647.

Parkin, D. M. 1989. "Cancers of the Breast, Endometrium and Ovary: Geographic Correlations." *European Journal of Cancer and Clinical Oncology* 25 (12): 1917–1925. doi:10.1016/0277-5379(89)90373-8.

Romeo, S., C. Viaggi, D. Di Camillo, A. W. Willis, L. Lozzi, C. Rocchi, M. Capannolo, et al. 2013. "Bright Light Exposure Reduces TH-Positive Dopamine Neurons: Implications of Light Pollution in Parkinson's Disease Epidemiology." *Scientific Reports* 3. doi:10.1038/srep01395.

Rosero-Bixby, L., M. W. Oberle, and N. C. Lee. 1987. "Reproductive History and Breast Cancer in a Population of High Fertility, Costa Rica, 1984-85." *International Journal of Cancer* 40 (6): 747–754. doi:10.1002/(ISSN)1097-0215.

Rybnikova, N., A. Haim, and B. A. Portnov. 2015. "Artificial Light at Night (ALAN) and Breast Cancer Incidence Worldwide: A Revisit of Earlier Findings with Analysis of Current Trends." *Chronobiology International* 32 (6): 757–773. doi:10.3109/07420528.2015.1043369.

Rybnikova, N. A., A. Haim, and B. A. Portnov. 2016. "Does Artificial Light-At-Night (ALAN) Exposure Contribute to the Worldwide Obesity Pandemic?" *International Journal of Obesity* 40: 815–823. doi:10.1038/ijo.2015.255.

Rybnikova, N. A., A. Haim, and B. A. Portnov. 2016a. "Is Prostate Cancer Incidence Worldwide Linked to Artificial Light at Night Exposures? Review of Earlier Findings and Analysis of Current Trends." *Archives of Environmental & Occupational Health* 1–12. doi:10.1080/19338244.2016.1169980.

Schernhammer, E., and K. Schulmeister. 2004. "Melatonin and Cancer Risk: Does Light at Night Compromise Physiologic Cancer Protection by Lowering Serum Melatonin Levels?" *British Journal of Cancer* 90: 941–943. doi:10.1038/sj.bjc.6601626.

Schernhammer, E. S., F. Laden, F. E. Speizer, W. C. Willett, D. J. Hunter, I. Kawachi, C. S. Fuchs, and G. A. Colditz. 2003. "Night-Shift Work and Risk of Colorectal Cancer in the Nurses' Health Study." *Journal of the National Cancer Institute* 95 (11): 825–828. doi:10.1093/jnci/95.11.825.

Stevens, R. G., and M. S. Rea. 2001. "Light in the Built Environment: Potential Role of Circadian Disruption in Endocrine Disruption and Breast Cancer." *Cancer Causes & Control* 12: 279–287. doi:10.1023/A:1011237000609.

Wallace Guy, G. M., D. F. Kripke, G. Jean Louis, R. D. Langer, J. A. Elliott, and A. Tuunainen. 2002. "Evening Light Exposure: Implications for Sleep and Depression." *Journal of the American Geriatrics Society* 50 (4): 738–739. doi:10.1046/j.1532-5415.2002.50171.x.

Zubidat, A. E., R. Ben-Shlomo, and A. Haim. 2007. "Thermoregulatory and Endocrine Responses to Light Pulses in Short-Day Acclimated Social Voles (Microtus Socialis)." *Chronobiology IntERNATIONAL* 24 (2): 269–288. doi:10.1080/07420520701284675.

Using multitemporal night-time lights data to compare regional development in Russia and China, 1992–2012

Mia M. Bennett and Laurence C. Smith

ABSTRACT

Multitemporal remotely sensed night-time lights data are often used as a proxy for population and economic growth, with China the most commonly researched area. Less is known about how lights respond to socioeconomic decline. Russia, a depopulating neighbour of China that experienced severe economic turmoil following the Soviet Union's disintegration in 1991, provides a useful case study to investigate the relationships between lights, depopulation, and economic contraction at national and provincial scales. We use the U.S. Air Force Defence Meteorological Satellite Program-Operational Linescan System (DMSP-OLS) V4 annual stable lights composites to compare changes in lights in Russia and China from 1992 to 2012. These two countries share a history of communist planning but have experienced divergent development patterns since the collapse of communism in the early 1990s. At the national scale, the total amount of lights in Russia declined between 1992 and 2012, while China's lights more than doubled. At the provincial scale, Russia exhibited an increase in inequality of lights per federal subject, while China's provinces became more equal to one another, particularly as Western China caught up to the more developed East Coast. To understand what may have driven these changes in lights, relationships with population and gross domestic product (GDP) are examined from 2000 to 2012 using panel regression models. While changes in population and GDP explain 81% of change over time in lights within China's provinces, they explain only 6% of change within Russia's provinces. The strong relationships found between changes in lights, population, and GDP in rapidly growing, urbanizing China appear to break down in areas undergoing depopulation and economic contraction.

1. Introduction

Since the late 1990s, remotely sensed night-time lights data captured by the U.S Air Force Defence Meteorological Satellite Program-Operational Linescan System (DMSP-OLS) have been used to estimate socioeconomic parameters (Sutton et al. 2001; Small, Pozzi, and Elvidge 2005; Imhoff et al. 1997; Elvidge et al. 2001). In the late 2000s, a related field of research emerged using multitemporal lights data to approximate changes in these parameters, with the most widely used dataset being the DMSP-OLS

Version 4 Annual Stable Lights composites produced by the National Oceanic and Atmospheric Association's National Geophysical Data Center (NOAA/NGDC). Changes in lights have been found to be significantly and positively associated with changes in population (Archila Bustos et al. 2015; Ceola, Laio, and Montanari 2014; Doll and Pachauri 2010), economic activity, often measured as gross domestic product (GDP) (Henderson, Storeygard, and Weil 2012; Chen and Nordhaus 2011; Weidmann and Schutte 2016), and urbanization (Zhang and Seto 2011; Gao et al. 2015; Tan 2015; Ma et al. 2012), to name some of the most widely studied trends.

If lights are assumed to reflect development, whose components we define for the purposes of our model as a combination of population and economic activity, then changes in lights may be able to approximate changes in levels of development. Given its supposedly global, standardized nature, lights data are arguably useful for examining trends in cross-border and comparative contexts (Henderson, Storeygard, and Weil 2012). But caution is warranted in assuming lights to be a globally standard predictor of population, GDP, or any other socioeconomic variable for that matter. This is due to three main reasons.

First, national-level institutions affect lights (Pinkovskiy 2013). A change in lights in China may not necessarily signal the same change in population or economic productivity as a change in lights in Russia. Second, assumptions about the responses of lights to socioeconomic variables may be biased by the fact that one of the most commonly researched areas is China (Bennett and Smith 2017; Huang et al. 2014), where urbanization and rapid economic growth have been the norm for much of the past two decades (e.g. Yi et al. 2014; Cao et al. 2014; Fan et al. 2014). Even outside of China, DMSP-OLS data are typically used to examine places that are largely growing rather than declining in lights (e.g. Keola, Andersson, and Hall 2015; Álvarez-Berríos, Parés-Ramos, and Mitchell Aide 2013; Zhou, Hubacek, and Roberts 2015). On one level, this makes sense since globally, lighted area increased between 1992 and 2009 (Cauwels, Pestalozzi, and Sornette 2014). Asia in particular has experienced some of the greatest changes in lights (Small and Elvidge 2013). Yet the relationships found between lights and socioeconomic variables in fast-developing places like China may break down in places where development is slower or altogether reversing.

This leads to the third reason why more research is needed into how lights respond across a variety of socioeconomic contexts: there may be limits to using lights to estimate a reversal of development that is characterized by depopulation and economic contraction. While Henderson, Storeygard, and Weil (2012) find that lights respond symmetrically to positive and negative changes in GDP, the dimming of lights in response to economic contraction appears lagged. More insight into the responses of lights to negative socioeconomic dynamics is necessary, for decreasing lights are less understood than increasing ones. A handful of studies specifically examine places where lights are decreasing for reasons like war (Li and Li 2014), light pollution control (Bennie et al. 2014), depopulation (Archila Bustos et al. 2015), and economic contraction (Li, Ge, and Chen 2013). But generally, the relative underuse of DMSP-OLS data for studying such trends represents an unusual gap in the research particularly since sensor issues like saturation can limit the data's practicality for studying brightly lit urban areas. DMSP-OLS cannot resolve further brightening in areas where lights have already reached the sensor's saturation point, when the observed digital number (DN) value reaches 63

(Small, Pozzi, and Elvidge 2005). Thus, DMSP-OLS data may actually prove more valuable for studying places where lights have not yet reached the saturation point, which are often non-urban areas, and places where lights are dimming rather than continuously brightening. However, countries that show unusual relationships between lights and socioeconomic variables like Russia are often simply labelled outliers and discarded from models without further analysis. This is the case in Zhang and Seto (2011)'s study of national-scale differences between normalized urban population and normalized total lights, where the two major outliers of Russia and Greenland are removed in order to substantially improve the model's fit.

Given the lack of attention to lights in areas of socioeconomic decline, this study has two goals. First, it attempts to examine and map how lights change in places known from other socioeconomic data to have declining population and/or economic activity. These negative socioeconomic trends occurred to varying degrees within Russia follow-ing the disintegration of the Soviet Union in 1991, right before the first year for which DMSP-OLS annual stable lights composites are available (1992). Post-Soviet Russia there-fore makes for a useful comparison with China, the most common case study in night lights research. Analysis of these two countries is made more compelling by the two countries' shared border and history of communist planning in the twentieth century. Additionally, although many multitemporal night lights studies compare national-scale dynamics, this study focuses on comparing provincial-scale dynamics between these two countries. Studies at this scale may provide insight into levels and trends of regional development, a topic that concerns the Russian and Chinese national governments. Both Moscow and Beijing have launched campaigns to reduce regional economic disparities partly out of concerns that failure to do so could endanger national stability. While there is a small subfield using lights to estimate regional development and economic activity (Hodler and Raschky 2014; Xu et al. 2015; Doll, Muller, and Morley 2006), few specifically examine these topics in former socialist and communist countries apart from China.

The second goal of this study is to assess whether changes in population and economic activity correspond similarly with lights in Russia and China. To do so, we take advantage of the availability of official provincial-scale population, GDP, and fixed capital investment records from the Russian and Chinese governments from 2000 to 2012 to predict lights using a panel regression model that controls for province and year fixed effects. These records are not available prior to 2000, which is why the panel does not cover the period 1992–1999 despite the availability of lights data.

This article is structured as follows. Section 2 presents a brief overview of political economy in Russia and China from 1992 to 2012. Section 3 reviews our data sources and methodology. Section 4 synthesises our results and discussion in two main subsections. The first main subsection (Section 4.1.) examines changes in lights at national, provincial, and pixel scales. At the national scale, descriptive statistics for lights in Russia and China in 1992 and 2012 are compared and evaluated. At the provincial scale, changes in the spatial distribution of lights are explored using a variety of measures, including changes in the Gini coefficient of provincial lights between 1992 and 2012 as a proxy for changes in regional development. Rates of change and path dependence in provincial lights are also examined both visually and statistically to assess whether relative development levels at the provincial scale within a country become 'locked in' at earlier stages. At the pixel scale, tritemporal night light imagery from 1992, 2002, and 2012 is compared with

high-resolution visible-light satellite imagery from Google Earth to illustrate correspondences between changes in lights and changes in daytime visible light imagery. The
second main subsection (Section 4.5.) discusses the results of the panel regression,
which estimates whether changes in population and GDP predict changes in lights
within provinces. We also examine interactions between regions and GDP to further
assess whether changes in lights reflect differences in regional development. Given the
existence of factors such as development strategies and endemic corruption concentrated at the scale of the region rather than the province, regional interactions with GDP
may prove important. In this section, we also draw on literature from the social sciences
to strengthen and extend our analysis. While this is perhaps unusual for a paper in this
journal, incorporation of this literature helps connect and contextualize remotely sensed
night light imagery with on-the-ground political and economic dynamics. Section 5
offers some conclusions and suggestions for future research, specifically in the way of
using lights data to explore regional development in countries that do not fit the 'China
model' of growth that consists of rapid economic expansion and urbanization.

2. Brief history of socioeconomic dynamics in Russia and China, 1992–2012

After the Soviet collapse, Russia struggled to maintain previously high levels of indus-
trialization during the rapid 'shock therapy' transition to a market economy. Demand for
industrial products and state subsidies to industry dropped substantially, with real
investment in industrial plants, equipment, and infrastructure falling 75% (Peterson
and Bielke 2002). The shrinking of public services hit Russia's northern and eastern
regions, which had been forcibly settled to varying degrees in previous decades,
particularly hard. Hundreds of thousands of people moved away from these areas,
some of which juxtapose China, to the country's relatively more prosperous western
economic centres like Moscow and St. Petersburg. Northern federal subjects (the high-
est-level administrative division within Russia) like Chukotka witnessed severe population declines. Between 1989 and 2002, Chukotka's population dropped 67% to 54,000
people (Thompson 2004). Overall, the national population dropped 3.6% from 1992 to
2012 (World Bank 2016). Although Russia's economy began to grow 7–8% annually
beginning in 2000, the crash in global energy prices during the 2007–2008 global
financial recession brought growth to a halt (Goldman 2008). Russia's post-Soviet
economic expansion has largely relied on the development of oil and gas and a select
few metropolitan powerhouses rather than a more spatially even style of regional
development as has occurred in China. From space, above the many Russian cities,
towns, and factories that have shrunk as the post-Soviet economy and population has
retreated to key urban centres and sites of resource production, DMSP-OLS can detect
declines in lights.

   As the Soviet Union was dissolving in the early 1990s, the Chinese state under the
leadership of Deng Xiaoping initiated economic reforms that ushered in average real
GDP growth of 9.9% per year according to China's National Bureau of Statistics, although
this figure is subject to dispute (Wu 2007). The reforms privileged the eastern coastal
region at the expense of the interior (Holbig 2004). To reduce this regional inequality, in
1999, the government launched a campaign aimed at advancing social and economic
development in China's interior and western regions called 'Open Up the West'

(Goodman 2004b). Two additional programmes called 'Uplifting of Central China' and 'Revitalizing the Northeast' were launched in subsequent years (Dunford and Li 2010). While the extent to which regional inequality has been reduced remains debated, since the mid-1990s, China's interior has received a greater share of project investment, labour-intensive industries have relocated away from the coast, and poverty reduction programmes have been established (Wei 2013). In addition, public investment in infrastructure has transformed the country's landscape, with DMSP-OLS data being used to effectively estimate rising steel use in China (Liang et al. 2014). Skyscrapers, hydroelectric power plants, and industrial facilities have sprung up all across China, emitting light into the sky that DMSP-OLS can capture from space.

3. Data and methodology

3.1. *Study area*

The study area covers the countries of Russia and China from 1992 to 2012 at the national, regional, and provincial scales. At the regional scale, Russia has eight federal districts, referred to in this article as regions. China has four economic planning regions in which various development schemes have been undertaken (Figure 1). At the provincial level, Russia has 83 federal subjects categorized as 46 *oblasts*, 21 republics, nine *krai*, four autonomous *okrugs*, two federal cities, and one autonomous *oblast*. China has

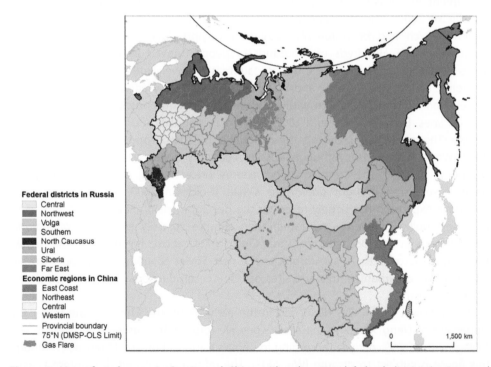

Figure 1. Map of study area in Russia and China with sub-national federal districts/regions and provincial-level administrative divisions illustrated. In this map, the region of Northeast China does not include Inner Mongolia, although five of its prefectures are included in the Chinese government's 'Revitalize the Northeast' regional development plan.

34 provincial-level administrative divisions categorized as 23 provinces (including Taiwan), four municipalities, five autonomous regions, and two special administrative regions (Hong Kong and Macau). For simplicity's sake, all provincial-level administrative divisions in both Russia and China are referred to herein as provinces. Administrative boundaries for Russia and China's subnational entities were downloaded from the Global Administrative Unit Layers (GAUL) 2015 Set developed by the Food and Agriculture Organization (FAO 2014). We include Hong Kong, Macau, and Taiwan within China's 34 provinces but exclude areas whose sovereignty the GAUL database categorizes as disputed (Arunachal Pradesh, Aksai Chin, and a handful of smaller areas). China's provinces vary in size from Macau (the smallest) to Xinjiang (the largest), while Russia's provinces vary from St. Petersburg (the smallest) to the Sakha Republic (the largest), which is slightly smaller than India. The two new federal subjects of Crimea and Sevastopol, which actually bring the total number to 85, are not included in our analysis of Russia as their sovereignty is also disputed. Regardless, Crimea was joined to Russia in 2014, after the end of the study period (2012). As a final note on the study area, portions of Russia's three northernmost provinces (Nenets, Krasnoyarsk, and Sakha) extend above the 75° N limit of DMSP-OLS. Yet since these northern areas account for a fraction of Russia's territory and are largely unpopulated and devoid of economic activity, their omission does not significantly affect our results.

3.2. Night-time lights data

DMSP-OLS was initially developed for daily detection of global cloud cover, but the high night-time sensitivity of its sensor permits detection of illuminated anthropogenic activities like city lights, gas flaring, and fishing fleets (Croft 1978). DMSP-OLS's nightly overpass occurs between 20:30 and 21:30 (Elvidge et al. 2001). Each swath is ~3000 km wide with a nominal resolution of 1 km, resampled from 2.8 km (Doll, Muller, and Morley 2006). During its night-time overpass, OLS can detect radiances down to 1.54×10^{-9} and up to 3.17×10^{-7} W cm^{-2} sr^{-1} μm^{-1} (Cinzano et al. 2000).

With this remotely sensed lights data, NOAA/NGDC produced a dataset of annual stable lights composites from 1992 to 2013 called the DMSP-OLS Version 4 (V4) annual stable lights composites, available freely online at http://ngdc.noaa.gov/eog/dmsp.html. This study uses the V4 annual stable lights composites from 1992 to 2012. The composites capture all persistent lighting emitted each year from −180° W to 180° E and −65° S to 75° N and exclude sunlit and moonlit data, glare, aurora, observations obscured by clouds, and ephemeral events like fires. Data values consist of DNs that range from 0 (no light) to 63 (the maximum value).

Lights data may generally be better suited to detecting negative socioeconomic changes than other remotely sensed data products like land cover. While a significant amount of research compares the two products' abilities to detect increases in land with human presence (Castrence et al. 2014; Small and Elvidge 2013; Xiao et al. 2014), as far as we know, a comparison of their abilities to detect decreases in land use has not been specifically studied. A land cover product would likely not detect any changes in a built-up area even if its associated population had moved away or economic activity had slowed because signatures from buildings, concrete, and asphalt would persist. Lights data, however, can detect a dimming of lights that may be associated with depopulation or deurbanization. Whereas lights detect fluctuations in radiance that can be associated with inward or outward flows of people, capital, or energy, land cover products tend to capture stocks of infrastructure that do not decline right away even if the flows typically associated with them have.

3.3. *Population, GDP, and fixed capital investment data*

Population, GDP, and fixed capital investment statistics for Russia were obtained from the online database of the Russian Federal State Statistics Service (Rosstat) (2014). At the federal subject level, population and fixed capital investment data are available from 1992 to 2012, while GDP data is only available from 1998 to 2012. Rosstat (Federal State Statistics Service 2014) reports GDP and fixed capital investment data in nominal roubles, which are not adjusted for inflation. These records were subsequently converted into 2005 US dollars using annual adjustment ratios from the World Bank (2015).

Population, GDP, and fixed capital investment data for China were obtained from the online database of the National Bureau of Statistics of China (2015). Regional-level data is available for all provinces except Taiwan, Hong Kong, and Macau, which is why they are excluded from our panel. GDP and fixed capital investment data at the provincial level are available from 1995 to 2012, while population data is only available from 2000 to 2012. As with Russia, Chinese GDP and fixed capital investment records were converted from nominal yuan into 2005 US dollars according to annual adjustment ratios from the World Bank (2015). Since population, GDP, and lights data are only available for all provinces in Russia and China from 2000 to 2012, this study's panel is limited to that period.

3.4. *Image processing and geostatistics*

To account for the fact that six different satellite pairs (F10, F12, F14, F15, F16, and F18) captured NTL imagery from 1992 to 2012, DMSP-OLS V4 composites were intercalibrated using an ordinary least squares (OLS) regression model developed by Elvidge et al. (2014). In principle, intercalibration allows for more accurate interannual comparison. The regression model applied is

$$\text{DNadjusted} = C_0 + C_1 \times \text{DN} + C_2 \times \text{DN}^2, \tag{1}$$

in which C_0, C_1, and C_2 are regression coefficients provided in Elvidge et al. (2014), DN is the value of the pixel in the uncalibrated composite, and DNadjusted is the adjusted pixel value. Elvidge et al. (2014) determine that F121999 has the brightest average values, so all other composites are intercalibrated to match its range. For our choice of an invariant area for the regression, since Russia and China cover approximately 17% of the world's land area, we follow Elvidge et al. (2014)'s global-scale regression in using Sicily.

In ArcGIS 10.3.1, the resulting intercalibrated rasters were clipped to Russia and China's spatial extents based on the GAUL boundaries. Areas with gas flaring, an externality associated with oil and gas production, were clipped from the spatial extents, as their inclusion can introduce inaccuracies into examinations of the relationship between lights and socioeconomic variables (Elvidge et al. 2009). Gas flares were removed based on shapefiles downloaded from NOAA at http://ngdc.noaa.gov/eog/interest/gas_flares.html. Gas flares were particularly prominent in the Russian oil and gas producing provinces of Khanty-Mansiysk, Yamalo-Nenets, and Nenets. It should be noted that these shapefiles do not exactly conform to gas flares in years before and after 2008, which may potentially introduce some inaccuracies into the results for areas known to have gas flaring whose levels have varied considerably over the years.

Next, each clipped raster was reprojected to the Asia North Albers Equal Area projection using the nearest neighbour resampling method and a 1 km cell size. Pixels whose DN <3 were reclassified to 0 and pixels assigned a DN >63 after intercalibration were reclassified to 63. Next, lights were summed for each province in Russia and China, generating total lights per province for every year from 1992 to 2012. For years in which two annual stable lights composites exist due to overlapping satellite records (1994, 1997, 1998, 1999, 2000, 2001, 2002, 2003, 2004, 2005, 2006, 2007, 2008, 2009), the mean total lights of the two composites was calculated as the value for that year. In the F142002 composite, many northern provinces had total lights values of 0 due to its apparent failure to include data for a large part of the Northern Hemisphere. The values from the F152002 value were thus retained as the provincial lights total rather than averaged with the zero values from the F142002 composite. This same procedure was applied for a handful of other instances where provinces were coded as having total lights values of 0, notably Nenets and Murmansk in F141997, F141998, F141999, and F142001. This issue may suggest some underlying defect with the F14 satellite in detecting lights near the northern polar latitudes that requires further investigation.

3.5. *Methods of analysis*

We used a combination of statistical analysis and visual inspection to determine national, provincial, and pixel-scale change in lights in Russia and China between 1992 and 2012. First, national-scale descriptive statistics were calculated to highlight changes in the intensity of lights in 1992 and 2012 across various categories of DNs: 0 (representing unlit area), dimly lit areas of DN = 4–6 and 6–10, brightly lit areas of DN = 11–20 and 21–62, and top-coded (saturated) areas of DN = 63. These follow the categories established in Henderson, Storeygard, and Weil (2012). Second, to visualize the differences in lights between the first, middle, and final years of the 1992–2012 time period, the F101992, F152002, and F182012 composites were combined into a single three-band raster in ArcGIS 10.3.1. This resulted in a pixel-scale tritemporal map of lights for Russia and China. This visualization method is based on Small and Sousa (2016), who use OLS composites from 1992, 2002, and 2012 to highlight change in lights in the Himalayas. In Figure 3(c), 1992 values were input into the blue channel, 2002 into the green channel, and 2012 into the red channel, with the raster displayed using a standard deviation stretch (n = 2.5). The resulting tritemporal map was visually inspected for Russia and China to obtain a sense of the geography of changes in lights in each country over the two-decade period.

To examine change in lights in Russia and China from 1992 to 2012, time-series plots of total annual lights at the national level and provincial scale were created. In the provincial lights time series, the natural log (ln) of total lights was used to facilitate comparison between provinces with widely varying amounts of light, particularly in

Russia. After each country's provinces with the largest and smallest log differences (a measure similar to finding the total percentage difference) in total light between 1992 and 2012 were identified, specific areas of decreasing, stable, and increasing lights were located in high-resolution Google Earth imagery. One image from as early as possible in this time period (2000-2007) was compared with an image from as late as possible in this time period (2011-2014). This provided insight into what decreasing, stable, or increasing lights represent on the ground. In future, further validation of the ability of DMSP-OLS V4 stable lights composites to detect areas experiencing socioeconomic decline, especially through comparison with daytime visible light imagery, is recommended.

At the provincial scale, various measures were examined to explore how lights changed in Russia and China between 1992 and 2012. First, using the *ineq* package in RStudio 0.99.902, the Gini coefficient of light inequality was calculated to estimate whether a country's provinces had grown more or less equal in lighting from 1992 to 2012. A decrease in the Gini coefficient corresponds to an increase in equality in total lights per province and suggests a trend of more spatially even national development. By contrast, an increase in the Gini coefficient corresponds to a decrease in equality in total lights per province and suggests a trend of more spatially uneven national development. Second, to assess path dependence in provincial lights, Pearson's *r* and Kendall's rank correlations were calculated for both logged and unlogged total lights per province in 1992 and 2012 in Russia and China.

Additionally at the provincial scale, using a variety of linear and panel regression models, the relationships between the dependent variable, lights per province, and the two independent variables, population and GDP, were examined from 2000 to 2012. All three variables were transformed using natural logarithms due to their positively skewed distributions across provinces in Russia and China. While other studies take lights per area as their unit of analysis (Hodler and Raschky 2014; Michalopoulos and Papaioannou 2014), we did not do this in our study since we only examine within-province change.

First, a multiple linear regression model using ordinary least squares (OLS) was tested to see if population and GDP affect lights. We also initially considered using fixed capital investment in our model, but it was shown to not add much explanatory power so the variable was discarded. The resulting model is written as

$$\text{Lights}_i = \beta_0 + \beta_1 P_{i1} + \beta_2 G_{i2} + \varepsilon_i, \tag{2}$$

in which 'Lights' is the dependent variable, logged lights, where i = province; β_0 is the intercept; P is logged population; G is logged GDP; β_1 and β_2 are coefficients for their respective independent variables, and ε_i is the standard error.

Next, using the *plm* package in R, a one-way fixed effects model was tested separately for Russia and China, which assumes that unobserved heterogeneity within provinces is time-invariant. The model is written as

$$\text{Lights}_{i,t} = \beta_1 P_{i,t} + \beta_2 G_{i,t} + a_i + u_{i,t}, \tag{3}$$

in which Lights$_{i,t}$ is the dependent variable, logged lights, where i = province and t = time; P is logged population; G is logged GDP; a_i is the unknown intercept for each province, β_1 and β_2 are coefficients for their respective independent variables; and $u_{i,t}$ is the error term plus the unobserved, time-invariant characteristics of regions (fixed effects). An F-test demonstrates that the fixed effects model offers an improvement over the multiple linear regression model ($p < 0.05$) for both Russia and China. Although simple and multiple linear regression models are a common method of exploring relationships between lights and socioeconomic variables (e.g. Sutton, Elvidge, and Ghosh 2007; Zhang and Seto 2011; Zeng et al. 2011; Ghosh et al. 2009; Shi et al. 2014), a fixed effects model more accurately controls for unobserved effects in multi-temporal data (Henderson, Storeygard, and Weil 2012; Chen and Nordhaus 2011). Furthermore, since checking for errors among the variables revealed that the data is heteroscedastic, the assumption of homogeneity of variance could be relaxed with the use of a fixed effects model.

Next, within-province variation over time in lights and socioeconomic variables was examined separately for Russia and China using the fixed effects model expressed in Equation (3). The panel data for China is balanced with 31 provinces and 403 observations. The panel for Russia is slightly unbalanced with 83 provinces and 1079 observations, as GDP data is missing for Chechnya from 2000 to 2004. Due to the poor fit of the model in Russia, it was run again separately for the 62 provinces that decreased in population between 2000 and 2012 and the 21 provinces that increased in population. As the model still showed a generally poor fit for both groups, the model was tested again on all observations in Russia and China with the addition of year fixed effects, which were introduced as a dummy variable into the model with 2000 as the reference year.

Finally, to explore whether there is a specific regional geography to the way in which GDP affects lights in Russia and China, time-invariant regional effects were interacted with GDP. The model that examines the effects of regional interactions with GDP on lights in addition to the original predictors of population and GDP is written as

$$\text{Lights}_{i,t} = \beta_1 P_{i,t} + \beta_2 G_{i,t} \times R + a_i + u_{i,t}, \tag{4}$$

in which $\text{Lights}_{i,t}$ is the dependent variable, logged lights, where i = province and t = time; P is logged population; G is logged GDP; a_i is the unknown intercept for each province; R represents the dummy variable for each region, β_1 and β_2 are coefficients for their respective independent variables; and $u_{i,t}$ is the error term plus the unobserved, time-invariant characteristics of provinces (fixed effects).

4. Results and discussion

4.1. National-level change in lights

Since the disintegration of the Soviet Union and the end of the Cold War in 1991, Russia and China have experienced dramatically divergent development patterns. This is readily apparent in analysis of night light imagery. In 1992, Russia had more total lights than China, but by 2002, China had more lights than Russia (Figure 2(a,b)), 9 years after its GDP in current US dollars surpassed Russia's (World Bank 2017). Strikingly, Russia actually had fewer lights in 2012 than it did in 1992, whereas China's lights more than doubled during this period. In other words, Russia's lights had not recovered by 2012 to their directly post-Soviet levels (Table 1).

Between 1992 and 2012, lights in China spread out across the country and intensified, whereas neither is particularly true for Russia. The percentage of unlit area in China dropped

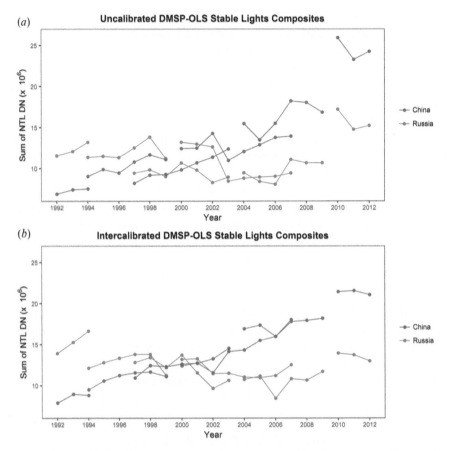

Figure 2. Annual total lights from 1992 to 2012 for Russia and China before intercalibration (a) and after intercalibration (b), following Equation (1).

from 92% in 1992 to 82% in 2012, exemplifying the spread of anthropogenic presence and pressure on land across Earth since the early 1990s (Geldmann, Joppa, and Burgess 2014). The percentage of lights in all other categories increased, too. In Russia, unlit area only declined slightly from 95% to 93%. The percentage of top-coded pixels in Russia increased, which may indicate growth in extent and intensity of lights in the primate city of Moscow and other major metropolitan centres including St. Petersburg, Yekaterinburg, and Novosibirsk, the latter two of which are located in Siberia (Figure 3(a–c)). The difference in the percentage of top-coded pixels between 1992 and 2012 in Russia, however, is much smaller than in China. Notably, Russia experienced a decline in the percentage of lights within the two brightest categories of pixels (DN = 11–20 and DN = 21–61/60[1]), which are typically associated with urban areas. This may explain why even though a larger extent of Russia's area was illuminated in 2012 compared to 1992, the country's total lights (in other words, the combined intensity of all lights) dropped. A substantial number of secondary cities like Khabarovsk were brighter in 1992 than in 2012 as depicted in Figure 3(c), which may reflect a national trend of urban shrinkage. According to Rosstat population data, eight of Russia's 13 cities with over 1 million people in 2002 have decreased in size since 1989 (Molodikova and Makhrova 2007). The apparent dimming of many of the country's cities

Table 1. Descriptive statistics for lights in Russia and China, 1992 and 2012.

	Russia		China	
Attribute	1992	2012	1992	2012
Proportion of country with 0 DN (Unlit area) (%)	94.95	92.66	91.09	82.12
4–5 DN (%)	0.09	3.48	3.93	5.94
6–10 DN (%)	1.77	3.35	2.06	6.87
11–20 DN (%)	2.01	0.90	1.14	2.25
21–61[a]/60[b] DN (%)	1.15	0.77	0.94	2.68
62[a]/61[b] DN (%)	0.02	0.05	0.02	0.12
Sum of all lights (DN)	13,933,462	13,021,075	7,942,838	21,089,640
Gini (provincial lights)	0.44	0.45	0.46 (0.44***)	0.40 (0.38***)
	1992 vs. 2012		1992 vs. 2012	
Pearson's r (lights)	0.95***		0.93*** (0.90***)	
Pearson's r (ln lights)	0.94***		0.90*** (0.90***)	
Kendall's rank	0.79***		0.71*** (0.70***)	

[a]Top-code for intercalibrated composite in 1992.
[b]Top-code for intercalibrated composite in 2012.
Number in parentheses is value with Macau excluded.
***$p < .001$.

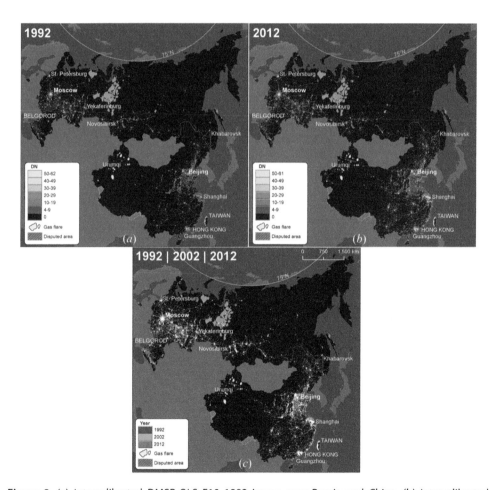

Figure 3. (a) Intercalibrated DMSP-OLS F10 1992 image over Russia and China. (b) Intercalibrated DMSP-OLS F18 2012 image over Russia and China. (c) Tritemporal composite showing F10 1992, F15 2002, and F18 2012 images over Russia and China. Gas flares were removed from all images following the method described in Section 3.4. Projection: Asia North Albers Equal Area Conic.

outside of Moscow and St. Petersburg along with the 1% increase in the country's rural population between 1992 and 2000 (Pivovarov 2003) may also help explain why changes in urban population in Russia do not correspond to changes in total lights in the same manner as most other countries' cities (Zhang and Seto 2011), though further research at the urban scale is necessary to explore this.

The tritemporal lights map visualizes differences in lights in Russia and China in 1992, 2002, and 2012 (Figure 3(a–c)). White pixels represent stable areas of relatively similar brightness in each of the 3 years. Blue pixels represent areas brighter in 1992 than in 2002 or 2012, suggesting decline. Green pixels represent areas brighter in 2002 than at the start or end period, suggesting growth around the new millennium that has since tailed off. Red pixels represent areas brightest in 2012, suggesting recent growth. Visual inspection of the tritemporal lights map highlights the geography of changes in lights within each country. The swath of green and red pixels along the coast between Beijing and Shanghai conveys recent growth at the outset of the twenty-first century. In southern China, the numerous concentrations of red pixels eventually converge on the Pearl River Delta, which includes cities such as Guangzhou and Hong Kong. Although the Pearl River Delta is the world's fastest growing region (Su et al. 2015), the conurbation appears relatively white, which may due to the sensor not being able to detect further brightening beyond the saturation level. The most obviously red pixels are in western China around cities like Hohhot and Ordos in Inner Mongolia and Urumqi in Xinjiang. A scattering of blue and purple pixels are concentrated northeast of Beijing in the provinces of Liaoning and Jilin, where the proliferation of decaying industries since the 1970s has led to pronouncements of a 'Rust Belt' (Li 1996). Areas represented by blue pixels were brightest in 1992, whereas purple ones may have recently recovered to near 1992 levels following a decrease in 2002.

Russia is characterized by an extensive spread of blue pixels from west to east, signifying the dimming of lights since 1992. These pixels follow the Trans-Siberian Railway and spread outwards from the transportation corridor to numerous surrounding settlements. The highest concentrations of green pixels, which indicate brightening during the early years of Russian President Vladimir Putin's administration, which commenced in 2000, appear within the Volga region, the southern Caucasus, and in north-west Russia around the Arctic port city of Murmansk. Red pixels are most obvious around the edges of Moscow and St. Petersburg, in the province of Belgorod, which has pursued intensive modernization of its agricultural sector since 2000 (Efendiev, Sorokin, and Kozlova 2014), and northwest Siberia, where the oil industry has expanded in recent years into places like the Yamal Peninsula. Lights in parts of the Northern Caucasus appear purple, suggesting that lights were dimmer in 2002 relative to 1992 but have since began to recover. The dimming of lights in 2002 in the conflict-ridden Caucasian republics of Dagestan and Ingushetia can possibly be explained by the First and Second Chechen Wars (1994–1995 and 1999–2000, respectively), which reduced cities like Grozny to rubble. Multi-billion dollar rebuilding efforts in recent years may likely explain the brightening apparent in 2012.

4.2. *Regional development and provincial-scale changes in lights*

The Russian and Chinese governments have sought to reduce disparities in regional development, with the latter more serious about tackling the issue than the former. In

China since 2004, interregional inequality has declined thanks to convergence in provincial economic growth rates (Fan and Sun 2008). Yet in Russia, the overall speed of regional convergence from 1998 to 2006 was very low by international standards, although high-income regions located in close proximity to one another exhibited strong regional convergence (Kholodilin, Oshchepkov, and Siliverstovs 2009). The Kremlin has announced policies of regional development over the years ranging from a federal programme called 'Reducing differences in socioeconomic development of the regions of the Russian Federation (2002–2010)' to more recent efforts to develop eastern Russia through the establishment in 2012 of the Ministry for Development of the Russian Far East. As Russia's relationships with Europe fracture, the government has set its sights on improving Russia's political and economic ties with Asia partly via the development of its regions closest to China, Japan, and South Korea. However, the majority of any related gains in the Russian Far East's development that could potentially be approximated by lights would occur after 2013, when production of DMSP-OLS V4 composites ceased. Night-lights based evaluation of the success of eastern Russia's development thus requires use of the newer Suomi National Polar-orbiting Partnership data, available since late 2011.

Existing findings of increased regional convergence in China and stagnant or decreasing regional convergence in Russia are corroborated by the results we generated using a variety of lights-based measures. The first is the Gini coefficient of lights, whose use to date appears limited to comparisons at the international level (Cauwels, Pestalozzi, and Sornette 2014; Henderson, Storeygard, and Weil 2012). In 1992, Russia and China (Macau excluded) had identical Gini coefficients of lights across provinces of .44, respectively. Results for China are examined both with and without Macau since the province is extremely small in size at ~31 km^2, densely populated, and highly developed, making it particularly subject to saturation. By 2012, Russia's Gini coefficient had increased slightly to .45. China's, however, had decreased to .38 (Macau excluded). This indicates an evening of light levels between Chinese provinces and a possible increase in spatially even national development. In Russia, by contrast, the increase in inequality in lights between provinces may be due to the intensification and outward extension of lights in Moscow and St. Petersburg and the dimming of lights in many of the country's peripheral provinces.

The lack of convergence in lights in Russia's provinces between 1992 and 2012 prompts an assessment of path dependence, or whether total lights in 1992 determined total lights in 2012. This was explored using various correlation measures of logged and unlogged lights. The estimated Pearson's r correlation coefficients vary slightly depending on whether the data is logged (Table 1). This point is worth mentioning because although transforming lights data can affect results, its implications have not been widely considered. Yet regardless of whether logged or unlogged data is use, Pearson's r correlation coefficients show that lights are more strongly correlated in Russia between 1992 and 2012 than in China.

A separate test that provides the same result for both logged and unlogged data is Kendall's rank correlation, which measures the similarity of how two quantities are ordered. Russia's provinces remain more similarly ordered between 2012 and 1992 than China's. Thus, both the Pearson's r and Kendall's rank correlations indicate a higher degree of provincial path dependence of lights in Russia than in China. This may mean that it is harder for Russian provinces to move up or down in relative development status. It could also suggest that the Chinese government's efforts to stimulate growth in its less developed provinces are actually helping certain areas to overcome initial low levels of development.

Scatterplots of the correlations of total lights per province in 1992 and 2012 show that in the unlogged data in both China and Russia, provinces with mid-range light levels appear to increase the most relative to the other provinces (Figure 4(a)). In the logged data (Figure 4(b)), the correlation appears fairly steady across provinces of varying amounts of total lights in both countries. Yet in Russia, the provinces with some of the lowest amounts of logged lights appear to underperform in 2012 compared to their 1992 levels. More robust time-series analysis of path dependence is necessary to test the hypothesis that relative levels of development among Russia's provinces are more determined by levels reached during the Soviet era than they are China prior to Deng Xiaoping's economic reforms.

Breaking down the national trend in Russia of decreasing lights to the provincial scale, Figure 5(a,b) illustrate how total lights in most of Russia's provinces stagnated or declined from 1992 to 2012. Over half (48/83) of Russia's provinces had fewer lights in 2012 than they did in 1992, suggesting weak levels of regional development in much of the country. In contrast, all of China's provinces except for Hong Kong experienced growth in lights during this period. Examination of rates of change in lights also shows how in China, provinces with lower levels of lights in 1992 appeared to catch up to more developed provinces, while the same cannot be said of Russia. Provinces in western and northern China, namely Tibet, Xinjiang, Qinghai, and Inner Mongolia, experienced some of the most rapid rates of change in lights. These initially less-developed western provinces grew more rapidly than coastal provinces like Hong Kong, Macau, Taiwan, Beijing, and Shanghai, which had the lowest rates of change. One exception is Hainan, an island province in southern China that experienced rapid growth in lights comparable to rates in western and northern China. Hainan's growth may be due to its historical lack

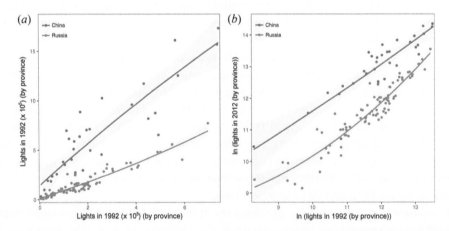

Figure 4. Comparison of scatterplots for correlation of (a) unlogged lights in 1992 and 2012 (b) logged lights in 1992 and 2012. Macau is excluded for the purposes of visualization since it emits very few total lights due to its small size. The grey buffer areas represent 95% confidence intervals for the fitted lines.

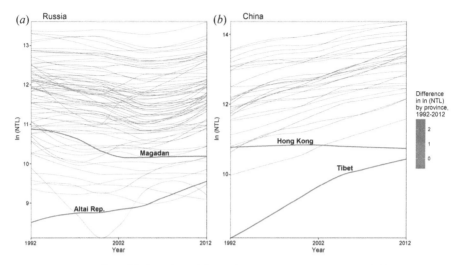

Figure 5. Time series of total lights plotted for each province in Russia (a) and China (b). Each time series has been smoothed to highlight general trends. In (a), Altai Republic and Magadan are the Russian provinces with the biggest increase and decrease in log lights between 1992 and 2012, respectively. In (b), Tibet and Hong Kong are the Chinese provinces with the biggest increase and decrease in log lights during this period, respectively.

of development and the recent promotion of sustainable development initiatives and tourism (Stone and Wall 2004). In China's poorer, less populated western provinces, the state-encouraged relocation of both capital and Han Chinese migrants (Becquelin 2004) may be driving them to catch up in total lights to the wealthier, more populous coastal provinces. Western China also holds significant deposits of resources like oil, gas, and coal, which the government began seeking to extract as part of its regional development strategy beginning in the early 2000s (Woodworth 2016). In comparison to western China and neighbouring Inner Mongolia and Heilongjiang, the northeast Chinese provinces of Jilin and Liaoning – home to the aforementioned 'Rust Belt' – experienced slower rates of change. China's provinces may be growing closer to one another in total lights, but regional inequalities remain.

4.3. Inferring development from decreasing, stable, and increasing lights

4.3.1. Decreasing lights in Magadan
Of all of Russia's provinces, lights in the northeast province of Magadan decreased the most. Total lights in 2012 equalled only 53% of total lights in 1992. Like much of the Russian North, this mining-intensive province has become depopulated through government programmes carried out with the assistance of the World Bank to resettle its residents in Russia's more temperate locations, where it is easier for the state to provide services (Round 2006). Comparison of the tritemporal map over Magadan City, which mostly consists of lights that dimmed since 1992, and the same location in Google Earth reveals encroachments of vegetation, demolished or abandoned infrastructure, and general urban shrinkage between 2002 and 2012 (Figure 6(a)).

4.3.2. *Stable lights in Hong Kong*

While 33/34 of China's provinces increased in lights between 1992 and 2012, Hong Kong's lights actually decreased to 91% of 1992 levels. Rather than reflecting a downturn in Hong Kong's level of development, however, this may be due to the sensor not capturing growth beyond the saturation level. As a highly developed and densely populated city-province, most of Hong Kong already appears close to saturation level in 1992. This would limit the ability of DMSP-OLS to detect further increases in brightness beyond a DN of 63, an issue which several studies attempt to remedy with various saturation corrections (Ma et al. 2014; Letu et al. 2012; Zhuo et al. 2015). Alternatively, Frolking et al. (2013) combine DMSP-OLS data with microwave scatterometer data to document the upward expansion of urbanization via the construction of taller buildings, which is common in Chinese cities. Hong Kong's growth could also simply be slower than in less developed or less urbanized provinces in places like western China. Indeed, little change is noticeable in a portion of the densely populated Kowloon Peninsula between 2000 and 2012 in Google Earth imagery (Figure 6(b)), and most of the city appears white (representing stable lights) in

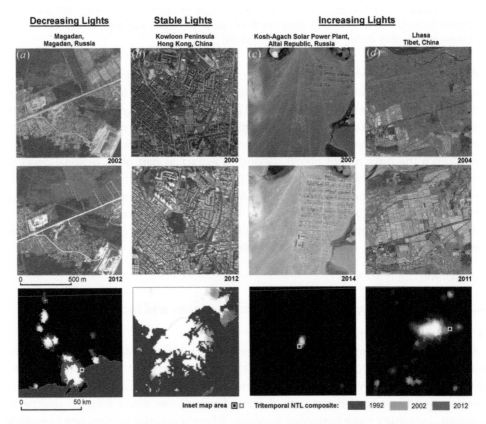

Figure 6. Comparison of Google Earth imagery and four key locations within the DMSP-OLS 1992–2002–2012 tritemporal composite. (a) Magadan, the province with the largest decrease in lights in Russia; (b) the Kowloon Peninsula in Hong Kong, the only province in China whose lights decrease but which is still relatively stable overall; (c) the Kosh-Agach Solar Power Plant in Altai, the province with the largest increase in lights in Russia; (d) Lhasa, capital of Tibet, the province with the largest increase in lights in China.

the tritemporal map. A similar 'flat lining' of lights is also apparent in the Russian city-province of Moscow, which already has many pixels at or close to saturation in 1992. Although its population has grown 28% since 1989 and its economy has continued to expand (Argenbright 2013), the stagnation of total light between 1992 and 2012 does not reflect this.

4.3.3. *Increasing lights in Altai Republic and Tibet*

The largest increase in lights in Russia is in the Altai Republic, a mountainous area that borders China. Lights in 2012 are 208% of 1992 levels. Some of the greatest changes in lights within Altai correspond to the recent completion of the Kosh-Agach solar power facility, whose construction is visible in Google Earth imagery between 2007 and 2014 (Figure 6(c)). Altai's increase in lights over this period resembles the general rate of increase among China's provinces, none of which bear any resemblance to Magadan's dramatic decrease in lights. Indeed, the similarity of the trend in lights between China's provinces and the Altai Republic, which borders China, Kazakhstan, and Mongolia, may be related to efforts by Chinese authorities to develop tourism and transportation infrastructure in the impoverished region (Nyíri and Breidenbach 2008). Over time, the diffusion of Chinese capital across its borders to develop its foreign hinterlands in places like Altai may come to represent an exception to Pinkovskiy (2013). This study identifies discontinuous increases in lights per capita upon crossing a border from a poorer to a wealthier country and particularly strong discontinuities along Asian borders such as the one between Russia and China.

In China, total lights in the southwestern province of Tibet increased the most of any province, with 2012 levels growing to 903% of 1992 levels. Xinjiang, Qinghai, Yunnan, and Inner Mongolia have the next highest growth rates, with 2012 lights all at least five times greater than 1992 levels. In recent decades, these provinces had very low levels of development compared to eastern China. Qinghai, for instance, had the worst economic performance (Goodman 2004a) and some of the lowest levels of road density of all of China's provinces (Fan and Zhang 2004). The dramatic increase in total lights in western China, especially in Tibet, may reflect state investment spurred by the national government's 'Open Up the West' campaign. Tibet has been the largest per capita recipient of subsidies from the central government. In 2004, $2 billion was invested on building infrastructure (Chansoria 2011). The large increase in lights around the Tibetan capital of Lhasa corresponds to an increase in built-up area visible in Google Earth between 2004 and 2011 (Figure 6(d)). A new, $4 billion public railway from the city of Golmud, Qinghai to Lhasa, Tibet, which opened in 2006, has brought increased numbers of tourists to the province, too (Su and Wall 2009). With the Chinese government seeking to connect Tibet by rail with neighbouring Xinjiang, Yunnan, and also Sichuan by 2020, growth in lights in these western provinces seems set to continue. Whether this type of development is improving the quality of life for ethnic Tibetans, however, remains contested (see, for example, Yeh 2013).

4.4. *Cross-border differences in lights*

Eastern Russia and northeast China share similar natural environments, peripheral locations relative to their national centres, and economies dominated by the state and agriculture (Ryzhova and Ioffe 2009). Whereas lights in northeast China substantially brightened between 1992 and 2012, however, they dimmed in eastern Russia (Figure 7).

Many of the dimming lights indicate the decline of infrastructure and villages along the Trans-Siberian Railway. The border clearly delineates differences in patterns of changes in light, illustrating the importance of provincial and national-level institutions in regional development. Eastern Russia's decline in the 1990s may have even accelerated northeast China's growth: the plummeting in production of consumer goods and food staples in Russian border provinces, for instance, incentivized Chinese production (Alekseev 2001). The one exception where Chinese growth appears to spill over into Russia is between the border cities of Heihe and Blagoveshchensk, possibly due to deliberate cooperation between the two municipalities to stimulate growth (Ryzhova and Ioffe 2009).

Northeast China's growth in lights, even if slower than in the southwest, may be due to the national government's 'Revitalize the Northeast' development programme. The government established this strategy in 2003 to stimulate growth in the 'Rust Belt' area that once formed one of China's key industrial bases but began declining in the late 1970s when the government switched its focus to coastal development. The programme aims to transform the region into one of the country's four economic engines by modernizing state-owned enterprises, stimulating manufacturing and services, and increasing trade with neighbouring countries. In contrast to China's efforts in its

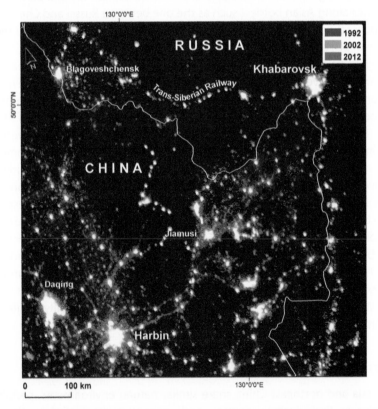

Figure 7. In this tritemporal composite, the Russia–China border sharply illustrates the divergence in the two countries' lights from 1992 to 2012, with lights in China generally brightening and lights in Russia generally dimming.

borderland regions, the Russian government made few concerted attempts to deliberately spur growth in the Russian Far East until 2012, when it created the new regional ministry for development.

4.5. *Predicting change in lights with change in population and GDP, 2000–2012*

The above examples demonstrate that lights are capable of capturing both positive and negative changes over time in regional development. They can also illustrate stark cross-border differences in lights that may be due to differences in national and regional administration and development policies. When examined systematically, however, it is unclear whether changes in lights respond similarly to positive *and* negative changes in population and economic activity.

Table 2 reports the fixed effect estimates regressing lights on population and GDP for province in China and Russia for every year between 2000 and 2012. All provinces in each country were included except for Macau, Hong Kong, and Taiwan in China, since no GDP or population data were available. We test the model with province and year fixed effects. When only province fixed effects are included, population and GDP each significantly predict lights over time within provinces in both Russia and China, but the coefficients vary. Within Russia, population has a larger effect on lights than GDP. In China, the opposite is true, with GDP having a larger effect than population. Unusually, population appears to have a negative effect, but this may be due to multicollinearity. When only population or only GDP is included as the predictor variable for lights, both coefficients are positive.

Since the variables are all logged, the coefficients can be reported as elasticities. Holding population constant in each country, the elasticity of provincial GDP growth with respect to lights growth is estimated to be 0.07 in Russia and a substantially higher 0.396 for China. Notably, the estimated elasticity between provincial GDP growth and lights growth for China is higher than the estimated global elasticity found between GDP and lights of 0.27 in Henderson, Storeygard, and Weil (2012)'s panel, which consists of 188 countries. Their model differs from ours, particularly since we include population. Yet their estimated within-country coefficient of determination (R^2) of .749 is close to the within-province R^2 of .81 we estimate for China. This indicates that in certain circumstances – perhaps in the context of rapid, urban-centric development – the relationship

Table 2. Fixed effect estimates regressing lights on population and GDP (constant 2005 US$) for provinces in Russia and China, 2000–2012.

	China	Russia	Russia	Russia	China	Russia
ln (population)	−0.648***	1.014***	0.772*	0.604	−0.622***	0.566***
	(0.132)	(0.204)	(0.456)	(0.373)	(0.122)	(0.186)
ln (GDP)	0.396***	0.074***	0.094***	0.050**	0.426***	0.214***
	(0.011)	(0.010)	(0.017)	(0.019)	(0.072)	(0.045)
No. of observations	403	1074	268	806	403	1074
Sample	All	All	Pop. increase (2000–2012)	Pop. decrease (2000–2012)	All	All
R^2	.810	.056	.217	.010	.855	.342
Year fixed effects	No	No	No	No	Yes	Yes

All specifications include province fixed effects. Standard errors are in parentheses.
*$p < .1$; **$p < .05$; ***$p < .01$.

between GDP, population, and lights may be scale-invariant. Further research is necessary to confirm or deny this hypothesis.

Overall, the ability of population and GDP to predict change in lights is much lower in Russia than in China. These two variables explain 81% of within-province change in lights in China but only 6% of within-province change in lights in Russia. Population and GDP therefore do not just have numerically different effects on lights in Russia than in China: they have little relationship with lights at all over time. This finding may due to a number of reasons explored in the following paragraphs.

The first possible reason is that lights may be coupled with GDP and population when they increase but not when they decrease. A number of studies have observed this tendency. Examining countries in Europe from 1992 to 2012, Archila Bustos et al. (2015) conclude that population decline is not always coupled with decline in lighted area. While ten countries steadily depopulate over this period, only two – Ukraine and Moldova – also experience steady declines in lighted area. Similarly, Elvidge et al. (2014) note that in several former Soviet countries (though not Russia), even as population declines or lags relative to GDP, lights continue to grow. In China's provinces between 2000 and 2012, population and GDP generally increased from one year to the next, so the possibility of lights decoupling from these variables when they become negative cannot be studied there at the provincial scale. Population did decrease in five provinces in China (Guizhou, Sichuan, Anhui, Guangxi, and Henan) between 1992 and 2012, but this number of observations is too small to fit to a model.

Yet in Russia's provinces, while reported GDP tended to increase annually for all provinces, population dropped for many. The possibility of decoupling can thus be explored. This is done by running the panel regression separately for the 62 provinces where population decreased between 2000 and 2012 and the 21 provinces where population increased over this period. While the model's fit improves to $R^2 = .22$ when only provinces that increased in population are included, population becomes a less significant predictor of lights ($p < .1$). GDP remains highly significant ($p < .01$), although the elasticity of change in GDP with respect to change in lights, 0.094, is still only approximately a quarter of its elasticity in China. This suggests that even in Russia's provinces that are reportedly increasing in both population and GDP, lights still do not respond in the same way to increases in lights as they do in China. The model's fit weakens even more to $R^2 = .02$ when only provinces that decreased in population are included, perhaps illustrating the erratic nature of the relationships between lights, population, and GDP in depopulating places.

The second possibility is that Russia may be experiencing aggregate, national-level shocks that are not controlled by the predictor variables of within-country GDP and population. This hypothesis is examined using year fixed effects. When these are included for Russia, the elasticity of change in GDP with respect to change in lights triples to 0.21, while the elasticity of change in population with respect to change in lights nearly halves to 0.57. All years are significant at $p < .001$ except for 2010, which is still significant at $p < .01$. The fit of the model for Russia improves substantially from $R^2 = .05$ to .34. For China in contrast, all but one year fixed effects are insignificant at $p < .01$. When year fixed effects are included, the effects of population and GDP hardly change, while the R^2 also does not improve much. This means that while population and

GDP sufficiently explain change in lights in China, they may not in Russia due to the existence of omitted variables at an aggregated, national, yearly level. In other words, although there appears to be a general 'background decay' of lights across Russia that appears to reduce the effect of GDP on lights, when this aggregate trend is controlled for with year fixed effects, the effect of GDP on lights within provinces is somewhat more comparable to the effect in China's provinces.

One possible omitted variable that could explain the 'background decay' of lights is infrastructure stock, which is deteriorating in many post-Soviet countries. However, including this variable in a model is difficult. As explained in Section 3.5, we attempted to use fixed capital investment to represent infrastructure, but it added little to the model. This is likely because the variable was collinear with GDP. In China, for instance, fixed capital investment is a key driver of GDP (Qin et al. 2006). Additionally, fixed capital investment figures may only account for addition to or enhancement of existing stock rather than its depreciation.

Aside from national-level year-fixed effects, a third reason for the poor fit of the model may be regional effects in Russia that influence the relationship between lights, GDP, and population. We interact the eight federal districts in Russia (Central, Far East, Northwest, North Caucasus, Siberia, Southern, Ural, and Volga) and four regional planning districts in China (East Coast, Central, Northeast, and Western) with GDP to see whether its effect on lights varies by region (Table 3). In China, regional interactions with GDP are insignificant in Central and Northeast China when compared to the East Coast reference region, which includes Beijing and Shanghai. However, GDP appears to have a significantly stronger effect on lights in western China, with an elasticity of 0.152 compared to the East Coast. This could mean that the government's 'Open Up the West' campaign is resulting in more lights generated per dollar invested in western China than in any other region in the country. Still, since the inclusion of regional interactions with GDP as predictors only slightly improves the model's fit, this means that the relationship between population, GDP, and lights in China can generally be sufficiently explained without incorporating regional variations into this relationship.

Contrastingly in Russia, regional interactions with GDP are significant in several cases and their addition substantially improves the overall fit of the model from $R^2 = .06$ to $R^2 = .14$. Regional effects are highly significant ($p < .01$) in the Far East and Northwest, where lights respond more positively to increases in GDP than in the reference region, Central Russia, which includes Moscow. These significant and positive interactions could reflect the federal government's ongoing efforts to develop the Far East and Northwest regions, both of which include Arctic areas where it is seeking to stimulate the economy and encourage infrastructure construction via federal transfers and increased investment (Russian Government 2014; Korolev 2016).

Regional interactions with GDP are also highly significant in the Volga Federal District ($p < .01$) and in the Southern and North Caucasus Federal Districts (both $p < .05$), yet GDP interacts with these three regions negatively compared to Central Russia. Notably, the geography of this effect corresponds to a 'southern belt' of corruption stretching from the Southern to Volga Federal Districts. These are two largely agricultural areas which once strongly supported communism (Dininio and Orttung 2005). Corruption is also a major concern among residents in the North Caucasus (Gerber and Mendelson, 2009). The negative effect of these three regions

Table 3. Fixed effect estimates regressing lights on population, GDP (constant 2005 US$), and interactions between GDP and regions for provinces in Russia and China, 2000–2012.

	ln (lights)	
	China	Russia
ln (population)	−0.318**	1.580***
	(0.151)	(0.221)
ln (GDP)	0.308***	0.078***
	(0.023)	(0.018)
ln (GDP) × Central	0.031	
	(0.029)	
ln (GDP) × Western	0.152***	
	(0.024)	
ln (GDP) × Northeast	0.020	
	(0.036)	
ln (GDP) × Far East		0.174***
		(0.030)
ln (GDP) × Northwest		0.090***
		(0.027)
ln (GDP) × North Caucasus		−0.066**
		(0.033)
ln (GDP) × Siberia		0.021
		(0.027)
ln (GDP) × Southern		−0.081**
		(0.034)
ln (GDP) × Ural		−0.012
		(0.036)
ln (GDP) × Volga		−0.076***
		(0.026)
No. of observations	403	1,074
Sample	All	All
R^2	.836	.136

All specifications include province fixed effects. Standard errors are in parentheses.
**$p < .05$; ***$p < .01$.

on GDP compared to the Central Federal District could limit the efficacy of investment-based regional development programmes. Future research could examine whether these negative regional interactions may be due to the presence of corruption, typically defined as using public office for private benefit (Taylor, 2007). If this were shown to be the case, then the finding of significant negative regional effects in Russia as opposed to China, where only the Western region shows a significant and positive interaction with GDP, could corroborate findings that corruption is less costly and destructive in China than in Russia (Sun 1999; Blanchard and Shleifer 2001).

A fourth and final reason for the relatively poor fit of the model in Russia compared to China may be that Russian GDP and/or population records are inaccurate. Although the fit of the model actually decreases slightly when only GDP ($R^2 = .03$) or only population ($R^2 = 0$) are included, the possibility of unreliable standard accounts could be further explored if another independent, provincial-level source of data were used. One possible dataset for future incorporation is Yale University's G-Econ data, which provide GDP at a 1° × 1° resolution at a global scale for four years (1990, 1995, 2000, and 2005) and have been used in a handful of lights studies (Chen and Nordhaus 2015, 2011). If the replacement of Russian GDP data with the G-Econ data substantially improved the model's fit, this could suggest that Russia's standard accounts are inaccurate and that the typical strong relationship found between GDP and lights could actually hold true in Russia. Then, the model could also be used to predict GDP or population within a certain area or scale of Russia for which statistics do not exist. Refinement of this model could also help better solve a debate as to whether the Russian economy actually 'collapsed' in the 1990s or whether economic statistics under communism were vastly inflated, thus making Russia's decline appear more calamitous statistically than it was in actuality (Åslund 2001).

5. Conclusion

The strong, positive relationships between change over time in lights, population, and economic activity found in rapidly growing places like China appear to break down when examined in places experiencing socioeconomic decline. Within China's provinces, changes in development, estimated using population and GDP records, are significant and account for over 80% of within-province change in lights. But within Russia's provinces, changes in population and GDP explain a mere 6% of change in lights over time even though they are significant predictors. The improvement of the model's fit in Russia with the addition of year fixed effects suggests that GDP actually has a stronger effect on lights than initially determined, but additional research is necessary to work out the unobserved variables that constitute these aggregate, yearly effects, which may indicate the broader deterioration of infrastructure that is occurring even as new investments in fixed capital are made. Until this can be done, our model using changes in population and GDP to explain changes in lights may not apply to areas of the world experiencing depopulation, economic decline, or other phenomena associated with a reversal of development such as deurbanization or deindustrialization.

While our study confirms existing research that shows strong relationships between GDP and lights over time in China at various scales (Zhao, Currit, and Samson 2011; Liu et al. 2012), we also show that the government's campaigns to reduce regional inequality, particularly in the relatively undeveloped region of Western China, appear to be making inroads that are visible in night lights imagery. China's provinces brightened from 1992 to 2012, while spatial inequality in lights between provinces declined. GDP also has a stronger effect on lights in Western China than in any of the other three economic planning regions. In contrast, over half of Russia's provinces, many of which are less developed and located in the country's periphery, dimmed while spatial inequality in lights between provinces increased. This suggests a negative and spatially unequal pattern of development since the dissolution of the Soviet Union. Additional lights-based research could help confirm whether globally, regional development is becoming more or less spatially uniform within the world's countries.

As far as we know, this study is the first to specifically use lights to examine development trajectories in Russia during the transition from communism to capitalism. Further study of this tumultuous period for post-Soviet countries is recommended, especially since DMSP-OLS V4 composites conveniently date back to 1992 – just one year after the global collapse of communism. While a model predicting change in lights in areas that confound the typical rural-to-urban development trajectory requires significant refinement, for now, lights can at least illustrate a more complex story of regional development that might be obscured by the more conventionally used national-level measures of population and GDP. As we illustrate with the use of interactions between GDP and regions, lights data also appear promising for identifying regions suffering from corruption, where despite invest-ment of public capital, light-emitting development is lower than predicted. A model that also better predicts decline in lights could be applied to other places experi-encing a reversal of development for various reasons, such as economically depressed Detroit or war-torn Syria. Greater research into the causes of changes in lights in rural, remote, distressed, and slowly developing regions is recom-mended, for these are often the places for which conventional socioeconomic indicators are least available and where lights data could prove most illuminating.

Note

1. Due to differences after intercalibration, the topcode for 1992 is 62, while the topcode for 2012 is 61.

Acknowledgements

We would like to thank Dr. David Rigby, Dr. Jamie Goodwin-White, and Dylan Connor of the UCLA Department of Geography for their assistance with this article.

Disclosure statement

No potential conflict of interest was reported by the authors.

Funding

This work was supported by the National Science Foundation under Grant [DGE-1144087]; and the UCLA Charles F. Scott Fellowship.

References

Alekseev, M. 2001. "Socioeconomic and Security Implications of Chinese Migration in the Russian Far East." *Post-Soviet Geography and Economics* 42 (2): 122–141. doi:10.1080/10889388.2000.10641154.

Álvarez-Berríos, N. L., I. K. Parés-Ramos, and T. Mitchell Aide. 2013. "Contrasting Patterns of Urban Expansion in Colombia, Ecuador, Peru, and Bolivia Between 1992 and 2009." *Ambio* 42 (1): 29–40. doi:10.1007/s13280-012-0344-8.

Archila Bustos, M. F., O. Hall, and M. Andersson. 2015. "Nighttime Lights and Population Changes in Europe, 1992–2012." *Ambio* 44 (7): 653–665. doi:10.1007/s13280-015-0646-8.

Argenbright, R. 2013. "Moscow on the Rise: From Primate City to Megaregion." *Geographical Review* 103 (1): 20–36. doi:10.1111/j.1931-0846.2013.00184.x.

Åslund, A. 2001. "Russia." *Foreign Policy* 20–25. doi:10.2307/3183323.

Becquelin, N. 2004. "Staged Development in Xinjiang." *The China Quarterly* 178 (June): 358–378. doi:10.1017/S0305741004000219.

Bennett, M. M., and L. C. Smith. 2017. "Advances in Using Multi-Temporal Night-Time Lights Satellite Imagery to Detect, Estimate, and Monitor Socioeconomic Dynamics." *Remote Sensing of Environment*192: 176–197. doi:10.1016/j.rse.2017.01.005.

Bennie, J., T. W. Davies, J. P. Duffy, R. Inger, and K. J. Gaston. 2014. "Contrasting Trends in Light Pollution Across Europe Based on Satellite Observed Night Time Lights." *Scientific Reports* 4 (January): 3789. doi:10.1038/srep03789.

Blanchard, O., and A. Shleifer. 2001. "Federalism With and Without Political Centralization: China Versus Russia." *IMF Staff Papers* 48 (1): 171–179. doi:10.2307/4621694.

Cao, X., J. Wang, J. Chen, and F. Shi. 2014. "Spatialization of Electricity Consumption of China Using Saturation-Corrected DMSP-OLS Data." International Journal of Applied Earth Observation and Geoinformation 28: 193–200. doi:10.1016/j.jag.2013.12.004.

Castrence, M., D. Nong, C. Tran, L. Young, and J. Fox. 2014. "Mapping Urban Transitions Using Multi-Temporal Landsat and DMSP-OLS Night-Time Lights Imagery of the Red River Delta in Vietnam." *Land* 3 (1): 148–166. doi:10.3390/land3010148.

Cauwels, P., N. Pestalozzi, and D. Sornette. 2014. "Dynamics and Spatial Distribution of Global Nighttime Lights." *EPJ Data Science* 3: 2. doi:10.1140/epjds19.

Ceola, S., F. Laio, and A. Montanari. 2014. "Satellite Nighttime Lights Reveal Increasing Human Exposure to Floods Worldwide." *Geophysical Research Letters* 41 (20): 7184–7190. doi:10.1002/2014GL061859.

Chansoria, M. 2011. "China's Infrastructure Development in Tibet: Evaluating Trendlines." *Manekshaw Papers* 32. 1–40. New Delhi: Centre for Land Warfare Studies.

Chen, X., and W. Nordhaus. 2011. "Using Luminosity Data as a Proxy for Economic Statistics." *Proceedings of the National Academy of Sciences* 108 (21): 8589–8594. doi:10.1073/pnas.1017031108.

Chen, X., and W. Nordhaus. 2015. "A Test of the New VIIRS Lights Data Set: Population and Economic Output in Africa." *Remote Sensing* 4937–4947. doi:10.3390/rs70404937.

Cinzano, P., F. Falchi, C. Elvidge, and K. Baugh. 2000. "The Artificial Night Sky Brightness Mapped from DMSP Operational Linescan System Measurements." *Monthly Notices of the Royal Astronomical Society* 657: 17. doi:10.1046/j.1365-8711.2000.03562.x.

Croft, T. 1978. "Nighttime Images of the Earth From Space." *Scientific American* 239 (1): 86–98. doi:10.1038/scientificamerican0778-86.

Dininio, P., and R. Orttung. 2005. "Explaining Patterns of Corruption in the Russian Regions." *World Politics* 57 (4): 500–529. doi:10.1353/wp.2006.0008.

Doll, C. N. H., J.-P. Muller, and J. G. Morley. 2006. "Mapping Regional Economic Activity from Night-Time Light Satellite Imagery." *Ecological Economics* 57 (1): 75–92. doi:10.1016/j.ecolecon.2005.03.007.

Doll, C. N. H., and S. Pachauri. 2010. "Estimating Rural Populations Without Access to Electricity in Developing Countries Through Night-Time Light Satellite Imagery." *Energy Policy* 38 (10): 5661–5670. Elsevier. doi:10.1016/j.enpol.2010.05.014.

Dunford, M., and L. Li. 2010. "Chinese Spatial Inequalities And Spatial Policies." *Geography Compass* 4 (8): 1039–1054. doi:10.1111/j.1749-8198.2010.00359.x.

Efendiev, A., P. Sorokin, and M. Kozlova. 2014. "Transformations in the Rural Life in Russian Belgorod Region in 2000-2013 Through 'modernization' theoretical perspective: increasing material well-being, growing individualism and Persisting Pessimism." *SSRN Electronic Journal.* doi:10.2139/ssrn.2537305.

Elvidge, C. D., D. Ziskin, K. E. Baugh, B. T. Tuttle, T. Ghosh, D. W. Pack, E. H. Erwin, and M. Zhizhin. 2009. "A Fifteen Year Record of Global Natural Gas Flaring Derived from Satellite Data." *Energies* 2 (3): 595–622. doi:10.3390/en20300595.

Elvidge, C. D., K. E. Baugh, F. C. Hsu, and T. Ghosh. 2014. "National Trends in Satellite-Observed Lighting." In *Global Urban Monitoring and Assessment through Earth Observation*, edited by Q. Weng, 97–118. Boca Raton: CRC Press.

Elvidge, C. D., M. L. Imhoff, K. E. Baugh, V. R. Hobson, I. Nelson, J. B. D. Jeff Safran, and B. T. Tuttle. 2001. "Night-Time Lights of the World: 1994-1995." *ISPRS Journal of Photogrammetry and Remote Sensing* 56: 81–99. doi:10.1016/S0924-2716(01)00040-5.

Fan, C. C., and M. Sun. 2008. "Regional Inequality in China, 1978-2006." *Eurasian Geography and Economics* 49 (1): 1–18. doi:10.2747/1539-7216.49.1.1.

Fan, J., M. Ting, C. Zhou, Y. Zhou, and X. Tao. 2014. "Comparative Estimation of Urban Development in China's Cities Using Socioeconomic and DMSP/OLS Night Light Data." *Remote Sensing* 6 (8): 7840–7856. doi:10.3390/rs6087840.

Fan, S., and X. Zhang. 2004. "Infrastructure and Regional Economic Development in Rural China." *China Economic Review* 15 (2): 203–214. doi:10.1016/j.chieco.2004.03.001.

FAO. 2014. The Global Administrative Unit Layers. 2016. Rome: EC-FAO Food Security for Action Programme funded by the European Commission.

Federal State Statistics Service. 2014. "Russia in Figures." Accessed 16 July, 2016 http://www.gks.ru/wps/wcm/connect/rosstat_main/rosstat/en/main.

Frolking, S., T. Milliman, K. C. Seto, and M. A. Friedl. 2013. "A Global Fingerprint of Macro-Scale Changes in Urban Structure From 1999 to 2009." *Environmental Research Letters* 8 (2): 024004. doi:10.1088/1748-9326/8/2/024004.

Gao, B., Q. Huang, H. Chunyang, and M. Qun. 2015. "Dynamics of Urbanization Levels in China From 1992 to 2012: Perspective from DMSP/OLS Nighttime Light Data." *Remote Sensing* 7 (2): 1721–1735. doi:10.3390/rs70201721.

Geldmann, J., L. N. Joppa, and N. D. Burgess. 2014. "Mapping Change in Human Pressure Globally on Land and Within Protected Areas." *Conservation Biology* 28 (6): 1604–1616. doi:10.1111/cobi.12332.

Gerber, T. P., and S. E. Mendelson. 2009. "Security Through Sociology: The North Caucasus And The Global Counterinsurgency Paradigm." *Studies In Conflict & Terrorism* 32 (9): 831-851. doi:10.1080/10576100903116656.

Ghosh, T., S. Anderson, R. L. Powell, P. C. Sutton, and C. D. Elvidge. 2009. "Estimation of Mexico's Informal Economy and Remittances Using Nighttime Imagery." *Remote Sensing* 1 (3): 418–444. Molecular Diversity Preservation International. doi:10.3390/rs1030418.

Goldman, M. 2008. *Petrostate: Putin, Power, and the New Russia*. Oxford: Oxford University Press.

Goodman, D. 2004a. "Qinghai and the Emergence of the West: Nationalities, Communal Interaction and National Integration." *The China Quarterly* 178 (June): 379–399. doi:10.1017/S0305741004000220.

Goodman, D.2004b. "The Campaign to 'Open Up the West': National, Provincial-Level and Local Perspectives." *The China Quarterly* 178: 317–334. doi:10.1017/S0305741004000190.

Henderson, J. V., A. Storeygard, and D. N. Weil. 2012. "Measuring Economic Growth From Outer Space." *American Economic Review* 102 (2): 994–1028. doi:10.1257/aer.102.2.994.

Hodler, R., and P. A. Raschky. 2014. "Regional Favoritism." *The Quarterly Journal of Economics* 129 (2): 995–1033. doi:10.1093/qje/qju004.

Holbig, H. 2004. "The Emergence of the Campaign to Open up the West: Ideological Formation, Central Decision-Making and the Role of the Provinces." *The China Quarterly* 178 (June): 335–357. doi:10.1017/S0305741004000207.

Huang, Q., X. Yang, B. Gao, Y. Yang, and Y. Zhao. 2014. "Application of DMSP/OLS Nighttime Light Images: A Meta-Analysis and a Systematic Literature Review." *Remote Sensing* 6 (8): 6844–6866. doi:10.3390/rs6086844.

Imhoff, M. L., W. T. Lawrence, D. C. Stutzer, and C. D. Elvidge. 1997. "A Technique for Using Composite DMSP/OLS 'City Lights' Satellite Data to Map Urban Area." *Remote Sensing of Environment* 61 (3): 361–370. doi:10.1016/S0034-4257(97)00046-1.

Keola, S., M. Andersson, and O. Hall. 2015. "Monitoring Economic Development From Space: Using Nighttime Light and Land Cover Data to Measure Economic Growth." *World Development* 66: 322–334. doi:10.1016/j.worlddev.2014.08.017.

Kholodilin, K. A., A. Y. Oshchepkov, and B. Siliverstovs. 2009. "The Russian Regional Convergence Process: Where Does it Go?" *SSRN Electronic Journal*. doi:10.2139/ssrn.1362355.

Korolev, A. 2016. "Russia's Reorientation to Asia: Causes and Strategic Implications." *Pacific Affairs* 89 (1): 53–73. doi:10.5509/201689153.

Letu, H., M. Hara, G. Tana, and F. Nishio. 2012. "A Saturated Light Correction Method For DMSP/ OLS Nighttime Satellite Imagery." *IEEE Transactions on Geoscience and Remote Sensing* 50 (2): 389–396. doi:10.1109/TGRS.2011.2178031.

Li, C. 1996. "China's Northeast: From Largest Rust Belt to Fourth Economic Engine." *China Leadership Monitor* 9: 1–15.

Li, X., L. Ge, and X. Chen. 2013. "Detecting Zimbabwe's Decadal Economic Decline Using Nighttime Light Imagery." *Remote Sensing* 5 (9): 4551–4570. doi:10.3390/rs5094551.

Li, X., and D. Li. 2014. "Can Night-Time Light Images Play a Role in Evaluating the Syrian Cisis?" *International Journal of Remote Sensing* 35 (18): 6648–6661. doi:10.1080/01431161.2014.971469.

Liang, H., H. Tanikawa, Y. Matsuno, and L. Dong. 2014. "Modeling in-Use Steel Stock in China's Buildings And Civil Engineering Infrastructure Using Time-Series of DMSP/OLS Nighttime Lights." *Remote Sensing* 6 (6): 4780–4800. doi:10.3390/rs6064780.

Liu, Z., H. Chunyang, Z. Qiaofeng, H. Qingxu, and Y. Yang. 2012. "Extracting the Dynamics of Urban Expansion in China Using DMSP-OLS Nighttime Light Data from 1992 to 2008." *Landscape and Urban Planning* 106 (1): 62–72. doi:10.1016/j.landurbplan.2012.02.013.

Ma, L., W. Jiansheng, L. Weifeng, J. Peng, and H. Liu. 2014. "Evaluating Saturation Correction Methods for DMSP/OLS Nighttime Light Data: A Case Study from China's Cities." *Remote Sensing* 6: 9853–9872. doi:10.3390/rs6109853.

Ma, T., C. Zhou, T. Pei, S. Haynie, and J. Fan. 2012. "Quantitative Estimation of Urbanization Dynamics Using Time Series of DMSP/OLS Nighttime Light Data: A Comparative Case Study From China's Cities." *Remote Sensing of Environment* 124: 99–107. doi:10.1016/j.rse.2012.04.018.

Michalopoulos, S., and E. Papaioannou. 2014. "National Institutions And Subnational Development in Africa." *The Quarterly Journal of Economics* 129 (1): 151–213. doi:10.1093/qje/qjt029.

Molodikova, I., and A. Makhrova. 2007. "Urbanization Patterns in Russia in the post-Soviet Era." *The Post-Socialist City* 53–70. Dordrecht: Springer Netherlands. doi:10.1007/978-1-4020-6053-3_4.

National Bureau of Statistics of China. 2015. Statistical Data. http://www.stats.gov.cn/english/

Nyíri, P., and J. Breidenbach. 2008. "The Altai Road: Visions of Development Across the Russian–Chinese Border." *Development and Change* 39 (1): 123–145. doi:10.1111/j.1467-7660.2008.00471.x.

Peterson, D. J., and E. Bielke. 2002. "Russia's Industrial Infrastructure: A Risk Assessment." *Post-Soviet Geography and Economics* 43 (1): 13–25.

Pinkovskiy, M. L. 2013. *Economic Discontinuities at Borders: Evidence from Satellite Data on Lights At Night,* Working Paper. Cambridge, MA: Massachusetts Institute of Technology Department of Economics.

Pivovarov, I. L. 2003. "The Urbanization of Russia in The Twentieth Century, Perceptions and Reality." *Sociological Research* 42 (2): 45–65. doi:10.2753/SOR1061-0154420245.

Qin, D., M. A. Cagas, P. Quising, and H. Xin-Hua. 2006. "How Much Does Investment Drive Economic Growth in China?" *Journal of Policy Modeling* 28 (7): 751–774. doi:10.1016/j.jpolmod.2006.02.004.

Round, J. 2006. "The Economic Marginalization of Post-Soviet Russia's Elderly Population and the Failure of State Ageing Policy: A Case Study of Magadan City." *Oxford Development Studies* 34 (4): 441–456. doi:10.1080/13600810601045791.

Russian Federal Government. 2014. Resolution No. 366. On Approval of the State Program of the Russian Federation for Social And Economic Development of the Russian Arctic for the Period Until 2020.April 21. http://www.rg.ru/2014/04/24/arktika-site-dok.html

Ryzhova, N., and G. Ioffe. 2009. "Trans-Border Exchange Between Russia and China: The Case of Blagoveshchensk and Heihe." *Eurasian Geography and Economics* 50 (3): 348–364. doi:10.2747/1539-7216.50.3.348.

Shi, K., Y. Bailang, Y. Huang, H. Yingjie, B. Yin, Z. Chen, L. Chen, and W. Jianping. 2014. "Evaluating The Ability of NPP-VIIRS Nighttime Light Data to Estimate the Gross Domestic Product and the Electric Power Consumption of China at Multiple Scales: A Comparison With DMSP-OLS Data." *Remote Sensing* 6 (2): 1705–1724. doi:10.3390/rs6021705.

Small, C., and C. D. Elvidge. 2013. "Night on Earth: Mapping Decadal Changes of Anthropogenic Night Light in Asia." *International Journal of Applied Earth Observation and Geoinformation* 22 (1): 40–52. doi:10.1016/j.jag.2012.02.009.

Small, C., F. Pozzi, and C. Elvidge. 2005. "Spatial Analysis of Global Urban Extent from DMSP-OLS Night Lights." *Remote Sensing of Environment* 96 (3–4): 277–291. doi:10.1016/j.rse.2005.02.002.

Small, C., and D. Sousa. 2016. "Humans on Earth: Global Extents of Anthropogenic Land Cover From Remote Sensing." *Anthropocene* 14: 1–33. doi:10.1016/j.ancene.2016.04.003.

Stone, M., and G. Wall. 2004. "Ecotourism and Community Development: Case Studies From Hainan, China." *Environmental Management* 33 (1): 12–24. doi:10.1007/s00267-003-3029-z.

Su, M. M., and G. Wall. 2009. "The Qinghai–Tibet Railway and Tibetan Tourism: Travelers' Perspectives." *Tourism Management* 30 (5): 650–657. doi:10.1016/j.tourman.2008.02.024.

Su, Y., X. Chen, C. Wang, H. Zhang, J. Liao, Y. Yuyao, and C. Wang. 2015. "A New Method for Extracting Built-Up Urban Areas Using DMSP-OLS Nighttime Stable Lights: A Case Study in The Pearl River Delta, Southern China." *GIScience & Remote Sensing* 52 (2): 218–238. doi:10.1080/15481603.2015.1007778.

Sun, Y. 1999. "Reform, State, and Corruption: Is Corruption Less Destructive in China Than In Russia?" *Comparative Politics* 32 (1): 1. doi:10.2307/422430.

Sutton, P. C., C. D. Elvidge, and T. Ghosh. 2007. "Estimation of Gross Domestic Product at Sub-National Scales Using Nighttime Satellite Imagery." *International Journal of Ecological Economics & Statistics* 8 (S07): 5–21.

Sutton, P. C., D. Roberts, C. Elvidge, and K. Baugh. 2001. "Census From Heaven: An Estimate of The Global Human Population Using Night-Time Satellite Imagery." *International Journal of Remote Sensing* 22 (16): 3061–3076. doi:10.1080/01431160010007015.

Tan, M. 2015. "Urban Growth and Rural Transition in China Based on DMSP/OLS Nighttime Light Data." *Sustainability* 7 (7): 8768–8781. Multidisciplinary Digital Publishing Institute. doi:10.3390/su7078768.

Taylor, B. D. 2007. "Putin's" Historic Mission:„ State-building And The Power Ministries In The North Caucasus." *Problems Of Post-communism* 54 (6): 3-16. doi:10.2753/PPC1075-8216540601.

Thompson, N. 2004. "Migration and Resettlement in Chukotka: A Research Note." *Eurasian Geography and Economics* 45 (1): 73–81. doi:10.2747/1538-7216.45.1.73.

Wei, Y. D. 2013. *Regional Development in China: States, Globalization and Inequality*. New York: Routledge.

Weidmann, N. B., and S. Schutte. 2016."Using Night Light Emissions for the Prediction of Local Wealth." *Journal of Peace Research*, May. doi:10.1177/0022343316630359.

Woodworth, M. D. 2016. "Disposable Ordos: The Making of an Energy Resource Frontier in Western China." *Geoforum*. doi:10.1016/j.geoforum.2016.04.007.

World Bank. 2015. "GDP at Market Prices (Constant 2005 US$)." http://data.worldbank.org/indicator/NY.GDP.MKTP.KD?display=fxnjwxlfd&locations=CN&order=wbapi_data_value_1983+wbapi_data_value&page=2&sort=asc

World Bank. 2016. "Russian Federation: Population, Total." http://data.worldbank.org/country/russian-federation

World Bank. 2017. "GDP (current US$)." http://data.worldbank.org/indicator/NY.GDP.MKTP.CD?locations=RU-CN&view=chart

Wu, H. X. 2007. "The Chinese GDP Growth Rate puzzle: How Fast Has the Chinese Economy Grown? *." *Asian Economic Papers* 6 (1): 1–23. doi:10.1162/asep.2007.6.1.1.

Xiao, P., X. Wang, X. Feng, X. Zhang, and Y. Yang. 2014. "Detecting China's Urban Expansion Over the Past Three Decades Using Nighttime Light Data." *Ieee Journal of Selected Topics in Applied Earth Observations and Remote Sensing* 7 (10): 4095–4106. doi:10.1109/Jstars.2014.2302855.

Xu, H., H. Yang, L. Xi, H. Jin, and D. Li. 2015. "Multi-Scale Measurement of Regional Inequality in Mainland China during 2005–2010 Using DMSP/OLS Night Light Imagery and Population Density Grid Data." *Sustainability* 7 (10): 13469–13499. doi:10.3390/su71013469.

Yeh, E. 2013. *Taming Tibet: Landscape Transformation and the Gift of Chinese Development*. Ithaca: Cornell University Press.

Yi, K., H. Tani, L. Qiang, J. Zhang, M. Guo, Y. Bao, X. Wang, and L. Jing. 2014. "Mapping and Evaluating the Urbanization Process in Northeast China Using DMSP/OLS Nighttime Light Data." *Sensors* 14 (2): 3207–3226. doi:10.3390/s140203207.

Zeng, C., Y. Zhou, S. Wang, F. Yan, and Q. Zhao. 2011. "Population Spatialization in China Based on Night-Time Imagery and Land Use Data." *International Journal of Remote Sensing* 32 (24): 9599–9620. doi:10.1080/01431161.2011.569581.

Zhang, Q., and K. C. Seto. 2011. "Mapping Urbanization Dynamics at Regional and Global Scales Using Multi-Temporal DMSP/OLS Nighttime Light Data." *Remote Sensing of Environment* 115 (9): 2320–2329. doi:10.1016/j.rse.2011.04.032.

Zhao, N., N. Currit, and E. Samson. 2011. "Net Primary Production and Gross Domestic Product in China Derived From Satellite Imagery." *Ecological Economics* 70 (5): 921–928. doi:10.1016/j.ecolecon.2010.12.023.

Zhou, N., K. Hubacek, and M. Roberts. 2015. "Analysis of Spatial Patterns of Urban Growth Across South Asia Using DMSP-OLS Nighttime Lights Data." *Applied Geography* 63: 292–303. doi:10.1016/j.apgeog.2015.06.016.

Zhuo, L., J. Zheng, X. Zhang, L. Jun, and L. Liu. 2015. "An Improved Method of Night-Time Light Saturation Reduction Based on EVI." *International Journal of Remote Sensing* 36 (16): 4114–4130. doi:10.1080/01431161.2015.1073861.

The short-term economic impact of tropical Cyclone Pam: an analysis using VIIRS nightlight satellite imagery

Preeya Mohan and Eric Strobl

ABSTRACT

Cyclones are relatively instantaneous shocks where arguably most of the important consequences take place in the first few weeks or months. In this article, we construct destruction proxies of wind exposure and storm surge damages and use satellite measures of nightlight intensity to investigate the short-term impact of tropical cyclones using the case study of Cyclone Pam, which struck the South Pacific Islands in March 2015. Using the unaffected islands as a control group, our regression analysis reveals that initially the storm reduced economic activity in the affected islands by as much as 111%, but by the seventh month there were positive boosts to nightlight intensity. By the ninth month this resulted in cumulative net increases in activities related to night-time electricity usage. More generally, our results suggest that there is likely considerable temporal heterogeneity in the response of areas affected by tropical cyclones and demonstrates the potential of using nightlight imagery to assess the short-term economic impact of tropical storms, and possibly other extreme event phenomena, in a relatively timely manner.

1. Introduction

Over the last few years there has been an increasing concern over the economic impact of tropical cyclones on countries, in part because of the possibility that the frequency or strength of storms may increase in certain regions due to climate change, see Walsh et al. (2015). Unsurprisingly, there are now a number of papers that have investigated the macroeconomic consequences of these storms. Although the evidence thus far has been mixed, most published studies suggest, if anything, a small negative effect. For instance, Rasmussen (2004), in a study of tropical storms in the Caribbean for the period 1970–2002, found a negative effect of 0.05% on GDP growth. Similarly, Strobl (2011) estimated a negative impact of 0.83% on the county income growth rates of USA. Importantly, all of the existing studies have examined the impact of tropical storms at a relatively low temporal frequency, i.e. annually or more long term. However, arguably, given that tropical storms are almost instantaneous shocks, much of the fundamental reaction to these storms occurs in the first few weeks or months. These short-term consequences are likely to be muddled in lower-frequency data, particularly if they are heterogeneous over time. For example, although one might expect an immediate negative impact on economic activity due to the destruction and indirect losses caused by cyclones, with time recovery through international aid, reconstruction, and investment may dominate any still-existing negative effects and eventually boost economic activity (Horwich 2000).

The main obstacle in trying to explore the short-term impact of cyclones has arguably been the lack of availability of appropriate temporally high-frequency economic activity data. However, researchers have over the last few years increasingly resorted to nightlight intensity imagery to measure local economic activity on a consistent basis when data collected by statistical agencies has not been sufficient; see, for instance, Chen and Nordhaus (2011), Henderson, Storeygard, and Weil (2012), and Rybnikova and Portnov (2014). These earlier studies generally used the Defense Meteorological Satellite Program (DMSP) images, which are publicly available annual composites of nightlight intensity at a local scale (30 arc-seconds) globally since 1992. However, more recently, the National Aeronautics and Space Administration (NASA) has also started providing satellite imagery data at higher frequency (monthly) and a higher resolution (15 arc-seconds). Specifically, a consistent series of monthly nightlight intensity images, known as the Visible Infrared Imaging Radiometer Suite (VIIRS) light data, has been available since January 2014 and updated with only a few months delay. Recent studies, such as Li et al. (2013), Ma et al. (2014), and Kyba et al. (2015), have confirmed their potential for measuring economic activity even at a very local scale.

The objective of this article is to investigate the impact of tropical cyclones measured by wind exposure and storm surge destruction indices on economic activity, proxied by VIIRS monthly nightlight data, using the case study of Cyclone Pam, which struck the South Pacific islands in May 2015. Cyclone Pam first formed in the Southern Pacific Basin in early March 2015, but by 12 March it mutated into the most powerful tropical cyclone recorded in the southern hemisphere in history, with estimated wind speeds of 250 km h^{-1} and wind gusts that peaked at around 320 km h^{-1} (Esler 2015; OCHA 2015a). It wrecked havoc mainly in the islands of Vanuatu, and to a lesser degree Solomon Islands and Tuvalu. The case study here provides an example of how monthly nightlight data can reveal the extent to which economic activity can be affected in the very short term by tropical cyclones.

2. Data

2.1. *Study region*

The country sample studied in this article consists of the South Pacific islands, which include American Samoa, Cook Islands, Fiji, Federated States of Micronesia, Guam, Kiribati, Marshall Islands, Northern Mariana Islands, New Caledonia, Norfolk Island, Niue, Nauru, Pitcairn, Palau, French Polynesia, Solomon Islands, Tokelau, Tonga, Tuvalu, Vanuatu, Wallis and Futuna, and Samoa. These are all small developing countries in terms of land area, population density, and gross domestic product (GDP), and rely mostly on tourism for income generation.

2.2. *VIIRS nightlight satellite imagery and economic activity*

There have been a number of studies that have used nightlight imagery to examine the economic impact of tropical cyclones, such as Elliott, Strobl, and Sun (2015) and

Bertinelli and Strobl (2013). These articles have all resorted to measures of night-time brightness generated from the DMSP satellites. However, these nightlight composites suffer from a number of disadvantages, including saturation at upper levels, and the inability to discriminate combustion sources from lights; see Elvidge et al. (2013). In contrast, the recently available VIIRS nightlight imagery collected and processed from the Suomi National Polar-Orbiting Partnership satellite offers a number of improvements, with particular relevance to the current context.

First, the VIIRS publicly available product provides monthly nightlight intensity data, whereas the published DMSP data is available only annually. Although the National Oceanic and Atmospheric Administration (NOAA) will sell monthly DMSP data available upon request, several observations are not usable due to the fact that the overpass of the source satellite is around 19h30 and thus when parts of the globe at certain times of the year are not dark yet. The VIIRS overpass, in contrast, is at 1h30 and hence captures nightlight all year round for most of the globe. Additionally, the VIIRS data does not have a saturation point, an aspect that can be important for urban cores. Moreover, VIIRS-Day/ Night Band image (DNB) images have a relatively low light detection limit of near 2×10^{-11} W cm^{-2} sr^{-1}, compared with the around 5×10^{-10} W cm^{-2} sr^{-1} light detection limit of the DMSP composites (Elvidge et al. 2013). Finally, the VIIRS provides intensity measures at a higher resolution (15 arc-seconds, around 750 m at the equator) than the DMSP product (30 arc-seconds, around 1 km at the equator).

To conduct our study, we have accessed the monthly composites from January 2014 to April 2016, i.e. we have 27 months of data, of which 13 are post Cyclone Pam. One should note that whereas normally light intensity is measured in radiance units, in the calibration of the data from the VIIRS, the effect of clear sky is taken into account by subtracting the estimates of this effect from the observed radiance values. Since this procedure does not take account of the contribution of airglow, the clear-sky offset tends to be too large. Thus, near the noise floor of the data, in which the values tend to be very small, there will be both small negative and positive values, resulting in what has been termed albedo radiance values; see Chen and Nordhaus (2015). One should note that the unit of the data as used throughout our analysis is in W cm^{-2} sr^{-1}. We show in Figure 1, as an example of nightlight intensity at the island levels, the distribution for parts of the Vanuatu island group in March 2014, where the level of lights ranges from low (yellow) to high (red). Figure 1 shows considerable differences across the islands as well as within islands, where the red areas correspond to the more populated areas, and hence economic activity, intense areas.

Some discussion is warranted as to what extent nightlight images are capturing economic activity, or at least what sort of economic activity they are likely to capture. There is certainly growing evidence that, in the face of a lack of alternative proxies, nightlight can serve as a reasonable proxy of countries' GDP; see, for instance, Chen and Nordhaus (2011) and Henderson, Storeygard, and Weil (2012). In terms of the VIIRS data, Li et al. (2013) also show that the derived nightlight images are highly correlated with regional GDP, capturing nearly 90% of their variation. However, at the same time it is likely that brightness at night may not be a good proxy for some types of economic activities. For instance, it is unlikely to capture agricultural cropland production, although this may be less of an issue in our South Pacific island example where the services sector, in particular tourism, rather than agricultural production is generally

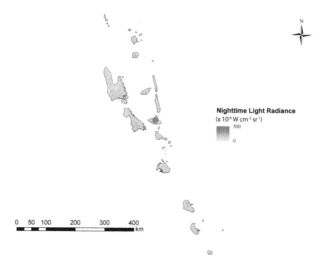

Figure 1. Nightlight intensity: Vanuatu (March 2014).

important in overall production. To further investigate this, we plot the nightlight intensity as measured by VIIRS and GDP km^{-2} for the islands in our sample in Figure 2. The variables show a clear positive relationship, with a significant correlation coefficient of 0.97 (*p*-value of 0.02).

2.3. *Typhoon destruction indices*

Destruction caused by tropical storms typically is due to damages due to strong winds, storm surge, and heavy rainfall. To capture the potential destruction due to strong wind exposure, we use an index in the spirit of Strobl (2012), which measures the wind speed experienced at a very localized level, taking account of the spatial heterogeneity of winds during a cyclone, and then use exposure weights to arrive at an island (group)-

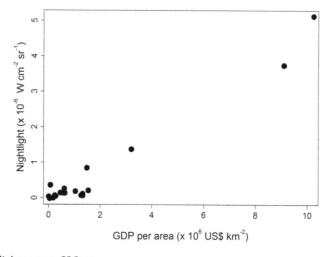

Figure 2. Nightlight versus GDP per area.

specific proxy. More specifically, for a set of locations, $i = 1, \ldots I$, in our case the centroids of the nightlight cells, in island j we define the destruction as

$$D_j = \sum_{i=1}^{I} w_{i,j} \left(W_{j,i}^{max} \right)^3 \quad W^{max} \geq W^*, \tag{1}$$

where D is the wind damage index, W^{max} is the maximum measured wind speed at point i during the storm, W^* is a threshold above which wind is damaging, and w is exposure weights in the month prior to the cyclone of locations, $i = 1, \ldots I$, which aggregates to 1 at the island j level. We set W^* equal to 119 km h^{-1}, which is the threshold that corresponds to the Saffir–Simpson Scale Level 1, i.e. the lowest wind speed at which a tropical storm is considered the equivalent of a tropical cyclone. One may want to note that we allow local destruction to vary with wind speed in a cubic manner, since, as noted by Emanuel (2011), kinetic energy from a storm dissipates roughly to the cubic power with respect to wind speed and this energy release scales with the pressure acting on a structure; see Kantha (2008) and ASCE (2006). From (1), our index D requires local wind speed, W, and exposure weights, w, as inputs in order to be operational.

To calculate the local wind speed, we use the Boose, Mayra, and David (2004) version of the well-known Holland (1980) wind field model. More specifically, the wind experienced due to the storm at any point $P = i$, i.e. W_i is given by

$$W_i = GF\left[M - S(1 - \sin(T_i))\frac{H}{2} \right] \left[\left(\frac{X}{R_i}\right)^B \exp\left(1 - \left[\frac{X}{R_i}\right]^B \right) \right]^{\frac{1}{2}}, \tag{2}$$

where M is the maximum sustained wind velocity anywhere in the storm, T is the clockwise angle between the forward path of the storm and a radial line from the storm centre to the pixel of interest, $P = i$, H is the forward velocity of the hurricane, X is the radius of maximum winds, and R is the radial distance from the centre of the storm to point $P = i$. The remaining variables in (2) consist of the gust factor G and the scaling parameters F, S, and B, for surface friction, asymmetry due to the forward motion of the cyclone, and the shape of the wind profile curve, respectively.

In terms of implementing (2), one should note that M can be obtained from storm track data, H can be directly calculated by following the storm's movements between locations along its track, and R and T are calculated relative to the point of interest $P = i$. All other parameters have to be estimated or assumed. We have no information on the gust wind factor G, but a number of studies, for instance Paulsen and Schroeder (2005), have measured G to be around 1.5, and we also use this value. For S we follow Boose, Mayra, and David (2004) and assume it to be 1. Although we also do not know the surface friction to directly determine F, Vickery et al. (2009) note that in open water the reduction factor is about 0.7 and reduces by 14% on the coast and by 28% further 50 km inland. We thus adopt a reduction factor that linearly decreases within this range as we consider points i further inland from the coast. Finally, to determine B we employ Holland's (2008) approximation method, and we use the parametric model estimated by Xiao et al. (2009) to estimate X. In terms of the implementation of (2), we use the best track data for Cyclone Pam as taken from the Joint Typhoon Warning Center, which provides information, amongst other things, on the maximum wind speed and the location of the storm eye at 6 hourly intervals. We interpolate these data to obtain hourly observations. For each hourly observation of the storm, we can then calculate the W for each nightlight cell centroid contained within our South Pacific islands and retain this value of at least 119 km h^{-1}.

To derive island-specific aggregate time-varying measures of destruction, we also want to take exposure into account using w. Ideally we would like to have time-varying information on the degree of dispersion of economic activity within islands at the most spatially disaggregated level as possible, given that wind speeds due to tropical storms can differ substantially across space. Since this is not available, we instead use the above-described nightlight imagery values at the cell level in the month (February 2015) before Cyclone Pam struck. As the calibration of the nightlight imagery induced negative values for some cells, we add a constant to all values equal to the absolute value of the largest negative value plus a small positive constant (0.00001). This enabled us to have positive weights for all nightlight cells and ensured that any weight, calculated as a percentage of an island's total, varied between 0 and 1.

Although the extent of wind damage due to cyclones is certainly correlated with the amount of storm surge, this is probably only very imperfectly so; see, for instance, Needham and Keim (2014). Storm surge should therefore ideally be modelled independently. Unfortunately, storm surge modelling requires detailed local data, such as bathymetry and surface roughness, which is unavailable for the South Pacific. We thus construct a rather crude index of storm surge-prone areas as the weighted share of the area of low-elevation coastal zone (L), in a similar spirit to Elliott, Strobl, and Sun (2015). More specifically, in order to identify L in the affected islands, we follow McGranahan, Balk, and Anderson (2007) and Brecht et al. (2012) and define land areas contiguous with the coastline up to a 10 m rise elevation using the Shuttle Radar Topography Mission (SRTM) 30 m elevation data set. We then isolate the share of nightlight in these areas to arrive at an island share of nightlight intensity in storm surge-prone areas. As before, we add a constant to all values equal to the absolute value of the largest negative value, along with a small positive constant (0.00001) to obtain positive weights. More precisely, we construct a measure of storm surge damages, S, as follows:

$$S_j = \sum_{i=1}^{I} w_{i,j} \frac{L_{i,j}}{A_j} \quad L_i = 1,0, \tag{3}$$

where L is an indicator of whether cell i lies within a low-elevation coastal zone (=1) or not (=0) and A is the area of island j. The estimated values of S of the three affected islands are shown in the last column of Table 1. Accordingly, the largest storm surge potential damages are in Tuvalu, followed by the Solomon Islands and then Vanuatu.

The final destructive aspect of tropical cyclones is that due to heavy rainfall. Unfortunately, we know of no data set that would provide us a local enough measure of rainfall on a monthly basis since 2014, covering all islands within our sample. We hence must rely on the somewhat scarce evidence that suggests that local rainfall during a tropical storm is considerably correlated with local wind exposure. For instance, it has been found that both winds and precipitation are the highest closer to the eye of

Table 1. Typhoon damage indices affecting nightlight intensity in the South Pacific Islands (dependent variable: nightlight intensity in W cm^{-2} sr^{-1}; January 2014 to April 2016 data; Panel Fixed Effects Regression).

Variable	Model 1 B[a]	Model 1 t[b]	Model 2 B[a]	Model 2 t[b]	Model 3 B[a]	Model 3 t[b]
S_t	−38.56	(−0.32)	20.20	(0.28)	9.13	(0.12)
S_{t-1}	20.00	(0.12)	−69.61	(−1.50)	−70.10	(−1.42)
S_{t-2}	−75.129	(−0.43)	41.54	(0.70)	32.73	(0.53)
S_{t-3}	202.53	(0.67)	11.12	(0.09)	43.39	(0.35)
S_{t-4}	−300.72	(−0.63)	−14.70	(−0.07)	−61.84	(−0.28)
S_{t-5}	497.01	(0.76)	78.09	(0.37)	133.32	(0.61)
S_{t-6}	−463.37	(−0.46)	170.17	(0.54)	96.12	(0.30)
S_{t-7}	841.23	(0.59)	−138.30	(−0.36)	9.53	(0.03)
S_{t-8}	−1354.04	(−0.62)	89.36	(0.14)	−118.96	(−0.18)
S_{t-9}	224.47	(0.67)	33.22	(0.03)	373.04	(0.34)
S_{t-10}	−3349.65	(−0.68)	−68.41	(−0.05)	−537.38	(−0.37)
S_{t-11}	4905.36	(0.65)	−98.66	(−0.04)	696.48	(0.29)
S_{t-12}	−7479.19	(−0.66)	14.25	(0.00)	−1170.439	(−0.32)
D_t	−0.06	(−3.07)***			−0.06	(−3.09)***
D_{t-1}	−0.08	(−4.67)***			−0.08	(−4.67)***
D_{t-2}	−0.09	(−4.96)***			−0.09	(−4.93)***
D_{t-3}	0.03	(1.01)			0.03	(1.00)
D_{t-4}	−0.11	(−6.16)***			−0.11	(−6.17)***
D_{t-5}	−0.07	(−3.15)***			−0.06	(−3.14)***
D_{t-6}	−0.01	(−0.35)			−0.01	(−0.35)
D_{t-7}	0.14	(3.73)***			0.14	(3.71)***
D_{t-8}	0.06	(1.69)*			0.06	(1.71)*
D_{t-9}	0.24	(6.39)***			0.23	(6.35)***
D_{t-10}	0.23	(9.35)***			0.23	(9.30)***
D_{t-11}	0.74	(9.83)***			0.73	(9.87)***
D_{t-12}	0.24	(8.89)***			0.24	(9.15)***
No. of Obs.	594		594		594	
R^2	0.14		0.08		0.14	
F-statistic	1.62***		1.24***		1.64***	

*Indicates a 0.1 two-tailed significance level; **Indicates a 0.05 significance level; ***Indicates a 0.01 significance level; [a]Unstandardized regression coefficient; [b]t-statistic; Time indicator variables are included but their coefficients are not reported.

the storm (Riehl 1954). It must nevertheless be kept in mind that at best our wind damage proxy is capturing both damages due to wind exposure and rainfall, and at worst that our analysis can only be interpreted in terms of capturing damages due to wind exposure and, to a limited extent, storm surge.

3. Methodology

3.1. *Graphical analysis*

We undertake two forms of graphical analysis. First we take the island of Vanuatu as an example to show how pixel-level night-time intensity might have changed after the typhoon. To this end, we extract all pixels within the islands' land surfaces for the composites March 2014, March 2015, and March 2016. We then subtracted the radiance values of March 2014 from March 2015 and the values of March 2015 from March 2016, in order to show possible changes in night-time intensity values that might have coincided with the storm. Second, we calculate the average of pixel values for each set of islands affected by the typhoon according to our damage indices D and S outlined above for each month of the composites. For those unaffected, we calculate the average across all islands. The monthly series of the three affected islands are then compared to the unaffected group.

3.2. Regression analysis

To statistically disentangle the effect of wind and storm surge destruction of Cyclone Pam on the South Pacific islands, we estimate the following:

$$N_{jt} = a + \sum_{k=0}^{12} \varphi_{t-k} S_j + \sum_{k=0}^{12} \beta_{t-k} D_j + \lambda_t + \mu_j + \varepsilon_{j,t}, \tag{4}$$

where N is the average nightlight intensity in island j at time t and D and S are, respectively, our wind and storm surge destruction proxies described above, with contemporaneous ($k = 0$) and lagged ($k > 0$) effects. One should note that prior to March 2015, i.e. prior to Typhoon Pam, D and S are zero. λ is a vector of time-specific indicator variables and is included in order to control for time-specific shocks common to all South Pacific islands. μ is a vector of island-specific time-invariant indicator variables. To account for these, we employ a panel fixed effects estimator, which essentially transforms the variables into deviations from their means, and thus purges μ (as well as a) from (4). ε is an i.i.d. error term, whereas a is a standard intercept term. To allow for spatial- and autocorrelation, we calculate Driscoll and Kraay (1998) standard errors. One may also want to note that after we control for island-specific time-invariant unobservables, μ, arguably D and S are exogenous since they can be considered as random realizations of storm occurrence. Thus one can, with reasonable confidence, interpret the coefficients of interest to be estimated, i.e. the φ's and β's, as capturing the causal effect of wind and storm surge destruction on average nightlight intensity.

3.3. Quantitative implications

One can use the estimated coefficients from (4) to gain insight into the quantitative significance of Cyclone Pam. More specifically, we can construct the following measure of cumulative impact, C, for each of our 12 months after the cyclone:

$$C_{j,s} = a + \sum_{k=0}^{12} \frac{\varphi_{t-k} S_j}{N_{j,t=-1}} 100 + \sum_{k=0}^{12} \frac{\beta_{t-k} D_j}{N_{j,t=-1}} 100 \ S = 1, ..., 12, \tag{5}$$

where $N_{j,t=-1}$ is the average nightlight intensity in island j in the month (February 2015) before Cyclone Pam. Importantly, one should note that as we calculate C for each period after the cyclone, we set β_{t-s} and φ equal to 0 when they are not significant. Moreover, for any k we set the additional C itself to 0 if an F-test of the coefficients, excluding the ones that are not significant, indicated that the null hypothesis that they are jointly equal to zero cannot be rejected at the 5% significance level. We are thus considering cumulative impacts as those where both marginal and total cumulative impacts were significant.

4. Results

Calculation of D and S from Equations (1)–(3) reveals that the Solomon Islands and Vanuatu were potentially affected by typhoon winds, with values of 0.00340 and 1.2389, respectively, whereas the Solomon Islands, Tuvalu, and Vanuatu were potentially affected by storm surge damage, with values of 1.309×10^{-6}, 0.0005, and 5.739×10^{-7}, respectively.

Our graphical pixel-level analysis in Figures 3 and 4, using Vanuatu as an example, shows that in Vanuatu compared with one year prior to the typhoon there were some visual decreases in night-time intensity, particularly where much of the intensity is concentrated, as suggested by Figure 1, whereas one year after the storm the intensity increased in this area again.

Figure 5 depicts the trends in average nightlight intensity for the three affected countries, i.e. parts (b), (c), and (d), as well as the average of all unaffected countries in part (a). For the three affected islands, there is a drop in intensity just after the storm. However, this fall in intensity seems to also have occurred on average for the unaffected groups. Moreover, it seems to coincide with the general seasonal patterns during this period of the year. This underlines the importance of trying to disentangle the potential effect of Typhoon Pam with regression analysis.

We estimate the regression Equation in (4) using all islands, affected and unaffected, the results of which are given as Model 1 in Table 1. As can be seen, after purging island-specific effects, μ, from (4) using a fixed-effects estimator, the control variables manage to explain 14% of the variation in nightlights. Considering the individual coefficients, one can see that wind exposure has a negative and significant impact in the month of the strike and the 5 months thereafter, except for the third month. From the seventh month onwards, the trend is reversed when a positive significant impact sets in, which continues until the end of our sample period, i.e. 12 months after Typhoon Pam first produced damage.

In contrast to wind exposure, the effect of storm surge is shown to be insignificant at the time of the cyclone strike as well as in the months thereafter. This may be because storm surge does not have a significant-enough or long-enough, if the impact lasts less

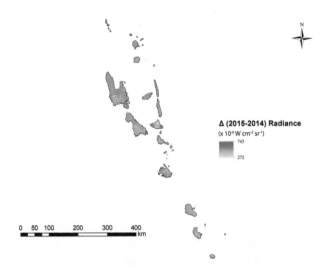

Figure 3. Nightlight intensity: Vanuatu (March 2014–March 2015).

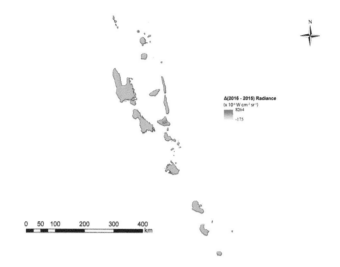

Figure 4. Nightlight intensity: Vanuatu (March 2015–March 2016).

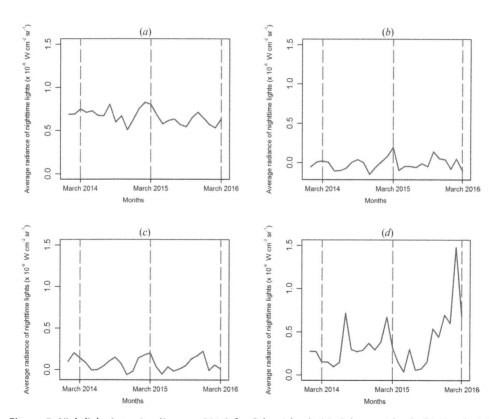

Figure 5. Nightlight intensity: (January 2014) for Other Islands (a), Solomon Islands (b), Tuvalu (c), and Vanuatu (d).

than a month, effect to show up in our nightlight measure of local economic activity. Alternatively, our 'modelling' of storm surge may be too simplistic to accurately capture its nature, hence inducing attenuation bias. Since the latter reason is likely to play at least some role, one must view our finding with regard to storm surge with at least some caution. Another reason may be that the wind destruction proxy is already capturing storm surge and hence there is a problem of multi-collinearity. Thus in Model 2 in Table 1, we excluded D and its lags from (4), but, as can be seen, this similarly produced insignificant coefficients on S and its lags and reduced the explanatory power of the model to an R^2 of 0.08.

Damage assessment reports suggest that Tuvalu experienced the most severe destruction from storm surge compared with Vanuatu and Solomon Islands in that there was inundation from storm surge and sea swells of 3–5 m in seven of its nine islands and the government had to declare a state of emergency (OCHA 2015c). We thus experimented with setting S equal to zero for Vanuatu and the Solomon Islands, but kept the value for Tuvalu unchanged. As depicted in Model 3 in Table 1, this did not produce any significant effect of S on island average nightlight intensity.

We next calculate the implied cumulative impact in (5) over time for wind exposure using the significant β's for the Solomon Islands and Vanuatu, but setting the ϕ's, in accordance with our regression results, to zero. The resultant values are shown in Figures 6 and 7, respectively. As can be seen, given that we are using the same β for both islands, the pattern of cumulative impact is identical. In this regard, one finds that the cumulative negative effect of Cyclone Pam increases slowly until the fifth month after the event, until it begins to fall as the marginal impact turns positive. By the eighth month, the overall cumulative impact is 0. Thereafter, as the positive effects in response to the storm accumulate, there is a starkly rising beneficial impact of the storm on nightlight intensity. Whereas the pattern of the impact is by construction the same across the two islands, the quantitative nature of this effect differs substantially. For the Solomon Islands, the total negative impact is never greater than two percentage points of the pre-storm level of intensity, whereas one year after the event the net cumulative positive effect has increased to a little over 5 percentage points. Hence, one can conclude that the overall net impact of Pam was relatively small for the Solomon Islands, at least measured in terms of nightlight intensity.

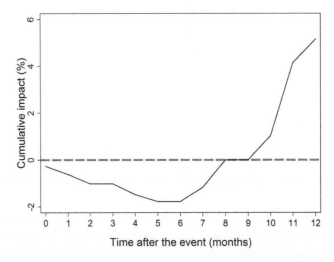

Figure 6. Cumulative impact of wind exposure destruction – Solomon Islands.

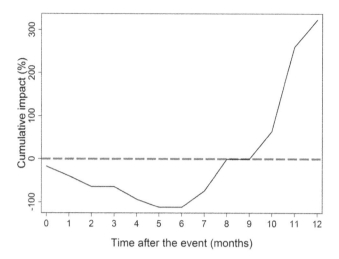

Figure 7. Cumulative impact of wind exposure destruction – Vanuatu.

Reports that assessed damage following Typhoon Pam similarly found that destruction was relatively lower for Solomon Islands. In Solomon Islands, the islands affected by strong winds and rainfall were among the least populated territories, thereby minimizing any negative impact. For instance, in Malaita and Temotu Province, just 30,000 people (5% of the total population) were affected by flooding caused by heavy rains (OCHA 2015c), whereas the islands of Anuta and Temotu were the most affected by strong winds and rainfall and are among the least populated territories (OCHA 2015d). The Solomon Islands also received some international support for relief and reconstruction immediately following the cyclone, perhaps accounting for the subsequent positive impact after the cyclone struck.

In contrast to the Solomon Islands, both within the first few months when effects were negative, as well as after 9 months when the cumulative effect turned positive, the impact in Vanuatu was fairly large. More precisely, our results suggest that the total net negative impact on local economic activity reached a loss of as much as 111% by the fifth month. Damage reports also found losses to be high and were estimated to be approximately US$ 449.4 million (64.1% of GDP), which is most likely an underestimation because the figure was based on the best available information at the time (Esler 2015). The vast destruction is likely because Typhoon Pam directly struck Vanuatu at category 5 strength and all six provinces were affected with the larger and more populated islands being more negatively affected (OCHA 2015b). Moreover, the centre of the storm passed east of the main island Efate, where the capital Port Vila was directly struck. The cyclone damaged or completely destroyed 17,000 buildings, including houses, schools, hospitals, and clinics (Esler 2015). However, our results show that once the storm started enhancing economic activity there was a large positive effect. As a matter of fact, as of the final month of available data, our regressions results suggest that local night-time brightness has increased by over 300% relative to prior to the storm.

Our finding of an initial large negative impact within the first 5 months after Cyclone Pam struck followed by a large positive impact is supported by damage assessment reports. Esler

(2015) stated that while Typhoon Pam was expected to reduce Vanuatu's GDP growth by 5.5 percentage points relative to the 2015 pre-cyclone forecast, the large scale of recovery and reconstruction activities along with international aid and funding received, which started to take place almost immediately following the storm, would allow for a large positive GDP growth of 1.4% in 2015, with further increases in 2016 and 2017. Furthermore, although tourism was one of the most negatively affected sectors, earnings in the industry were only negatively affected for 3–6 months while hotels remained closed for clean up and reconstruction (Esler 2015). Additionally, government spending in quick response for recovery and reconstruction and international funds received, along with policies to increase tax exemptions and commercial bank lending, might have significantly increased economic activity within months after Typhoon Pam struck (Esler 2015).

Just days after Typhoon Pam struck, the government of Vanuatu led response efforts with support from the Pacific Humanitarian Team and the Vanuatu Humanitarian Team and worked with various development partners for recovery and reconstruction activities. The government spent US$ 2.21 million from its Emergency Relief Fund to support immediate humanitarian response and redeployed additional funds from their 2015 fiscal budgets (Esler 2015). Furthermore, US$ 3.28 million was received from the European Union (EU), a US$ 1.84 million insurance payout was received from the World Bank for the 2015 recurrent fiscal budget to finance recovery-related expenditures, and grant funding of US$ 6.09 million was received from donors for recovery operations (Esler 2015). As of 26 March, just about 12 days after Typhoon Pam struck, US$ 18 million was already received (OCHA 2015e). Overall, recovery and reconstruction were estimated at US $ 316 million, of which US$ 95 million was for short-term needs (Esler 2015).

5. Discussion

We investigate the short-term economic impact of Cyclone Pam using monthly composites of nightlight intensity and tropical cyclone destruction indices. Our regression analysis reveals that initially the storm reduced economic activity in the affected islands. However, by the seventh month, positive boosts to nightlight intensity, possibly due to reconstruction activities and government programmes, began to counteract any negative effects and by the ninth month it resulted in cumulative net increases in activities related to night-time electricity usage. More generally, our results suggest that there is likely considerable temporal heterogeneity in the response of areas affected by tropical cyclones. Moreover, the analysis demonstrates the potential of using nightlight imagery to assess the short-term economic impact of tropical storms, and possibly other extreme event phenomena, in a relatively timely manner.

There are of course a number of shortcomings of our study that future research could address. Most obviously, waiting for the availability of more monthly images would allow one to explore what the impact of tropical cyclones are beyond one year from their occurrence. From a methodological point of view, clearly a better storm surge destruction proxy, as well as one capturing the damages due to heavy rainfall destruction, would provide a more accurate picture of the economic consequences of these storms. In addition, although very convenient in terms of spatial and temporal resolution, further investigation into what sort of economic activity nightlight images are capturing would be beneficial. Finally, since we find different patterns of impact as time passes since the event, a further understanding of what is driving these patterns would be insightful and aid policymakers in post event management strategies.

Disclosure statement

No potential conflict of interest was reported by the authors.

References

ASCE. 2006. *Minimum Design Loads for Buildings and Other Structures, ASCE/SEI 7-05, American Society of Engineers*. Reston, VA: American Society of Civil Engineers.

Bertinelli, L., and E. Strobl. 2013. "Quantifying the Local Economic Growth Impact of Hurricane Strikes: An Analysis from Outer Space for the Caribbean." *Journal of Applied Meteorology and Climatology* 52: 1688–1697. doi:10.1175/JAMC-D-12-0258.1.

Boose, E., I. S. Mayra, and R. F. David. 2004. "Landscape and Regional Impacts of Hurricanes in Puerto Rico." *Ecological Monograph* 74 (2): 335–352. doi:10.1890/02-4057.

Brecht, H., S. Dasgupta, B. Laplante, S. Murray, and D. Wheeler. 2012. "Sea-Level Rise and Storm Surges: High Stakes for a Small Number of Developing Countries." *The Journal of Environment and Development* 21 (1): 120–138. doi:10.1177/1070496511433601.

Chen, X., and W. Nordhaus. 2011. "Using Luminosity Data as a Proxy for Economic Statistics." *Procedings of the National Academy of Sciences* 108 (21): 8589–8594. www.pnas.org/cgi/doi/10.1073/pnas.1017031108.

Chen, X., and W. Nordhaus. 2015. "A Test of the New VIIRS Lights Data Set: Population and Economic Output in Africa." *Remote Sensing* 7: 4937–4947. doi:10.3390/rs70404937.

Driscoll, J. C., and A. C. Kraay. 1998. "Consistent Covariance Matrix Estimation with Spatially Dependent Panel Data." *Review of Economics and Statistics* 80 (4): 549–560. doi:10.1162/003465398557825.

Elliott, R., E. Strobl, and P. Sun. 2015. "The Local Impact of Typhoons on Economic Activity in China: A View from Outer Space." *Journal of Urban Economics* 88: 50–66. doi:10.1016/j.jue.2015.05.001.

Elvidge, C., K. Baugh, M. Zhizhin, and F. Hsu. 2013. "Why VIIRS Data are Superior to DMSP for Mapping Nighttime Lights." *Proceedings of the Asia-Pacific Advanced Network* 35: 62–69. doi:10.7125/APAN.35.

Emanuel, K. 2011. "Global Warming Effects on U.S. Hurricane Damage." *Weather Climate and Society* 3: 261–268. doi:10.1175/WCAS-D-11-00007.1.

Esler, S. 2015. "Vanuatu Post-Disaster Needs Assessment Tropical Cyclone Pam." Government of Vanuatu https://www.gfdrr.org/sites/default/files/publication/PDNA_Cyclone_Pam_Vanuatu_Report.pdf.

Henderson, V., A. Storeygard, and D. N. Weil. 2012. "Measuring Economic Growth from Outer Space." *American Economic Review* 102 (2): 994–1028. doi:10.1257/aer.102.2.994.

Holland, G. 2008. "A Revised Hurricane Pressure–Wind Model." *Monthly Weather Review* 136: 3432–3445. doi:10.1175/2008MWR2395.1.

Holland, J. 1980. "An analytic Model Of The Wind And Pressure Profiles In Hurricanes." *Monthly Weather Review* 108 (8): 1212-1218. doi:10.1175/1520-0493(1980)108<1212:AAMOTW>2.0.CO;2.

Horwich, G. 2000. "Economic Lessons of the Kobe Earthquake." *Economic Development and Cultural Change* 48 (3): 521–542. doi:10.1086/452609.

Kantha, L. 2008. "Tropical Cyclone Destructive Potential by Integrated Kinetic Energy." *American Meterological Society* 89: 219–221. doi:10.1175/BAMS-89-2-219.

Kyba, C., S. Garz, H. Kuechly, A. de Miguel, J. Zamorano, J. Fischer, and F. Hoelker. 2015. "High-Resolution Imagery of Earth at Night: New Sources, Opportunities and Challenges", *Remote Sensing* 7: 1–23.

Li, X., H. Xu, X. Chen, and C. Li. 2013. "Potential of NPP-VIIRS Nighttime Light Imagery for Modeling the Regional Economy of China." *Remote Sensing* 5: 3057–3081. doi:10.3390/rs5063057.

Ma, T., C. Zhou, T. Pei, S. Haynie, and J. Fan. 2014. "Responses of Suomi-NPP VIIRS-Derived Nighttime Lights to Socioeconomic Activity in China's Cities"." *Remote Sensing Letters* 5: 165–174. doi:10.1080/2150704X.2014.890758.

McGranahan, G., D. Balk, and B. Anderson. 2007. "The Rising Tide: Assessing the Risks of Climate Change and Human Settlements in Low Elevation Coastal Zones." *Environment and Urbanization* 19 (1): 17–37. doi:10.1177/0956247807076960.

Needham, H., and B. Keim. 2014. "Correlating Storm Surge Heights with Tropical Cyclone Winds at and before Landfall." *EARTH Interactions* 18: 1–26. doi:10.1175/2013EI000558.1.

OCHA. 2015a. "Vanuatu: Severe Tropical Cyclone Pam Situation Report No. 1 (As of 15 March 2015)." http://reliefweb.int/report/vanuatu/vanuatu-severe-tropical-cyclone-pam-situation-report-no-1-15-march-2015.

OCHA. 2015b. "Vanuatu: Severe Tropical Cyclone Pam Situation Report No. 2 (As of 16 March 2015)." http://reliefweb.int/report/vanuatu/vanuatu-severe-tropical-cyclone-pam-situation-report-no-2-16-march-2015.

OCHA. 2015c. "Vanuatu: Tropical Cyclone Pam Situation Report No. 4 (As of 18 March 2015)." http://reliefweb.int/report/vanuatu/tropical-cyclone-pam-situation-report-no-4-16-march-2015.

OCHA. 2015d. "Vanuatu: Tropical Cyclone Pam Situation Report No. 7 (As of 21 March 2015)." http://reliefweb.int/sites/reliefweb.int/files/resources/OCHA_VUT_TCPam_Sitrep7_20150321.pdf.

OCHA. 2015e. "Vanuatu: Tropical Cyclone Pam Situation Report No. 12 (As of 26 March 2015)." http://reliefweb.int/sites/reliefweb.int/files/resources/OCHA_VUT_TCPam_Sitrep12_20150326.pdf.

Paulsen, B. M., and J. L. Schroeder. 2005. "An Examination of Tropical and Extratropical Gust Factors and the Associated Wind Speed Histograms." *Journal of Applied Meteorology and Climatology* 44 (2): 270–280. doi:10.1175/JAM2199.1.

Rasmussen, T. N. 2004. "Macroeconomic Implications of Natural Disasters in the Caribbean." Working Paper 04/224, Washington, DC: The International Monetary Fund. doi:10.5089/9781451875355.001.

Riehl, H. 1954. *Tropical Meteorology*. New York: McGraw-Hill Book Company.

Rybnikova, N., and B. Portnov. 2014. "Mapping Geographical Concentrations of Economic Activities in Europe Using Light at Night (LAN) Satellite Data." *International Journal of Remote Sensing* 35: 7706–7725. doi:10.1080/01431161.2014.975380.

Strobl, E. 2011. "The Growth Impact of Hurricane Activity: Evidence from US Coastal Counties." *Review of Economics and Statistics* 93 (2): 575–589. doi:10.1162/REST_a_00082.

Strobl, E. 2012. "The Macroeconomic Impact of Natural Disasters in Developing Countries: Evidence from Hurricane Strikes in the Central American and Caribbean Region." *Journal of Development Economics* 97 (1): 130–141. doi:10.1016/j.jdeveco.2010.12.002.

Vickery, P. J., D. Wadhera, M. D. Powell, and C. Yingzhao. 2009. "A Hurricane Boundary Layer and Wind Feld Model for Use in Engineering Applications." *Journal of Applied Meteorology* 48 (2): 381–405. doi:10.1175/2008JAMC1841.1.

Walsh, K. J. E., J. L. McBride, P. J. Klotzbach, S. Balachandran, J. Camargo, G. Holland, T. R. Knutson, et al. 2015. "Tropical Cyclones and Climate Change." *WIRES Climate Change* 7 (1): 65–89.

Xiao, Q., X. Zhang, C. A. Davis, J. D. Tuttle, G. J. Holland, and P. J. Fitzpatrick. 2009. "Experiments of Hurricane Initialization with Airborne Doppler Radar Data for the Advanced- Research Hurricane WRF (AHW) Model." *Monthly Weather Review* 137: 2758–2777. doi:10.1175/2009MWR2828.1.

Port economics comprehensive scores for major cities in the Yangtze Valley, China using the DMSP-OLS night-time light imagery

Chang Li ⓘ, Guangping Chen, Jing Luo, Shice Li and Jia Ye

ABSTRACT

In this article, the Defence Meteorological Satellite Program Operational Linescan System (DMSP-OLS) night-time light remotely sensed data on a small scale is proposed to evaluate port economic comprehensive scores (PECS) for the ports of the major cities, including Shanghai, Nanjing, Wuhan, and Chongqing, in the Yangtze River Valley, China. First, the concept and calculation method of port night-time light intensity (PNLI) are proposed. Second, an estimation method of PECS is proposed using factor analysis. Third, two regression models (i.e. first- and second-order polynomials) between PNLI and PECS are built and tested. The goodness-of-fit of the two models are compared, both with and without outliers. Finally, experimental results show that the proposed methods for evaluating PECS in the Yangtze River Valley, China are feasible. Wuhan has the strongest correlation ($R^2 = 0.925$) and passes the F-test, $F_t = 55.429 > 4.26 = F_{t\,(0.05)}$ (2, 9); Shanghai has the weakest correlation (R^2: 0.688) but still passes as well, $F_t = 9.944 > 4.26 = F_{t(0.05)}$ (2, 9). Factors confounding the correlation for Shanghai are discussed. Overall, this study not only proposes a new set of methods to evaluate the performance of port economies but also provides a feasible way to use DMSP-OLS data to study the geographic problems of small scales.

1. Introduction

The port economy is a type of regional economy that is typically composed of shipping, port-neighbouring industries, trade, tourism, and other relevant industries in a certain area. The study of port economies has become an important component of economic geography and regional economics. However, the factors that affect the development of the port economy are complex, and it is often difficult to effectively evaluate the performance of a port economy. Many international organizations and scholars are committed to building a set of methods and techniques for evaluating the condition of port economies based on various economic indicators. In the 1980s, the United Nations Conference on Trade and Development (UNCTAD, 1983) developed a statistical index system of port performance in a prescriptive manual that can be used to assess the level of comprehensive port development. By comparing the economic operation of 20 ports and given certain standards, the data envelopment analysis (DEA) method was also used to evaluate port performance (Roll and Hayuth 1993). Some studies have found that port economic activities, resource abundance in the hinterland, and port linkage relationships are key elements of assessments of the competitiveness of a port economy (De Martino and Morvillo 2008). Others considered the available indicators of major Brazilian ports from 2006 to 2009 and created a model for obtaining the performance of container terminals based on a multicriteria and multivariate methodology to evaluate port performance (Madeira et al. 2012). What these studies have in common is that traditional socioeconomic factors form the basis of the evaluation. However, the traditional socioeconomic data is not sufficient for the calculation of port economics comprehensive scores (PECS) because of the infrequency of data updates, the poor timeliness of data acquisition, and the strong subjectivity inherent in a qualitative analysis. Moreover, many key factors are often unavailable because the relevant statistics department does not record them. Therefore, there exists a need for a more scientific, effective and convenient evaluation method.

Compared to the traditional socioeconomic data, the Defence Meteorological Satellite Program Operational Linescan System (DMSP-OLS) night-time light remotely sensed data, which can be downloaded from the website of the National Centres for Environmental Information (NCEI), National Oceanic and Atmospheric Administration (NOAA), have certain advantages, e.g. objective reflection, dynamic updating and wide coverage. Moreover, the data acquisition is easy and convenient. For these reasons, it has been successfully used in various fields to monitor temporal and spatial distributions of night-time light. At present, it has mainly been applied on macroscales and used in the study of regional economic development (Elvidge et al. 1997; Li, Ge, and Chen 2013), population migration and distribution (Sutton et al. 1997; Lo 2001; Zhuo et al. 2009), urbanization and urban spatial expansion (He et al. 2006; Small and Elvidge 2013; Yu et al. 2014), event evaluation (Li and Li 2014), energy consumption (Letu et al. 2010; He et al. 2014), and ecological environment effects (Kuechly et al. 2012; Bauer et al. 2013). However, few theories and methods have been proposed that utilize DMSP-OLS data in small-scale studies, and thus additional research in this area is needed. Obviously, it would be desirable to build a set of application techniques, methods and theoretical systems for the use of DMSP-OLS data on small scales through a rigorous empirical research process.

Theoretically, in fact, the process of utilizing DMSP-OLS imagery on smaller scales is a kind of scaling transformation in geography. According to the laws of geography, geographic objects and their properties have the characteristics of spatial correlation, heterogeneity and symmetry, and the application of small-scale and large-scale mechanisms have universality and individuality in some aspects (Meentemeyer 1989; Fotheringham 1989; Atkinson 1997; Marceau and Hay 1999; Raffy 1992). In terms of geographical spatial range of scales, port areas belong to the category of small scale. A large number of studies have successfully demonstrated the correlation between night-time light and urban development (e.g. urban spatial expansion and economy) in mechanism and practice (Elvidge et al. 1997; He et al. 2006; Small and Elvidge 2013; Yu et al. 2014). The port, as a 'subset' of the city, plays an important role in the city by greatly supporting the city's economic development. If the city's night lights and economy are closely correlated, it stands to reason that the 'subset' (i.e. city's port night lights) and the city's port economy might also be correlated. The scaling must allow the full information of geographic objects to be retained and ensure that the original geographic information will not be lost (Woodcock and Strahler 1987; Raffy 1994; Moody and Woodcock 1995; McCabe and Wood 2006); on that basis, we chose the appropriate spatial and statistical analysis. Hence, using the small-scale DMSP-OLS data to study a port economy is feasible.

In this article, we use the DMSP-OLS data on a small scale to study the port economy and propose a method that evaluates port economics using remotely sensed data. Moreover, the work and contributions of this article are as follows:

(1) A small-scale extraction technique of the DMSP-OLS data is proposed, which contributes a new method and technological reference for data processing.
(2) A comprehensive evaluation method is established between the DMSP-OLS data and the port economic multiple indicators by factor analysis and regression analysis, and a varimax rotation is used to simplify the expression of a particular sub-space in terms of just a few major items.
(3) Linear regression models for forecasting PECS in major cities of the Yangtze Valley are built and tested, which will be helpful to future studies of the whole Yangtze River Valley, China and even in other ports.

2. Study area and data

2.1. *Study area*

The Yangtze River is the world's third longest river and the largest in Asia; it stretches approximately 6300 km and the area of its watershed is approximately 1.8×10^6 km²; it is a world's famous river with an important role in China's economy. In recent years, the construction of the Yangtze River economic belt has become an important component of China's national strategy, and as such, development of the major ports in the Yangtze River Valley is underway.

The Yangtze River economic belt is the largest set of inland industrial and manufacturing districts in the world; it plays an important and strategic role in China's social and economic development. In addition, it has great potential for continued economic development because of its unique geographic location and natural resources. It is critical to select suitable ports as the objects of study because the Yangtze River Valley has many ports, both large and small. According to a report released by the Chinese Association of Development Strategy Studies (CADSS) (2011) and Beijing Jiaotong University (BJTU), Shanghai and Wuhan ports are two of the major ports in China; in addition, Nanjing and Chongqing ports are also important ports in the Yangtze River Valley. An evaluation of these four ports will be of great significance, for the ports belong to four of the largest cities in China. In this study, Shanghai, Nanjing, Wuhan, and Chongqing ports form the study area; the locations of these ports are shown in Figure 1.

2.2. *Study data*

2.2.1. *DMSP-OLS night-time light data*
The annual DMSP-OLS data was downloaded from the website of the NCEI, NOAA (http://www.ngdc.noaa.gov/eog/dmsp/downloadV4composites.html). The data include

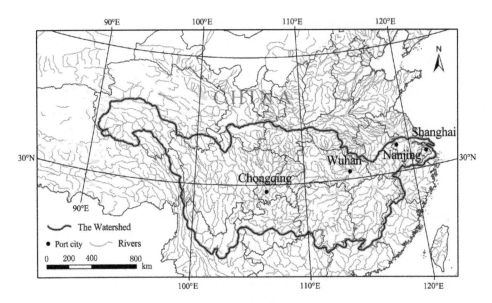

Figure 1. The four main ports city in Yangtze River Valley, China.

average visible light, cloud-free coverage and stable light average data from 1992 to 2013. The stable light average data represent the light from cities, towns, and other sites with persistent lighting, including the light from gas flares. The data values range from 1 to 63. We selected these particular datasets because some major outliers (such as those from fires) had already been discarded. Figure 2 shows the DMSP-OLS data for the four main port cities in the Yangtze River Valley, China.

Pre-processing of the DMSP-OLS data mainly included three steps:

(a) *Reprojection and resampling on raster imagery.* The projection coordinate system of the imagery is Geographic (latitude/longitude), and the spheroid is WGS 1984. In this study, the projection coordinate system was converted into Lambert Conformal Conic, the spheroid was converted into Clarke 1866 and the data were then resampled at a 1 km resolution in order to make it convenient to clip imagery and calculate areas.
(b) *Clipping the imagery.* We clipped out the DMSP-OLS stable light average data of the main ports in Yangtze River Valley from 2001 to 2013.
(c) *Imagery intercalibration.* In consideration of the algorithm accuracy, stability and efficiency, based on our previous study (Li et al. 2016), the LMedS (i.e. the least-median-of-squares method) is used to intercalibrate the radiometric problem.

2.2.2. *Port economic indicator data*

There is no unified standard for port economic indicator selection. As shown in Table 1, we cannot build a port economic indicator system without considering three important factors (Roll and Hayuth 1993; Madeira et al. 2012). The first is port infrastructure (e.g. berth number, pier length, berthing capacity, loading and unloading machine, warehouse capacity, storage yard area and capacity). The second is economic scale (e.g. cargo

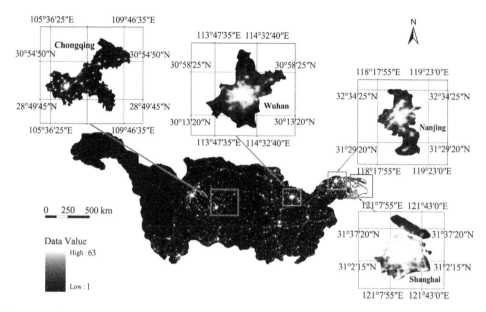

Figure 2. The DMSP-OLS data of the Shanghai, Nanjing, Wuhan and Chongqing in Yangtze River Valley, China (F18-2013).

Table 1. The general indicator system of port economy evaluation.

Target layer	First-class indicator	Second-class indicator	Code
		Number of berths	A_1
		Pier length	A_2
		Loading and unloading machine	A_3
	Port infrastructure (PI)	berthing capacity	A_4
		The warehouse area	A_5
		The warehouse capacity	A_6
		Storage yard area	A_7
		Storage yard capacity	A_8
		Cargo throughput	A_9
		Ore	A_{10}
		Coal	A_{11}
		Oil	A_{12}
The comprehensive performance of port economy	Economies of scale (ES)	Container throughput	A_{13}
		Automobile throughput	A_{14}
		Passenger throughput	A_{15}
		Revenue Passenger Kilometres (RPKs)	A_{16}
		Fixed investments	A_{17}
		Profits	A_{18}
	Investment and efficiency (IE)	Loading, unloading cost	A_{19}
		Production efficiency	A_{20}
		The city's GDP	A_{21}
	City's diffusion effect (DE)	The proportion of secondary industry in GDP	A_{22}
		The city's population	A_{23}
	Preferential policies (PP)	Tax policy	A_{24}
		Free-port policy	A_{25}

GDP: gross domestic product.

Table 2. The comprehensive evaluation indicator system of main ports in Yangtze River Valley, China.

Port	Indicator code	Number of indicators	Indicator integrity(%)
Shanghai	$A_1, A_2, A_4, A_5, A_7, A_9, A_{10}, A_{11}, A_{12}, A_{13}, A_{15}, A_{16}, A_{17}, A_{18}, A_{21}, A_{22}, A_{23}, A_{24}, A_{25}$	19	76
Nanjing	$A_1, A_2, A_5, A_6, A_7, A_8, A_9, A_{10}, A_{11}, A_{12}, A_{13}, A_{17}, A_{21}, A_{22}, A_{23}, A_{24}, A_{25}$	17	68
Wuhan	$A_1, A_2, A_4, A_9, A_{10}, A_{11}, A_{12}, A_{13}, A_{14}, A_{15}, A_{17}, A_{18}, A_{21}, A_{22}, A_{23}, A_{24}, A_{25}$	17	68
Chongqing	$A_1, A_2, A_5, A_7, A_9, A_{10}, A_{11}, A_{12}, A_{13}, A_{14}, A_{15}, A_{16}, A_{17}, A_{21}, A_{22}, A_{23}, A_{24}, A_{25}$	18	72

throughput, container throughput, passenger throughput). The third is the investment and efficiency (e.g. fixed investments, annual income, loading, unloading cost, and production efficiency). Some additional factors, such as the city's diffusion effect (e.g. the GDP, second industry ratio, and urban population) and preferential policies (e.g. comprehensive bonded policy), have been used to evaluate port economies (Bichou and Gray 2004; Peter, Michiel, and Martijn 2007).

In this study, the port economic statistics are mainly drawn from the *Chinese Ports Year Book* (http://cyfd.cnki.com.cn/N2013100053.htm), and some of the data were provided by the port statistics department (http://english.chinaports.org/). We considered the economic indicators' availability, along with their representativeness regarding the development of the main ports in the Yangtze River Valley, and selected the most satisfactory indicators; Table 2 shows the comprehensive economic indicator system of each port after selection, and the indicator codes in Table 2 correspond to the indicators in Table 1.

2.2.3. Auxiliary data

Our auxiliary data includes a 1:40,00,000 grid map of vector polygon data, which was provided by the National Geomatics Centre of China (NGCC, http://ngcc.sbsm.gov.cn/), along with the extracted vector data of the main ports from Google Earth in different years from 2001 to 2013.

3. Methodology

In this study, the vector polygon data were extracted from an overlay of the Google Earth high-resolution imagery. We overlaid the vector polygon data on the DMSP-OLS data and then clipped out the port imagery. Port night-time light intensity (PNLI) is calculated based on the port area. PECS is then calculated using factor analysis. By establishing a linear regression model, we demonstrate the temporal correlations between PNLI and PECS of the main ports in the Yangtze River Valley and evaluate the regression correlation with the usual parameters. Finally, we obtain the results of the experiment. The process flow of our study is illustrated in Figure 3.

3.1. Extraction of PNLI

Since the DMSP-OLS night-time light imagery is not a high-resolution dataset, a single pixel may be much larger than the port area. Therefore, it is useless to obtain the PNLI by clipping the port imagery on that small scale directly. To acquire the digital number

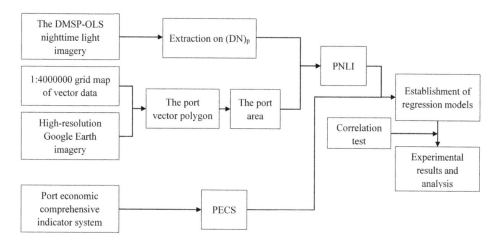

Figure 3. Research process flow chart.

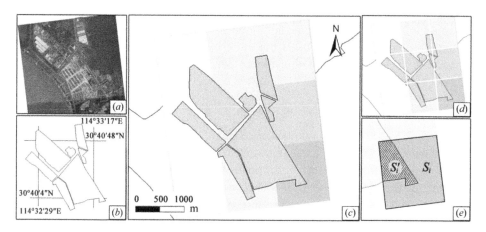

Figure 4. Extraction process of port area and digital number values (a case of Yang-Luo port); the pixel of DMSP-OLS data represented by lattice grey shade; the port vector borders represented by the polygon. (a) Extraction imagery in port area; (b) the port vector polygon; (c) the DMSP-OLS night-time light imagery of port; (d) the rasterized DMSP-OLS night-time light imagery of port; (e) Calculation on (DN)′.

in the port area (DN)′, the port vector polygon data are rescaled such that all the polygons cover the relevant pixels. The vector polygon data is then overlaid onto the DMSP-OLS raster imagery, and then the port night-time light imagery is clipped out (approximately). Take Yangluo port, Wuhan as an example. As shown in Figures 4(a)–(e), the process of extracting S (i.e. the area of port) and (DN)′ includes five steps:

(a) Extraction of the Google Earth imagery in the port area from 2001 to 2013.
(b) The port vector polygon.
(c) The DMSP-OLS night-time light imagery of the port.
(d) The rasterized DMSP-OLS night-time light imagery of the port.
(e) Calculation on (DN)′.

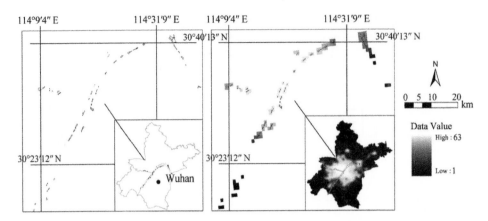

Figure 5. Spatial distribution of Wuhan port and its DMSP-OLS data (F18-2013).

Similarly, others can be extracted by this method. Figure 5 shows the spatial distribution of Wuhan port and its DMSP-OLS data acquired by sensor F18 in 2013.

Changes in DMSP-OLS night-time light intensity can reflect changes in socioeconomic activities. The digital number (DN) is the intensity quantified, and PNLI is formatted as (DN) per unit area. By using the equivalence between the area (i.e. S) and (DN) of each port, PNLI can be calculated by

$$S_i'/S_i = (DN)_i'/(DN)_i,$$ (1)

where $i = 1, 2, \ldots, n$; n is the number of pixels (i.e. the DMSP-OLS data represented by lattice grey shade) in the port area, as shown in Figure 4(e), S_i denotes the area of raster pixel i; S_i' denotes the area of port vector polygon in pixel i, here, $0 < S_i' \leq S_i$; $(DN)_i$ denotes the digital number of pixel i; $(DN)_i'$ denotes the digital number of the port in pixel i, here, $1 \leq (DN)_i' \leq (DN)_i \leq 63$. According to the above description, the number of pixels in the Yangluo port area is 9, and so $i = 1, 2, \ldots, 9$. Using this relationship, we have

$$PNLI = (DN)_p/S_p = \sum_{i=1}^{n}(DN)_i'/S_p$$
$$= \sum_{i=1}^{n}\left[(DN)_i S_i'/S_i\right]/S_p$$ (2)

where $(DN)_p$ denotes the total digital number of the port; S_p denotes the total area of the port vector region. The PNLI were calculated by this method and the results are presented in Table 3.

3.2. The model of calculating PECS

There are a number of indicators commonly used to evaluate port economies, but they often have some degree of information overlap; thus, we used factor analysis to standardize the indicators for multi-variable data processing. This allowed us to use several common factors instead of all the indicators to reflect most of the information from the comprehensive indicator system. These factors were given different weights. Suppose we have n samples; each sample has p variables, and after the standardization

Table 3. The PNLI extraction results of main ports in Yangtze River Valley, China from 2001 to 2013.

Year	PNLI			
	Shanghai	Nanjing	Wuhan	Chongqing
2001	3.394	6.537	4.049	1.258
2002	5.985	8.217	4.415	1.431
2003	5.862	7.894	3.997	1.303
2004	6.228	8.767	4.608	1.790
2005	5.336	8.415	4.746	2.200
2006	5.822	8.757	4.815	2.397
2007	4.080	10.281	5.379	2.349
2008	4.074	10.075	5.440	2.250
2009	3.982	8.349	4.372	2.032
2010	4.559	11.459	5.597	2.580
2011	4.648	11.183	5.572	2.350
2012	4.209	10.253	5.519	2.938
2013	4.618	11.406	5.637	2.921
Sum	62.796	121.593	64.146	27.800

process, the mean of standardized variables is 0, and the variance is 1. Combining with the selected indicators in Table 2, we have the port economic indicators M_1, M_2, ..., M_p, and $\{M_1, M_2, \cdots, M_p\} \subset \{A_1, A_2, \cdots, A_{25}\}$. The common factors are expressed by F_1, F_2, ..., F_m, and $0 < m < p \leq 25$, we have:

(a) $X = (M_1, M_2, ..., M_p)^{\mathsf{T}}$ is the vector of economic indicators for a given port, the mean vector $E(X) = 0$, the covariance matrix $Cov(X) = \Sigma$, and it is equal to the correlation matrix R.

(b) $F = (F_1, F_2, ..., F_m)^{\mathsf{T}}$ $(m < p)$ is the standardized vector indicator, and its mean vector $E(F) = 0$, the covariance matrix $Cov(F) = 1$, that is, the vector of each component is independent.

(c) $\varepsilon = (\varepsilon_1, \varepsilon_2, ..., \varepsilon_p)^{\mathsf{T}}$ and F are independent of each other, and $E(\varepsilon) = 0$, the covariance matrix of ε is a diagonal matrix, that is, each component ε is independent. The factor analysis model can be expressed as follows:

$$M_1 = a_{11}F_1 + a_{12}F_2 + \ldots + a_{1m}F_m + \varepsilon_1$$
$$M_2 = a_{21}F_1 + a_{22}F_2 + \ldots + a_{2m}F_m + \varepsilon_2$$
$$\cdots$$
$$M_p = a_{p1}F_1 + a_{p2}F_2 + \ldots + a_{pm}F_m + \varepsilon_p,$$

(3)

where F_1, F_2, ..., F_m are the common factors, ε_1, ε_2, ..., ε_p are the specific factors of M_p, and each specific factor and all the common factors are independent. The orthogonal factor model is

$$X = AF + \varepsilon.$$

(4)

Here, we have

$$X = \begin{bmatrix} M_1 \\ M_2 \\ \vdots \\ M_p \end{bmatrix}, A = \begin{bmatrix} a_{11} & a_{12} & \cdots & a_{1m} \\ a_{21} & a_{22} & \cdots & a_{2m} \\ \vdots & \vdots & & \vdots \\ a_{p1} & a_{p2} & \cdots & a_{pm} \end{bmatrix}, F = \begin{bmatrix} F_1 \\ F_2 \\ \vdots \\ F_m \end{bmatrix}, \varepsilon = \begin{bmatrix} \varepsilon_1 \\ \varepsilon_2 \\ \vdots \\ \varepsilon_p \end{bmatrix},$$

(5)

where (1) $m \le p$; (2) $\text{Cov}(\mathbf{F}, \boldsymbol{\varepsilon}) = \mathbf{0}$, that is $\mathbf{F}$ and $\boldsymbol{\varepsilon}$ are independent; (3) $E(\mathbf{F}) = \mathbf{0}$, $D(\mathbf{F}) = \mathbf{I}_m$, that is $F_1, F_2, ..., F_m$ are independent and the variances are 1. $\mathbf{A} = (a_{pm})$ is the component matrix; the greater of the absolute value of a_{pm}, the stronger interdependence between M_p and F_m, or the greater factor loading of the common factor F_m for M_p.

To make it easier to explain the meaning of each common factor (i.e. to acknowledge the variables represented by the common factors), we use varimax rotation to obtain the satisfied common factors; then, the factor loading squares are normalized (in 0–1), thus making the factor contribution disperse. The purpose is to eliminate all variables for the common factor dependent on different effects.

Each factor's score F_m is calculated by factor analysis; we take the rotated cumulative proportion for the coefficient and then calculate the comprehensive score after weighing all the factors. The calculation equation of PECS is expressed as follows:

$$\text{PECS} = [(CPP)_1 F_1 + (CPP)_2 F_2 + \cdots + (CPP)_m F_m] / \sum_{i=1}^{m} (CPP)_i, \tag{6}$$

where $(CPP)_i$ is the contribution of the ith common factor, F_i is the scores of the ith common factor, and $1 < m < p$.

3.3. Regression model and test method

In this study, we assume that PNLI and PECS have a relationship, that PNLI is the independent variable and PECS is the dependent variable, and that their functional relationship is either approximately linear or nonlinear. We establish the first- and second-order polynomials of the regression model as follows:

(1) The first-order polynomials (FOPs) model equation is as follows:

$$\text{PECS} = \hat{b}_0 + \hat{b}_1(\text{PNLI}), \tag{7}$$

where $\hat{b}_1$ is the slope; $\hat{b}_0$ is the intercept, which denotes the initial value of the dependent variable PECS when the independent variable (PNLI) is 0.

(2) The second-order polynomials (SOPs) model equation is as follows:

$$\text{PECS} = \hat{b}_0 + \hat{b}_1(\text{PNLI}) + \hat{b}_2(\text{PNLI})^2, \tag{8}$$

where $\hat{b}_0, \hat{b}_1, \hat{b}_2$ are constants, $\hat{b}_2 \neq 0$.

The goodness-of-fit of the model is represented by the coefficient of determination R^2, and an F-test (F_t) is used to determine whether the goodness-of-fit of the SOPs model is significant. When evaluating the regression between the explained variable PECS and a set of explanatory variables $\hat{b}_1, \hat{b}_2$, the null and alternative hypotheses are $H_0 : \hat{b}_1 = \hat{b}_2 = 0$ and $H_1 : \hat{b}_i (i = 1, 2)$ are not both equal to 0, respectively. Rejection of H_0 indicates that there is a relationship between the explanatory variables and PECS. If the result of F_t is to accept H_0, it shows that 2 explanatory variables do not have regression relationships with PECS.

4. Experimental results and analysis

4.1. *Results of the extraction and analysis of PNLI*

As shown in Table 3, the PNLI results of the selected main ports in Yangtze River Valley, China from 2001 to 2013 are calculated by Equations (1) and (2).

Given the consistency of the data sources and processing methods, the PNLI are comparable between ports. Figure 6 shows that the PNLI of every port increases overall from 2001 to 2013, but not without fluctuation and somewhat of a 'trough' in 2009. Specifically, the PNLI of Shanghai is low before rising to approximately 6.0 in 2004 and then remains steady; the PNLI of Nanjing rises sharply but fluctuates wildly; Wuhan and Chongqing are similar, in which the PNLI rises gradually but growth is not obvious.

4.2. *Results of the calculation and analysis of PECS*

As to the selection and weighing of indicators, some methods, such as Delphi, Analytic Hierarchy Process (AHP), etc., are subjective. However, the method (i.e. Factor analysis) we used in this study is objective. Because of the differences of selected indicators, the weights used are not the same at all ports. Given the different weights of common factors, the PECS of selected ports are calculated by Equation (6). Take Wuhan port as an example. According to the principle of eigenvalues greater than 1 (i.e. eig >1), we select three common factors (i.e. cargo throughput, fixed investments and port infrastructure) and accumulate the proportion in ANOVA (i.e. analysis of variance). The factor loading is computed by PCA (i.e. Principal components analysis), and the factor loading after rotation is computed by varimax. The factor analysis model of Wuhan port is then expressed as follows:

$$\begin{cases} M_1 = 0.249F_1 + 0.018F_2 + 0.884F_3 \\ M_2 = 0.109F_1 + 0.046F_2 + 0.734F_3 \\ \quad \cdots \\ M_{17} = 0.385F_1 + 0.210F_2 + 0.023F_3 \end{cases} . \tag{9}$$

As is shown in Table 4, the economic meaning of factor loading is more obvious after rotation than before. The first common factor reflects on M_4, M_7, M_8, M_9, M_{10}; the second

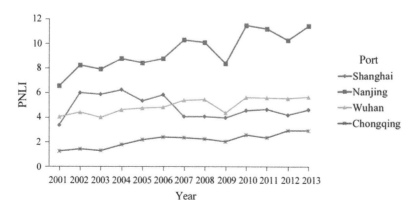

Figure 6. The PNLI time-changing trend chart of main ports in Yangtze River Valley, China from 2001 to 2013.

Table 4. The factor loading before and after rotation.

Factor	Factor loading before rotation				Factor loading after rotation			
	F_1	F_2	F_3	Communalities	F_1	F_2	F_3	Communalities
M_1	0.267	−0.092	0.896	0.898	0.249	0.018	0.884	0.893
M_2	0.128	0.062	0.790	0.757	0.109	0.046	0.734	0.753
M_3	−0.641	−0.245	0.875	0.823	0.321	0.041	0.868	0.851
M_4	0.905	0.430	0.239	0.891	0.889	0.450	0.205	0.903
M_5	−0.289	0.597	−0.237	0.653	−0.165	0.589	0.077	0.653
M_6	−0.218	0.619	−0.011	0.765	−0.102	0.612	0.008	0.763
M_7	0.648	0.106	0.210	0.639	0.624	0.104	0.192	0.639
M_8	0.834	0.244	0.327	0.877	0.808	0.192	0.268	0.875
M_9	0.767	0.388	0.272	0.852	0.745	0.397	0.261	0.850
M_{10}	0.723	0.405	0.094	0.749	0.713	0.370	−0.154	0.749
M_{11}	0.302	−0.239	0.707	0.718	0.348	−0.185	0.692	0.718
M_{12}	0.274	0.806	0.317	0.915	0.253	0.782	0.312	0.915
M_{13}	0.317	0.524	−0.162	0.631	0.306	0.512	−0.107	0.633
M_{14}	0.290	0.411	0.218	0.401	0.303	0.407	0.220	0.404
M_{15}	0.116	0.243	0.092	0.238	0.124	0.219	0.063	0.213
M_{16}	0.182	0.194	0.021	0.217	0.205	0.187	0.017	0.209
M_{17}	0.364	0.201	0.030	0.378	0.385	0.210	0.023	0.378
Cumulative proportion (%)	50.174	75.402	82.015		48.280	71.143	82.015	

common factor reflects on M_5, M_6, M_{12}, M_{13}; the third common factor reflects on M_1, M_2, M_3, M_{11}; the commonalities of M_{14}, M_{15}, M_{16}, M_{17} are low.

Each factor score is calculated by the factor score coefficient from the component score coefficient matrix and the standardized value of original factors. Weighted by the cumulative proportion, then PECS is calculated by

$$PECS = (48.280F_1 + 22.863F_2 + 10.872F_3)/82.015. \qquad (10)$$

The results of Wuhan are shown in Table 5. The PECS of others main ports in the Yangtze River Valley, China were calculated in a similar fashion and are shown in Table 6.

The results show that the PECS of the selected main ports are generally on the rise from 2001 to 2013, which is shown in Figure 7. The PECS of Chongqing rises most obviously, mainly because the annual cargo throughput, the main factor, grows rapidly; the PECS of Shanghai rises steadily, mainly because the number of port berths and cargo throughput (mainly including domestic and foreign trade) grows steadily in this period; it can also reflect that the port infrastructure and the domestic and foreign trade activities have a pronounced

Table 5. The results of factor comprehensive score.

Year	Common factor			PECS
	F_1	F_2	F_3	
2001	−1.354	−0.319	0.314	−0.844
2002	−1.221	−0.175	0.349	−0.721
2003	−1.148	−0.552	−0.003	−0.830
2004	−0.799	−0.230	0.470	−0.473
2005	−0.602	0.381	0.420	−0.192
2006	−0.224	0.298	−1.979	−0.311
2007	0.449	1.640	−1.462	0.528
2008	0.531	0.420	0.085	0.441
2009	0.914	1.360	2.008	1.183
2010	1.066	0.267	−0.217	0.673
2011	1.223	−0.948	0.357	0.503
2012	1.165	−2.143	−0.340	0.043
2013	1.303	−0.877	0.521	0.591

Table 6. The PECS results of main ports in Yangtze River Valley, China from 2001 to 2013.

	PECS			
Year	Shanghai	Nanjing	Wuhan	Chongqing
2001	−2.132	−2.926	−0.844	−5.279
2002	−1.629	−2.795	−0.721	−4.130
2003	−1.312	−2.259	−0.83	−3.189
2004	−0.992	−1.128	−0.473	−2.538
2005	−0.535	−0.052	−0.192	−2.069
2006	−0.285	0.517	−0.311	−0.922
2007	−0.188	0.868	0.528	−0.019
2008	−0.214	1.058	0.441	1.033
2009	0.196	2.250	1.183	0.940
2010	0.989	1.322	0.673	2.172
2011	1.790	1.453	0.503	3.300
2012	1.901	1.028	0.043	4.516
2013	2.411	1.364	0.591	5.075
Sum	0.000	0.700	0.591	−1.109

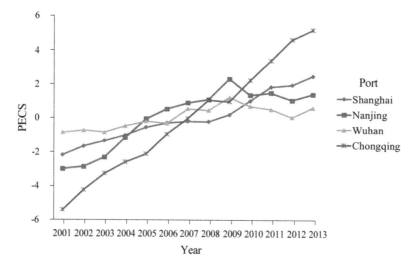

Figure 7. The PECS time-changing trend chart of main ports in Yangtze River Valley, China from 2001 to 2013.

effect on the economic development of Shanghai; the situations of Wuhan and Nanjing are similar in that they rise in the early years and peak in 2009, then decline before a rebound in 2012. However, the reasons are not the same, as Wuhan port is mainly affected by the cargo throughput and profit, and Chongqing is mainly affected by the investment in fixed assets.

In terms of the average annual growth rate (AAGR), namely the arithmetic mean of a series of growth rates, of PECS, the AAGR of each port is calculated based on the results of Table 6: that of Chongqing port is 0.797; Shanghai is 0.349; Nanjing is 0.330; and Wuhan is 0.110. Therefore, in terms of the speed of economic development, Chongqing grows fastest, followed by Shanghai, Nanjing, and Wuhan.

4.3. Regression analysis and correlation test

The two regression models (i.e. first- and second-order polynomials) were built based on Equations (7) and (8). However, after a preliminary inspection, we found that the thirteen samples of each selected port have outliers. Shanghai port has two outliers (i.e. a: 2012 and b: 2013); Nanjing port and Wuhan port each have an outlier in 2009; and Chongqing port has an outlier in 2011.

The outliers were eliminated by the random sample consensus (RANSAC) algorithm in case they were affecting the experimental results. The goodness-of-fit of the two regression models were then compared in both cases, either retaining or removing outliers.

4.3.1. Shanghai port

Figure 8 shows the difference of R^2 between the two regression models before outlier removal.

As is shown in Figures 9(a) and (b), the goodness of fit of FOPs are small after removing the outliers, so the independent variable (i.e. PNLI) is unable to predict the dependent variable (i.e. PECS). However, the goodness of fit of SOPs becomes much better, and the goodness-of-fit of removing the 2012 outlier is better than that of removing the 2013 outlier. Therefore, the relationship of PNLI and PECS in Shanghai better conforms to SOPs after removing the 2012 outlier.

4.3.2. The other three ports

As is shown in Figures 10–12, whether the outliers are removed or not, the PNLI and PECS of Nanjing, Wuhan and Chongqing are positively correlated. After removing the outliers of Nanjing and Wuhan in 2009 and of Chongqing in 2011, the goodness of fit of these two regression models are further improved, indicating that PECS can be estimated by the PNLI. However, the goodness of fit of SOPs are much better than of FOPs, which suggests that the relationship of PNLI and PECS conforms more to SOPs after removing the outliers.

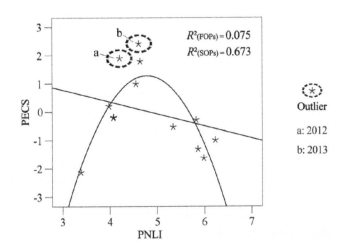

Figure 8. Comparison of the goodness of fit (i.e. R^2) between the two regression models of Shanghai before removing the outliers (i.e. a: 2012 and b: 2013). First-order polynomials (FOPs); and second-order polynomials (SOPs).

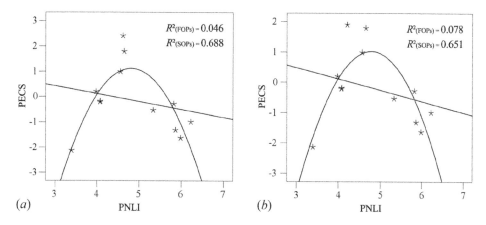

Figure 9. Comparison on the goodness-of-fit between the two regression models of Shanghai after removing the different outliers. Removing the outlier a: 2012 in (*a*); removing the outlier b: 2013 in (*b*).

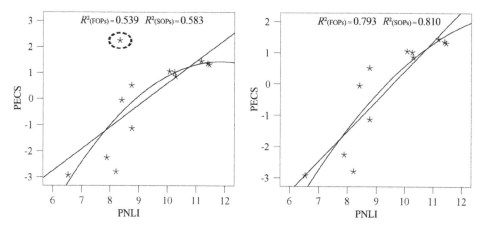

Figure 10. Comparison on the goodness-of-fit between the two regression models of Nanjing after removing the outlier in 2009.

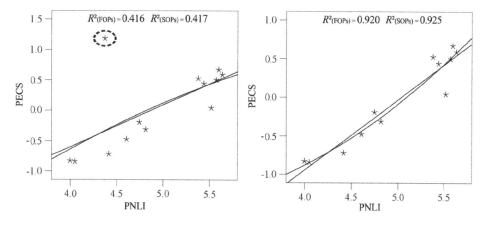

Figure 11. Comparison on the goodness-of-fit between the two regression models of Wuhan after removing the outlier in 2009.

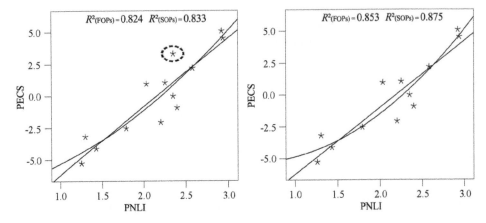

Figure 12. Comparison on the goodness-of-fit between the two regression models of Chongqing after removing the outlier in 2011.

On the basis of the coefficient of determination R^2, the goodness-of-fit of each of the two models are evaluated so that the best model can be selected as the regression prediction model of the PECS (seen in Table 7).

Take Wuhan port as an example. The regression results are calculated as follows:

$\hat{b}_0 = -0.839$; $\hat{b}_1 = -0.653$; $\hat{b}_2 = 0.161$; Correlation coefficient $R = 0.962$; Coefficient of determination $R^2 = 0.925$; $F_t = 55.429$.

Moreover, the regression model is expressed as follows:

$$PECS = 0.161(PNLI)^2 - 0.653(PNLI) - 0.839. \tag{11}$$

Actually, the correlation coefficient is $0.962 \square (0.8, 1]$, from which we can conclude that PNLI and PECS are strongly correlated. In addition, the determination coefficient is 0.925, indicating a high goodness of fit. Given the significance level $a = 0.05$, we have

Table 7. A comparison on the two regression models of main ports in Yangtze River Valley, China.

Port	Year	Outlier Removed/Not removed	Type	F_t	R^2	Model test(Yes/No)	Selection
	–	Not removed	FOPs	0.897	0.075	No	
			SOPs	10.275	0.673	Yes	
Shanghai	2012	Removed	FOPs	0.482	0.046	No	
			SOPs	9.944	0.688	Yes	Selected
	2013		FOPs	0.842	0.078	No	
			SOPs	8.410	0.651	Yes	
Nanjing	–	Not removed	FOPs	12.874	0.539	Yes	
			SOPs	7.001	0.583	Yes	
	2009	Removed	FOPs	38.412	0.793	Yes	
			SOPs	19.197	0.810	Yes	Selected
Wuhan	–	Not removed	FOPs	7.827	0.416	No	
			SOPs	3.575	0.417	No	
	2009	Removed	FOPs	114.907	0.920	Yes	
			SOPs	55.429	0.925	Yes	Selected
Chongqing	–	Not removed	FOPs	51.659	0.824	Yes	
			SOPs	24.989	0.833	Yes	
			FOPs	58.055	0.853	Yes	
	2011	Removed	SOPs	31.608	0.875	Yes	Selected

Table 8. The PECS regression prediction models of main ports in Yangtze River Valley, China based on PNLI.

Port	Regression prediction model
Shanghai	$PECS = -1.579(PNLI)^2 + 15.193(PNLI) - 35.426$
Nanjing	$PECS = -0.095(PNLI)^2 + 2.718(PNLI) - 17.126$
Wuhan	$PECS = 0.161(PNLI)^2 - 0.653(PNLI) - 0.839$
Chongqing	$PECS = 1.691(PNLI)^2 - 1.551(PNLI) - 5.163$

$F_t = 55.429 > 4.26 = F_{t\ (0.05)}(2, 9)$ and F_t is calculated from Equation (12), rejecting the null hypothesis. If $\hat{b}_0, \hat{b}_1, \hat{b}_2$ are not all 0, then the FOPs model makes sense at the level of 95%. Hence, the explaining variables have a significant effect on the explained variables, and thus the model is verified.

The above results show that the PNLI and PECS of Wuhan are strongly correlated, and we can use the prediction models of SOPs to estimate the basic trend of economic development in the future. Similarly, we get the best prediction models for Shanghai, Nanjing and Chongqing, which are listed in Table 8.

4.4. Discussion of some potential problems

4.4.1. The saturation problem of PNLI

In this study, we selected four rapidly developing port cities in the Yangtze River Valley, China. According to Table 9, there are few saturated pixels (i.e. the digital number of brightness is 63) in these cities, and the proportion of the saturated areas is small. Among the four selected cities, Shanghai is the most developed city in China, and has the largest number of saturated pixels. Therefore, we take it as an example. As shown in Table 10, the total pixels of Shanghai are 9979. In the years of 2001, 2005, 2010, 2013, the number of saturated pixels is 479, 750, 1134, 1694, and correspondingly, the proportion of saturated pixels is 4.80%, 7.52%, 11.36%, 16.98%, respectively. Figure 13 shows the comparison of saturation region of Shanghai DMSP/OLS night-time light imagery at different times.

The main ports of Shanghai are located away from the saturated regions. As shown in Figure 14, the saturated pixels mostly occur in the urban area, whereas the three parts of Shanghai port are located in the suburban area (part 2 is located on the edge of the

Table 9. The Saturation proportion of the four selected cities.

City	Shanghai	Nanjing	Wuhan	Chongqing
Total pixels	9979	9358	11821	109331
The number of saturated pixels	1694	141	286	21
Saturation proportion (%)	16.98	1.51	2.42	0.02

Table 10. The number of saturated pixels and its proportion in four times (a case of Shanghai port).

Year	2001	2005	2010	2013
Sensor	F15	F15	F18	F18
Total pixels		9979		
The number of saturated pixels	479	750	1134	1694
Saturation proportion (%)	4.80	7.52	11.36	16.98

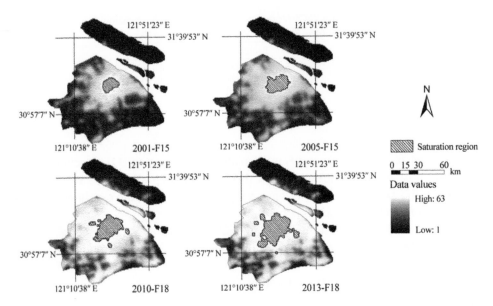

Figure 13. The saturation region of DMSP-OLS night-time light imagery in four times (A case of Shanghai).

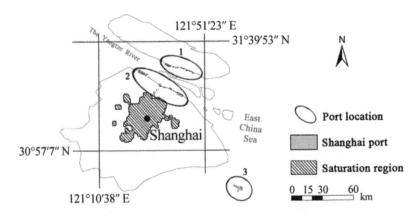

Figure 14. The saturation region and spatial distribution of Shanghai port.

saturated region). Furthermore, the digital numbers of brightness in Shanghai port are less than 63, and most are between 30 and 55. This means that there is no saturation within the Shanghai port area. Therefore, our study results will not be affected by the interference of saturation.

4.4.2. Shanghai problem of correlation between PNLI and PECS

We found that the PNLI and PECS of Shanghai port are negatively correlated; this is mainly because there were five 'Unusual cases' (i.e. the values in 2002, 2003, 2004, 2005, 2006) in which the value of PNLI is high and the value of PECS is low (as shown in Figure 15). These 'Unusual cases' can be reasonably explained as follows:

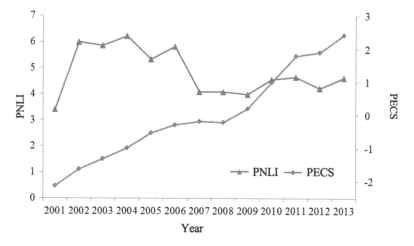

Figure 15. The PNLI and PECS time-changing trend chart of Shanghai from 2001 to 2013.

On the one hand, the main port location changes from 2002 to 2005. According to the Shanghai city master plan (1999–2020), as well as the related port development planning, a 'Retreat into the three' (i.e. a development policy limiting the secondary industries and promoting the tertiary industries in the industrial spatial structure adjustment) development strategy has been implemented. The ports, which are located in the main urban area near Huangpu River, are gradually being moved to locations along the Yangtze River and the coast. For example, the fourth and fifth period project of Waigaoqiao Port are located in the southern part of the Yangtze river estuary, and they were successively completed from 2002 to 2004. In addition, the Yangshan Deep Water Port, which is located at the coast, was officially put into operation in December 2005.

Due to the interference from the ambient light, the changes of port location will inevitably result in changes of PNLI value. Once the Shanghai main port location has changed, the PNLI will be less affected by the light from the main urban area. Therefore, the PNLI of Shanghai has experienced a 'big change' in 2006.

On the other hand, the port's economic development was in an initial stage during 2001–2006. As shown in Figure 16, the berth number and the cargo throughput are the two main factors in the port economic indicators, and during 2001–2013, the berth number increased from 1087 to 1253 and the cargo throughput increased from 220.992 million tons to 775.75 million tons.

According to the trends of these two main factors in this period, we find that it is not 'linear' growth, and it can be roughly divided into two phases: (a) the first is in 2001–2006. Shanghai port economic development is in the early stage, and its basic features are that it starts low but develops fast; (b) the second is in 2007–2013. It is in the stage of steady development, and the basic features are that it develops faster and its development is at a higher level. Overall, Shanghai port has experienced rapid development from 2001 to 2013.

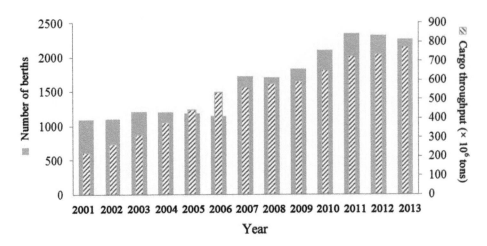

Figure 16. The statistical data of the two common factors (i.e. Number of berths and cargo throughput) of Shanghai from 2001 to 2013.

5. Conclusions and future work

Based on the DMSP-OLS data, this article puts forward a hypothesis that the PNLI and PECS are correlated; we then build regression models for the main ports in the Yangtze River Valley, China to estimate the PECS, and the models are tested. The experimental results show that the PNLI and PECS are closely correlated; it also verified the feasibility of using the PNLI to estimate PECS and the extraction method for small-scale DSMP-OLS data. From the results, the following conclusions can be drawn:

(1) This study proves that the DMSP-OLS data can be applied to a certain small-scale (e.g. the port area) research by using the proposed methods.
(2) The total PECS values of main ports in Yangtze River Valley, China are consistent with the total PNLI values, which basically satisfy the relationship: Nanjing>Wuhan>Shanghai>Chongqing.
(3) In terms of the average annual growth rate (AAGR) of PNLI from the years 2001 to 2013, we have Chongqing > Nanjing > Shanghai >Wuhan, and of the PECS, we have Chongqing > Wuhan > Shanghai > Nanjing.
(4) The PNLI and PECS of main ports in Yangtze River Valley, China have a linear correlation, and the SOPs regression model is the best model to describe the relationships after removing the outliers. Among them, the PECS of Shanghai port can only be estimated by the SOPs regression model.

In this study, the DMSP-OLS data is used to study the port economy, and the two regression models of PNLI and PECS are compared and discussed based on their correlation. In view of remote sensing technology, a small-scale extraction method of the night-time light data is first proposed. Perhaps this method is worth promoting in emerging research areas (e.g. the port economy), and in future studies, perhaps we might use the whole Yangtze River Valley, China or even ports in other areas, as the study area for more comprehensive research.

Acknowledgements

The authors thank the National Natural Science Foundation of China (NSFC) (grant nos. 41101407 and 41371183) and the Natural Science Foundation of Hubei Province (grant nos. 2014CFB377 and 2010CDZ005), China, the self-determined research funds of CCNU from the colleges' basic research and operation of MOE (grant nos. CCNU15A02001 and CCNU16JCZX09) for supporting this work, Wuhan Youth Science and technology plan (Grant No. 2016070204010137), and the National undergraduate training programmes for innovation and entrepreneurship (Grant No. 201410511037). We also thank Xudong Xing and Guangyao Zhang for their help in data collection and pre-processing. We are grateful for the comments and contributions of the anonymous reviewers and the members of the editorial team.

Disclosure statement

No potential conflict of interest was reported by the authors.

Funding

The authors thank the National Natural Science Foundation of China (NSFC) [Grant Numbers. 41101407 and 41371183] and the Natural Science Foundation of Hubei Province [Grant Numbers. 2014CFB377 and 2010CDZ005], China, the self-determined research funds of CCNU from the colleges' basic research and operation of MOE [Grant Numbers CCNU15A02001 and CCNU16JCZX09] for supporting this work, Wuhan Youth Science and technology plan [Grant Number 2016070204010137], and the National undergraduate training programmes for innovation and entrepreneurship [Grant Number 201410511037].

ORCID

Chang Li ⓘ http://orcid.org/0000-0003-2988-798X

References

Atkinson, P. M. 1997. "Selecting the Spatial Resolution of Airborne MSS Imagery for Small-Scale Agricultural Mapping." *International Journal of Remote Sensing* 18 (9): 1903–1917. doi:10.1080/014311697217945.

Bauer, S. E., S. E. Wagner, J. Burch, R. Bayakly, and J. E. Vena. 2013. "A Case Referent Study: Light at Night and Breast Cancer Risk in Georgia." *International Journal of Health Geographics* 12 (1): 1–10. doi:10.1186/1476-072X-12-23.

Bichou, K., and R. Gray. 2004. "A Logistics and Supply Chain Management Approach to Port Performance Measurement." *Maritime Policy & Management* 31 (1): 47–67. doi:10.1080/0308883032000174454.

Chinese Association of Development Strategy Studies. 2011. *The Blue Book of the Urban Space Value Assessment of Chinese Port Cities.* Guang Zhou: Social Science Association.

De Martino, M., and A. Morvillo. 2008. "Activities, Resources and Inter-Organizational Relationships: Key Factors in Port Competitiveness." *Maritime Policy & Management* 35 (6): 571–589. doi:10.1080/03088830802469477.

Elvidge, C. D., K. E. Baugh, E. A. Kihn, H. W. Kroehl, E. R. Davis, and C. W. Davis. 1997. ""Relation between Satellite Observed Visible-Near Infrared Emissions, Population, Economic Activity and Electric Power Consumption." *International Journal of Remote Sensing* 18 (6): 1373–1379. doi:10.1080/014311697218485.

Fotheringham, A. S. 1989. "Scale-Independent Spatial Analysis." In *Accuracy of Spatial Databases*, edited by M. F. Goodchild and S. Gopal. London: Taylor and Francis.

He, C., J. Li, J. Chen, P. Shi, J. Chen, Y. Pan, J. Li, L. Zhuo, and I. Toshiaki. 2006. "The Urbanization Process of Bohai Rim in the 1990s by Using DMSP/OLS Data." *Journal of Geographical Sciences* 16 (2): 174–182. doi:10.1007/s11442-006-0205-0.

He, C., Q. Ma, Z. Liu, and Q. Zhang. 2014. "Modeling the Spatiotemporal Dynamics of Electric Power Consumption in Mainland China Using Saturation-Corrected DMSP/OLS Nighttime Stable Light Data." *International Journal of Digital Earth* 7 (12): 993–1014. doi:10.1080/17538947.2013.822026.

Kuechly, H. U., C. C. M. Kyba, T. Ruhtz, C. Lindemann, C. Woltera, J. Fischer, and F. Hölker. 2012. "Aerial Survey and Spatial Analysis of Sources of Light Pollution in Berlin, Germany." *Remote Sensing of Environment* 126: 39–50. doi:10.1016/j.rse.2012.08.008.

Letu, H., M. Hara, H. Yagi, K. Naoki, G. Tana, F. Nishio, and O. Shuhei. 2010. "Estimating Energy Consumption from Night-Time DMPS/OLS Imagery after Correcting for Saturation Effects." *International Journal of Remote Sensing* 31 (16): 4443–4458. doi:10.1080/01431160903277464.

Li, C., J. Ye, S. Li, G. Chen, and H. Xiong. 2016. "Study on Radiometric Intercalibration Methods for DMSP-OLS Night-Time Light Imagery." *International Journal of Remote Sensing* 16 (37): 3675–3695. doi:10.1080/01431161.2016.1201232.

Li, X., L. Ge, and X. Chen. 2013. "Detecting Zimbabwe's Decadal Economic Decline Using Nighttime Light Imagery." *Remote Sensing* 5 (9): 4551–4570. doi:10.3390/rs5094551.

Li, X., and D. R. Li. 2014. "Can Night-Time Light Images Play a Role in Evaluating the Syrian Crisis?" *International Journal of Remote Sensing* 35 (18): 6648–6661. doi:10.1080/01431161.2014.971469.

Lo, C. P. 2001. "Modeling the Population of China Using DMSP Operational Linescan System Nighttime Data." *Photogrammetric Engineering and Remote Sensing* 67 (9): 1037–1047.

Madeira, A. G. Junior, M. M. Cardoso Junior, M. C. N. Belderrain, A. R. Correia, and S. H. Schwanz. 2012. "Multicriteria and Multivariate Analysis for Port Performance Evaluation." *International Journal of Production Economics* 140 (1): 450–456. doi:10.1016/j.ijpe.2012.06.028.

Marceau, D. J., and G. J. Hay. 1999. "Remote Sensing Contributions to the Scale Issue." *Canadian Journal of Remote Sensing* 25 (4): 357–366. doi:10.1080/07038992.1999.10874735.

McCabe, M. F., and E. F. Wood. 2006. "Scale Influences on the Remote Estimation of Evapotranspiration Using Multiple Satellite Sensors." *Remote Sensing of Environment* 105: 271–285. doi:10.1016/j.rse.2006.07.006.

Meentemeyer, V. 1989. "Geographical Perspectives of Space, Time, and Scale." *Landscape Ecology* 3 (3/4): 163–173. doi:10.1007/BF00131535.

Moody, A., and C. E. Woodcock. 1995. "The Influence of Scale and the Spatial Characteristics of Landscapes on Land-Cover Mapping Using Remote Sensing." *Landscape Ecology* 10 (6): 363–379. doi:10.1007/BF00130213.

Peter, D. L., N. Michiel, and V. D. H. Martijn. 2007. "New Indicators to Measure Port Performance." *Journal of Maritime Research* 4 (1): 23–36.

Raffy, M. 1992. "Change of Scale in Models of Remote Sensing: A General Method for Spatialisation of Models." *Remote Sensing of Environment* 40: 101–112. doi:10.1016/0034-4257(92)90008-8.

Raffy, M. 1994. "Change of Scale Theory: A Capital Challenge for Space Observation of Earth." *International Journal of Remote Sensing* 15 (12): 2353–2357. doi:10.1080/01431169408954249.

Roll, Y., and Y. Hayuth. 1993. "Port Performance Comparison Applying Data Envelopment Analysis (DEA)." *Maritime Policy & Management* 20 (2): 153–161. doi:10.1080/03088839300000025.

Small, C., and C. D. Elvidge. 2013. "Night on Earth: Mapping Decadal Changes of Anthropogenic Night Light in Asia." *International Journal of Applied Earth Observation and Geoinformation* 22: 40–52. doi:10.1016/j.jag.2012.02.009.

Sutton, P., D. Roberts, C. D. Elvidge, and H. Melj. 1997. "A Comparison of Nighttime Satellite Imagery and Population Density for the Continental United States." *Photogrammetric Engineering and Remote Sensing* 63 (11): 1303–1313.

UNCTAD. 1983. *Manual on a Uniform System of Port Statistics and Performance Indicators*. Geneva: UNCTAD.

Woodcock, C. E., and A. H. Strahler. 1987. "The Factor of Scale in Remote Sensing." *Remote Sensing of Environment* 21: 311–332. doi:10.1016/0034-4257(87)90015-0.

Yu, B. L., S. Shu, H. X. Liu, W. Song, J. P. Wu, L. Wang, and Z. Q. Chen. 2014. "Object-Based Spatial Cluster Analysis of Urban Landscape Pattern Using Nighttime Light Satellite Images: A Case Study of China." *International Journal of Geographical Information Science* 28 (11): 2328–2355. doi:10.1080/13658816.2014.922186.

Zhuo, L., T. Ichinose, J. Zheng, J. Chen, J. P. Shi, and X. Li. 2009. "Modeling the Population Density of China at the Pixel Level Based on DMSP/OLS Non-Radiance Calibrated Nighttime Light Images." *International Journal of Remote Sensing* 30 (4): 1003–1018. doi:10.1080/01431160802430693.

Urban mapping using DMSP/OLS stable night-time light: a review

Xuecao Li and Yuyu Zhou

ABSTRACT

The Defense Meteorological Satellite Program/Operational Linescane System (DMSP/OLS) stable night-time light (NTL) data showed great potential in urban extent mapping across a variety of scales with historical records dating back to 1990s. In order to advance this data, a systematic methodology review on NTL-based urban extent mapping was carried out, with emphases on four aspects including the saturation of luminosity, the blooming effect, the intercalibration of time series, and their temporal pattern adjustment. We think ancillary features (e.g. land surface conditions and socioeconomic activities) can help reveal more spatial details in urban core regions with high digital number (DN) values. In addition, dynamic optimal thresholds are needed to address issues of different exaggeration of NTL data in the large scale urban mapping. Then, we reviewed three key aspects (reference region, reference satellite/year, and calibration model) in the current intercalibration framework of NTL time series, and summarized major reference regions in literature that were used for intercalibration, which is critical to achieve a globally consistent series of NTL DN values over years. Moreover, adjustment of temporal pattern on intercalibrated NTL series is needed to trace the urban sprawl process, particularly in rapidly developing regions. In addition, we analysed those applications for urban extent mapping based on the new generation NTL data of Visible/Infrared Imager/Radiometer Suite. Finally, we prospected the challenges and opportunities including the improvement of temporally inconsistent NTL series, mitigation of spatial heterogeneity of blooming effect in NTL, and synthesis of different NTL satellites, in global urban extent mapping.

1. Introduction

Although the Defense Meteorological Satellite Program/Operational Linescane System (DMSP/OLS) was originally developed for the purpose of detecting the global distribution of clouds and cloud top temperature, it has become a predominate source for observing a series of faint emission sources since 1970s, such as city lights, shipping fleets, industrial sites, gas flares, and fires (Croft 1978; Elvidge et al. 1997b; Imhoff et al. 1997; Huang et al. 2014). The DMSP/OLS sensor contains two spectral bands (visible/near-infrared – VNIR, 0.4 – 1.1 μm and thermal infrared – TIR, 10.5 – 12.6 μm) with a swath of ~3000 km (Doll 2008). In addition, this data set has a near global coverage (spanning from −180° to 180° in longitude and −65° to 65° in latitude), and it spans two decades (1992–2013). Through geolocation processing, the nominal resolution of DMSP/OLS data set is 30 arc second, which equals to 1 km in equator.

According to the latest *World Urbanization Prospects* (United Nations 2015), the percentage of global urban population has exceeded 54% in 2014, and this proportion is estimated to reach 66% by 2050. More importantly, most of the newly increased population in the near future is likely to occur in developing regions (Africa and Asia), which will lead to a series of environmental or ecological issues related to the rapid urbanization process (Li and Gong 2016b). Therefore, acquiring the historical record of urban sprawl or urban population change, as well as predicting its future trajectories, is of great importance to sustainable urban development. The DMSP/OLS night-time light (NTL) data provide a particular perspective with a unique data set to study urban expansion and relevant sociodemographic activities across a variety of spatial scales, such as population density (Zhuo et al. 2009; Sutton, Elvidge, and Obremski 2003; Sutton et al. 2001; Lo 2002; Amaral et al. 2006), physical urban extent mapping (Elvidge et al. 1997b, 2007; Zhou et al. 2014; Small, Pozzi, and Elvidge 2005), energy consumption (Doll and Pachauri 2010; Letu et al. 2010), socioeconomic activities (Chen and Nordhaus 2011; Zhao and Samson 2012), and environmental changes (e.g. light pollution) (Davies et al. 2013; Falchi et al. 2016). Presently, there are three categories of DMSP/OLS data sets, including the stable lights, the calibrated radiance, and the average digital number (DN) (Elvidge et al. 1999; Doll 2008; Elvidge et al. 2009 2009a). Among them, the stable NTL dataset is the most widely used one for regional or global urban studies (Huang et al. 2014) because (1) the radiance calibrated dataset is only available for specific years without continuous time series; and (2) the average DN dataset may contain other emissions sources (e.g. fires and other background noise) in addition to city light.

One of the most important applications of DMSP/OLS stable NTL dataset is mapping urban extent (or boundary) and its temporal dynamics at the regional or global scales (Elvidge et al. 2007; Huang et al. 2014; Zhou et al. 2014; Elvidge et al. 1997a). Although a wide range of relevant studies have been carried out, most of them focus on particular local or regional areas using varying ancillary datasets or mapping approaches (Huang et al. 2014; Liu and Leung 2015; Ma et al. 2014; He et al. 2006; Liu et al. 2012; Yi et al. 2014; Milesi et al. 2003). Potential challenges are still remaining for pursing a globally consistent mapping of urban area using the DMSP/OLS stable NTL dataset. These challenges include the sensitivity of threshold for obtaining urban clusters (Liu and Leung 2015; Zhou et al. 2014), saturated DN values in urban core regions (Zhang, Schaaf, and Seto 2013; Cao et al. 2009), temporally inconsistency of NTL dataset over years (Zhao, Zhou, and Samson 2015; Elvidge et al. 2009b), and complicated urban sprawl patterns with different development levels (Zhang and Seto 2011; Ma et al. 2012). Few efforts have been made to summarize the difference of current approaches or comparison of different mapping results, although there are some general works on meta-analysis or summary of specific applications of NTL data (Huang et al. 2014; Li, Zhao, and Xi 2016a).

Hence, a systematic methodology review on these topics is urgently needed, for achieving a globally consistent mapping of urban dynamics with NTL datasets over past 20 years (Elvidge et al. 2007; Zhou et al. 2015).

This article aims to provide a comprehensive review on methodologies of urban mapping using DMSP/OLS NTL data. The reminder of this article is organized as follows. In Section 2, we discussed the challenges and reviewed current studies in urban mapping using DMSP/OLS stable NTL data. Thereafter, a brief introduction of the upgraded Visible/Infrared Imager/Radiometer Suite/Day Night Band (VIIRS/DNB) data was presented in Section 3. At the end, we prospected future opportunities of spatiotemporal urban extent mapping using NTL data in Section 4.

2. NTL-based urban mapping and challenges

The definition of 'urban extent' is different when referring to different cases (Liu et al. 2014). For NTL relevant studies, the commonly used terms include impervious surface, human settlement, urban clusters, population density, human population, and urban boundary (Elvidge et al. 1997a; Sutton et al. 1997; Elvidge et al. 1999; Sutton et al. 2001; Henderson et al. 2003; Elvidge et al. 2007; Zhou et al. 2015). In this review, we covered three groups of studies in NTL-based urban mapping, including population density, urban extent, and impervious surface area (Figures 1(a)–(f)). The first is population density mapping in a perspective of land use (Figure 1(a)), by linking NTL data with census (e.g. demographic) data (Figures 1(d) and (e)). The second one is urban extent (Figure 1(b)), which indicates the boundary that separates urban areas from surrounding rural areas based on NTL images (Figure 1(e)). The third one refers to impervious surface mapping (Figure 1(c)), which excludes other land cover types (e.g. water, vegetation, and bare land) within the urban domain (Figures 1 (e) and (f)). We included population density mapping in this review because (1) NTL datasets are often conjunctively used with demographic inventory (or census data), and the output of them can be used as an intermediate to map the urban extent (Elvidge et al. 2007; Lu and Weng 2006; Martinuzzi, Gould, and Ramos González 2007); and (2) population density essentially is a crucial indicator to describe the urban extent (Angel et al. 2005; Schneider, Friedl, and Potere 2010; Lo 2002).

Presently, studies on NTL-based urban mapping mainly focus on two domains as shown in Figure 2. Both spatial and temporal dimensions of NTL data have been extensively explored for urban mapping. At the spatial dimension, the inherent deficiencies of NTL dataset, that is, the saturated DN values in the urban core region and blooming effects on the urban–rural boundary, limit its application in urban mapping at a large extent (Zhang, Schaaf, and Seto 2013; Elvidge et al. 2007). At the temporal dimension, due to the lack of on-board calibration, additional processes on the annual composites of stable NTL data, such as intercalibration or temporal pattern adjustment, are needed to investigate the urban dynamics (Elvidge et al. 2009b; Zhang and Seto 2011). Consequently, a wide range of studies have been carried out to address these issues for consistent urban mapping at the regional or global scales. In this review, we discussed these issues in the following sections with more details.

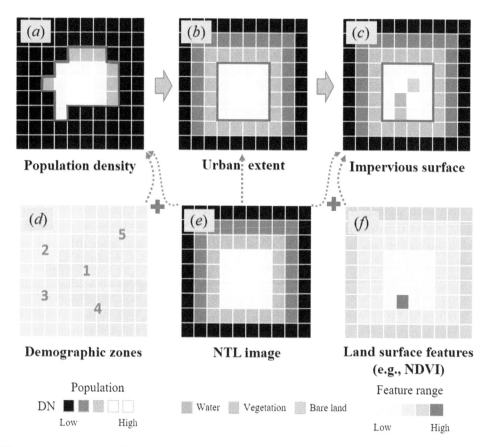

Figure 1. Night-time light (NTL)-based urban mapping: (*a*)–(*c*) are contents of NTL-based urban mapping; (*d*)–(*f*) illustrate necessary inputs for generating these maps.

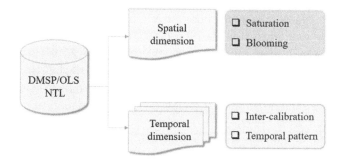

Figure 2. Research domains on NTL-based urban mapping.

2.1. *Spatial dimension*

2.1.1. *Saturation of NTL luminosity*

There exists a notable saturation effect of luminosity (i.e. the same or similar DN values in urban core area) in the DMSP/OLS NTL data because (1) the nominal resolution of

1 km is resampled from the 2.7 km native resolution (Doll 2008); and (2) the limit of DMSP/OLS sensor sensitivity is 6 bit (i.e. DN value ranges from 0 to 63). Although DMSP/OLS radiance data with on-board gain setting is an accurate way to differentiate the saturated luminosity in the urban core region, it is still limited for dynamic urban mapping because the implementation of radiance calibration is difficult, and this data set is only available for limited years (Elvidge et al. 1999; Doll 2008; Elvidge et al. 2001; Letu et al. 2012). A variety of attempts have been made to mitigate this saturation effect by using ancillary data to retrieve the heterogeneity within the urban extent. In general, there are two widely used ancillary datasets, land surface features and census population data, for conjunctively use with NTL data to map urban extent.

The saturation effect of luminosity in urban core region can be mitigated by incorporating land surface features as an intermediate output to map urban impervious surface. For instance, vegetation cover is a useful variable to reduce the saturation effect of NTL, which has been confirmed by a variety of studies (Li and Gong 2016a; He et al. 2014; Zhang, Schaaf, and Seto 2013; Liu et al. 2015a; Zhou et al. 2014). Lu et al. (2008) proposed a human settlement index (HSI) that incorporated Moderate Resolution Imaging Spectroradiometer (MODIS) normalized difference vegetation index (NDVI) with NTL data for settlement mapping in the southeastern China. Zhang, Schaaf, and Seto (2013) proposed a vegetation-adjusted NTL urban index (VANUI), which is simple but efficient in revealing the heterogeneity in regions with saturated DN values (Ma et al. 2014; Shao and Liu 2014; Li et al. 2016b). Liu et al. (2015a) combined both NDVI and normalized difference water index (NDWI) with NTL to reduce the pixel saturation in HSI and VANUI with a new indicator of normalized urban areas composite index (NUACI). Through incorporating remotely sensed land surface index, regions belong to non-urban but with high DN values can be recognized and removed in further processing. In addition, land-use/land-cover (LULC) datasets at a finer spatial resolution (e.g. 30 m) is able to provide more details in saturated regions in NTL data (Zhou et al. 2014; Liu et al. 2012), which can be served as a fraction of urban area when aggregated them to the same resolution as NTL data, or a statistic of total urban area over a particularly region. These land surface features have been used in NTL-based urban mapping together using classification (e.g. Support Vector Machine (SVM), random forest or spatially adaptive regression) or threshold methods (Cao et al. 2009; Xiao et al. 2014; Liu and Leung 2015; Shao and Liu 2014; Li et al. 2016b; Huang, Schneider, and Friedl 2016).

Demographic features can also help mitigate the saturation issue in NTL data in urban core areas by incorporating additional socioeconomic information to get the density of population (Sutton et al. 1997, 2001; Lo 2002; Sutton, Elvidge, and Obremski 2003; Amaral et al. 2006). In general, these datasets are associated with specific census unit, which can be used to differentiate DN values that are saturated but have different demographic levels (e.g. population or density) (Zhuo et al. 2009). Spatially explicit demographic information (e.g. demographic level or zone) can be introduced to group saturated pixels in the raw NTL data for further applications (e.g. population density mapping). It is worth noting that this mitigation of saturation effect in NTL luminosity depends on scales (or resolutions) of census data (Sutton, Elvidge, and Obremski 2003). More sophisticated approach with incorporation of demographic features and land surface factors can improve the performance in differentiating saturated DN values. Zhuo et al. (2009) performed a polynomial regression to calibrate the relationship between NTL and population at the county level in China, and then allocated the estimated population density to each pixel with the consideration of natural habitable condition (e.g. vegetation). The relationships between population and NTL vary among cases, and the spatial unit and level (e.g. state, province, county, and city) are a crucial factor influencing the relationship. In addition, NTL-derived population (or density) estimation can be used to delineate urban extent (Elvidge et al. 2007; Lu and Weng 2006; Martinuzzi, Gould, and Ramos González 2007).

2.1.2. Blooming effect in NTL

The blooming effect in the NTL data we discussed here specifically refers to the fact that outside of the actual urban extent, the DN values of NTL are still significantly above zeros. The blooming effect in NTL data increases difficulties to separate urban from its surrounding non-urban regions (Liu et al. 2015a; Zhang, Schaaf, and Seto 2013). A number of studies have been performed to address this issue for urban extent mapping (Henderson et al. 2003; Gallo et al. 2004; He et al. 2006; Cao et al. 2009; Liu et al. 2012). Among these studies, the approaches can be grouped roughly into two categories: (1) threshold based and (2) classification based.

Because of the blooming effect in NTL data, threshold-based approaches have been extensively used to extract urban extent from NTL data (see Figure 3) (Elvidge et al. 1997a; Imhoff et al. 1997; Henderson et al. 2003; He et al. 2006; Zhou et al. 2014; Liu et al. 2015a). Commonly, the status of urban and non-urban is determined by the threshold, that is, if the DN greater than the threshold, then it will be assigned as urban; otherwise it is classified as non-urban (see Figure 3, yellow rectangles). Essentially, the extracted urban extent is very sensitive to the threshold, and an optimal one is needed to

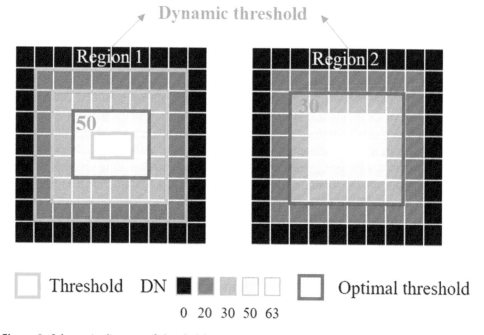

Figure 3. Schematic diagram of threshold approach using NTL.

maximally separate urban and non-urban regions using the NTL data (see Figure 3, red rectangles) (Zhou et al. 2014; Liu et al. 2012). Given that the spatial heterogeneity of urbanization features (e.g. urbanization level and urban size) over different regions, the optimal thresholds (see Figure 3, blue texts) vary across space and a scheme of dynamic (spatial and temporal) thresholds is required for large-scale and temporal dynamic urban extent mapping (Zhou et al. 2014; Elvidge et al. 1997b; Imhoff et al. 1997; Small, Pozzi, and Elvidge 2005; Elvidge et al. 2009b; Cao et al. 2009). Previous attempts on threshold approaches focused on NTL data only. For example, the 'light picking' approach was proposed to estimate the threshold for a local window based on the background information (Elvidge et al. 1997b), and urban shape (e.g. area or perimeter) was used to find the 'sudden jump' point through searching continuous thresholds (Imhoff et al. 1997; Liu and Leung 2015). More attentions have been given to determine those dynamic thresholds using ancillary information (He et al. 2006; Cao et al. 2009; Zhou et al. 2014; Liu et al. 2015a). For instance, He et al. (2006) iteratively searched optimal thresholds to match the statistical urban area at the province level. In a similar manner, statistical information of urban area of the region or city has been used to derive the optimal thresholds at these levels (Liu et al. 2012; Milesi et al. 2003; Yu et al. 2014). In addition, classified LULC data at a finer spatial resolution have been used to derive optimal thresholds over the conterminous space (Liu et al. 2015a; Li, Gong, and Liang 2015). Using the aggregated LULC information, Zhou et al. (2014) developed a method to derive dynamic optimal thresholds to map urban extent for each urban cluster, which was generated using a segmentation algorithm. This method was then extended to map urban extent at the global level (Zhou et al. 2015). Similarly, there are other site-based studies to estimate the empirical threshold based on the collected referred dataset (e.g. existing land-use cover data or impervious surface information) and the modified NTL indices (e.g. VANUI or NUACI) (Li et al. 2016b; Liu et al. 2015a).

Classification-based methods have always been used to extract the urban extent from NTL data with additional features such as NDVI and NDWI (Huang, Schneider, and Friedl 2016; Cao et al. 2009; Xiao et al. 2014). Cao et al. (2009) proposed a SVM-based region-growing algorithm to extract urban area using NTL data and Satellite Pour l'Observation de la Terre (SPOT) NDVI. Urban training samples were initially selected as seeds and thereafter they were iteratively updated through using newly classified urban pixels within a 3 × 3 window of these seeds to composite a new training set. This method outperforms those results derived from global-fixed or local-optimized approaches (Cao et al. 2009; Xiao et al. 2014). Huang, Schneider, and Friedl (2016) used a Random Forest regression model to estimate the urban percentage from stacked time series of NTL and MODIS NDVI data. In this method, urban percentage aggregated from Landsat-based land-cover data were used in the model training.

In addition to these two prevailing branches of methodology, there are other approaches to mitigate the blooming effect in NTL data. For instance, Townsend and Bruce (2010) developed an Overglow Removal Model (ORM) to correct the diffusion of NTL based on the empirical relationship between the light strength (sum of the total DN value) and the dispersion distance. But this relationship needed to be calibrated in advance with additional information (e.g. electricity use and population of each city). Su et al. (2015) adopted a neighbourhood statistics approach to detect the spatial differ-ence of NTL data between urban and associated non-urban regions in the Pearl River

Delta (China). Pre-defined thresholds are not needed in this method, but the mapped urban extent is sensitive to the neighbourhood morphology (e.g. configuration and size) and NTL magnitude within the neighbourhood (e.g. maximum and minimum), which needs be examined when being applied in other regions. Tan (2016) developed a method to generate inside buffers based on the empirical relationship between surveyed urban area and lit area of NTL data for mitigating the blooming effect. These methods are similar with the approach of dynamic optimal thresholds in determining the buffers to separate urban and non-urban regions. However, it should be cautious when applying them in a large area with high spatial heterogeneity.

2.2. Temporal dimension

2.2.1. Intercalibration of annual NTL data

Due to the absence of on-board calibration, the DMSP/OLS stable NTL annual composites product derived from multiple sensors (F12–F16) and different years (1992–2013) are not comparable directly (Doll 2008). Therefore, intercalibration of annual NTL composites product is highly needed to investigate urban dynamics using the NTL data. Elvidge et al. (2009b) built the framework of intercalibration for annual NTL composites product, which is the most widely used framework currently (Elvidge et al. 2014; Ma et al. 2014; Liu and Leung 2015; Zhao, Zhou, and Samson 2015; Huang, Schneider, and Friedl 2016; Li et al. 2016b; Zhang, Pandey, and Seto 2016; Tan 2016; Yi et al. 2014). This proposed framework includes three procedures: (1) selection of the reference region; (2) determination of the reference satellite and year for calibration; and (3) model development for intercalibration. Currently, most works requiring intercalibration of NTL series followed these procedures.

The reference regions vary among different studies for particular applications at the regional or global scales. There are two criteria in selecting reference regions: (1) small changes in lighting over years and (2) covering a wide range of DN values (Elvidge et al. 2009b; Wu et al. 2013). Therefore, in addition to Sicily Island selected by Elvidge et al. (2009b) for an early calibration work, many other reference regions have been used to intercalibrate the annual NTL composites product for urban dynamics analyses. We surveyed literature on NTL-based intercalibration and summarized the hotspot map of reference regions (Figure 4). Presently, the collected reference regions in Figure 4 include different countries (Italy, USA, China, Japan, and India), covering both mainland and islands (Puerto Rico, Mauritius, and Okinawa) (Wu et al. 2013). Although these regions were selected for different purposes, they showed potential for intercalibration of NTL dataset at the global level. In addition, apart from those reference regions that contain a wide range of DN values, there are also some attempts using automatically or manually collected sites (or points) as references for intercalibration (Yi et al. 2014; Zhang, Pandey, and Seto 2016; Li et al. 2013). For instance, Li et al. (2013) used a linear regression model to iteratively filter out pixels that may be experienced a change of DN value to collect the referenced sites for intercalibration. This method is more appropriate for local applications because the iteration process is time consuming. Liu et al. (2015b) set a simple rule (i.e. DN >30) for sample collection in New York for multi-temporal NTL data intercalibration. However, it should be noted that those pixels involved in the

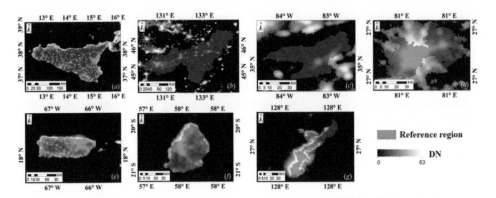

Figure 4. Reference regions used for intercalibration of annual NTL composites product. (*a*) Sicily island (Italy) (Elvidge et al. 2009 2009a); (*b*) Jixi county (China) (Liu et al. 2012); (*c*) Swain county (USA) (Li et al., 2016b); (*d*) Lucknow and Nawabganj (India) (Pandey, Joshi, and Seto 2013); (*e*), (*f*), and (*g*) are Puerto Rico (USA), Mauritius (an Indian Ocean island), and Okinawa (Japan) (Wu et al. 2013). The background NTL image is derived from F121999.

calibration process are very sensitive to the calibrated results (Zhang, Pandey, and Seto 2016).

The reference satellite and year are always determined based on the criterion that the sum (or averaged) of DN values in the reference region or the whole study area is the highest (Elvidge et al. 2009b; Pandey, Joshi, and Seto 2013; Ma et al. 2012). Other criterion in reference year/satellite selection is based on time-series of the NTL data, which aims to choose the year/satellite that lies in the middle of series for minimizing the effect of NTL change in the long time period (Zhang, Pandey, and Seto 2016). Once the reference year/satellite is selected, other NTL data were calibrated for achieving a comparable series over time. There are a variety of calibration models developed, such as six-order polynomial model (Bennie et al. 2014), second-order regression model (Elvidge et al. 2009b), simplified first-order regression model (Liu et al. 2015b) and power function (Wu et al. 2013). Among them, the second-order regression model has been extensively used to intercalibrate annual NTL composites product (Elvidge et al. 2009b; Zhao, Zhou, and Samson 2015; Ma et al. 2012; Liu et al. 2012; Liu and Leung 2015; Pandey, Joshi, and Seto 2013; Zhang, Pandey, and Seto 2016), and its formula can be expressed as Equation (1):

$$V_{adjust} = C_0 + C_1 \times V + C_2 \times V^2, \tag{1}$$

where V_{adjust} is the calibrated DN value, V is the original value, C_0, C_1, and C_2 are the coefficients, which were derived from the second-order regression model between DN values of reference image and others to be calibrated.

2.2.2. Temporal pattern adjustment

It is critical to evaluate the temporal pattern of the annual NTL data in terms of its consistency for tracing the urban sprawl process, particularly in rapidly developing regions (e.g. China and India) (Liu et al. 2012; Ma et al. 2014). Although it is a somewhat subjective modification of the intercalibrated NTL series, it is still needed because (1) the

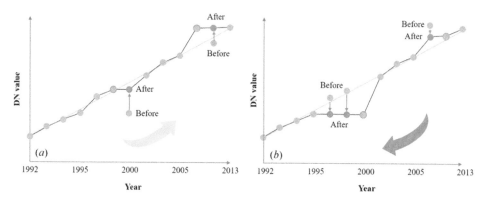

Figure 5. Temporal pattern adjustment of intercalibrated NTL time series: (a) 1992–2013 and (b) 2013–1992.

intercalibration is likely to introduce errors for some sites with abnormal NTL sequences that are not consistent over time; and (2) the pathway of urban expansion in rapidly developing regions is more certain with continuously expansion and increasing lit areas, whereas the obtained NTL series may not follow this trajectory (Liu et al. 2012; Li, Gong, and Liang 2015; Mertes et al. 2015; Zhao, Zhou, and Samson 2015). Liu et al. (2012) proposed an inter-annual series correction to modify the abnormal pixels (see Figure 5 (a)). In their study, based on the NTL series, temporally neighboured DN values are compared. Inconsistent pixels in the series were modified to achieve a continuously increasing pattern (see red and green circles in Figure 5(a)). Similar approaches can be found in Huang, Schneider, and Friedl (2016). Furthermore, to reduce the possible system errors caused by the initial year (e.g. 1992 in Figure 5(a)), Liu and Leung (2015) proposed a two-way modification of NTL series to combine sequences of 1992–2013 (i.e. green arrow in Figure 5(a)) and 2013–1992 (i.e. red arrow in Figure 5(b)). The mean of these two adjusted sequences was used in their studies based on the assumption that the positive and negative errors were offset (Zhao, Zhou, and Samson 2015; Liu and Leung 2015).

The adjustment of temporal pattern on the intercalibrated NTL series is needed for urban dynamics analyses in regions with rapid development while it may be not necessary for all areas. The natural pattern of NTL series may reflect multiple pathways (or archetypes) of urbanization, e.g. constant urban activity, earlier urban growth, de-urbanization, constant urban growth, and recent urban growth (Zhang and Seto 2011; Ma et al. 2012). Although most of these archetypes show temporally increasing total DN values, an opposite trajectory is also seen due to crisis such as war (e.g. Syria war) (Li and Deren 2014) or population migration due to poverty (Zhao, Zhou, and Samson 2015). The adjustment of intercalibrated NTL series is helpful in analysing dynamics of urban expansion, whereas the knowledge of the study area is needed for designing reasonable adjustment rules (i.e. linearly changed or not). Given that the land cover change from urban to non-urban rarely occurred (Li, Gong, and Liang 2015; Mertes et al. 2015), the temporal pattern adjustment is efficient for most urban lit areas on the planet. Nevertheless, it is still challenging to distinguish those pseudo changes from the actual expansion based on the calibrated NTL time series.

3. Successor of DMSP/OLS: VIIRS/DNB

The new generation of NTL, VIIRS, carried on the Suomi National Polar-orbiting Partnership (NPP) satellite (http://npp.gsfc.nasa.gov) was launched in 2011. Compared to the DMSP/OLS, the sensor DNB in the VIIRS is more advanced in (1) on-board calibration; (2) spatial resolution (about four times finer than DMSP); and (3) radiometric resolution (14 bit) (Miller et al. 2012; Elvidge et al. 2013). As a consequence, VIIRS is able to provide more details in terms of the detected night-time light (Small, Elvidge, and Baugh 2013). However, due to the short period since the VIIRS data were available, studies on urban mapping using VIIRS data are relatively limited currently. In addition, most of them centred around the comparison with DMSP/OLS data using similar approaches as we documented earlier, to enhance the benefits of VIIRS with improved spatial details. For example, Shi et al. (2014) evaluated the performance of VIIRS NTL data for extracting urban areas using the thresholds calibrated from statistical data based on 12 cities in China, and they found that the obtained accuracies were higher than that using DMSP/OLS data. Guo et al. (2015) integrated the VIIRS data with MODIS NDVI data to map the impervious surface area in China using the regression model. This procedure was similar to the approaches discussed in Section 2.1.1 to mitigate the saturation effect in NTL data with DMSP/OLS replaced by VIIRS. Sharma et al. (2016) made a similar attempt to estimate the thresholds at the global scale using data such as MODIS-derived 'Urban Built-up Index (UBI)' to estimate the thresholds. The thresholds for urban extent delineation in their work were determined based on region-specific values in each 10° × 10° tile for the whole globe. In addition, NTL-based observations with high spatial resolution (1 m) are emerging now, such as the Israeli EROS-B satellite (Levin et al. 2014), which is of great value in urban studies at the local scale.

4. Discussion and future opportunities

The DMSP/OLS NTL data showed great potential in urban extent mapping across a variety of scales with historical records dating back to 1990s. This article provides a systematic review on NTL-based urban mapping, including the saturation of luminosity, the blooming effect of NTL data, the intercalibration of NTL series, and adjustment of intercalibrated temporal patterns. Although NTL data are useful in urban extent mapping over large areas, it is worth to note that it is limited to its spatial resolution (1 km) and could be influenced by other light disturbance (e.g. gas) (Zhang and Seto 2013). The urban extent from NTL data may omit small city and include pseudo lit areas. However, the DMSP/OLS NTL data are highly recommended for global urban mapping studies. Compared to urban mapping using other datasets (e.g. MODIS, Landsat, and Orthophoto) (Schneider, Friedl, and Potere 2010; Gong et al. 2013; Small, Pozzi, and Elvidge 2005; Henderson et al. 2003; Zhou and Wang 2008), although they can provide more details of urban structure or extent, NTL data show advantages in generating a global consistent urban map series because of (1) more direct observations of night-time city light; and (2) less data volume requirement with globally consistent measurements (Zhou et al. 2015; Elvidge et al. 2007). However, due to the challenges discussed, there still lacks multi-temporal urban products based on a consistent mapping scheme from regional to global levels. These challenges also provide opportunities and open future research avenues in temporal dynamic urban mapping from regional to global levels using NTL data.

(1) Improvement of temporally inconsistent NTL series. After Elvidge et al. (2009b) proposed the general framework for intercalibration of global inconsistent NTL series, few attempts have been made for multi-year global urban extent mapping. Recently, Zhang, Pandey, and Seto (2016) improved the intercalibration with carefully selected reference pixels to improve the initial NTL DN values. This study will undoubtedly promote the global mapping studies over multiple years. However, there are still two concerns to be addressed in this calibration framework in the future work. One is the notably disturbance of DN values through implementing the calibration model for almost all the pixels. Another is the shift of the initial pattern of NTL series over time after calibration (Wu et al. 2013). Novel methods are needed to reduce the uncertainty introduced from the calibration by detecting systematic errors of images of different satellites and years.

(2) Mitigation of blooming effect and its spatial heterogeneity in NTL. Presently, mapping approaches using NTL dataset at the global scale is still under development. The first global impervious surface map was built using a linear relationship and population data (Elvidge et al. 2007). However, the spatial heterogeneity of local socio-economic development was not well considered in this method. Zhou et al. (2015) used a logistic-model to estimate the optimal threshold for each urban clusters derived from NTL dataset for global urban extent mapping. Although spatial heterogeneities have been considered for each urban cluster, the finer resolution land cover data used in threshold estimation were merely based on two representative regions: China and USA These challenging issues still exist in the global urban extent mapping using the NTL data, and more efforts are needed in the future.

(3) Synthesis of DMSP/OLS and VIIRS NTL datasets. The temporal coverage of DMSP/OLS NTL dataset is 1992–2013, and the continuing project of VIIRS is ongoing. The new satellite and sensor make it possible to detect more details of night-time city lights, whereas the inconsistent setting of sensors and resolutions between DMSP/OLS and VIIRS raise challenges to combine these two data sources for continuously monitoring of global urban expansion since 1990s. Although there are several studies have been carried out for comparison between VIIRS and DMSP/OLS datasets (Small, Elvidge, and Baugh 2013; Shi et al. 2014; Guo et al. 2015), few attempts have been made to integrate the DMSP/OLS and VIIRS/DNB for a consistent observation, which is of great importance to understand the dynamics of long-term urban expansion. More efforts are required to take advantage of them for a continuing mapping of urban dynamics at the global scale.

Acknowledgements

This work was supported by the NASA ROSES LULC Program 'NNH11ZDA001N-LCLUC'. We thank three anonymous reviewers and editor for their valuable comments to improve this manuscript.

Disclosure statement

No potential conflict of interest was reported by the authors.

Funding

This work was supported by the NASA ROSES LULC Program 'NNH11ZDA001N-LCLUC'.

References

Amaral, S., A. M. V. Monteiro, G. Câmara, and J. A. Quintanilha. 2006. "Dmsp/Ols Night-Time Light Imagery for Urban Population Estimates in the Brazilian Amazon." *International Journal of Remote Sensing* 27 (05): 855–870. doi:10.1080/01431160500181861.

Angel, S., S. Sheppard, D. L. Civco, R. Buckley, A. Chabaeva, L. Gitlin, A. Kraley, J. Parent, and M. Perlin. 2005. *The Dynamics of Global Urban Expansion*. Washington DC, USA: Citeseer.

Bennie, J., T. W. Davies, J. P. Duffy, R. Inger, and K. J. Gaston. 2014. "Contrasting Trends in Light Pollution across Europe Based on Satellite Observed Night Time Lights." *Scientific Reports* 4: 3789. doi:10.1038/srep03789.

Cao, X., J. Chen, H. Imura, and O. Higashi. 2009. "A Svm-Based Method to Extract Urban Areas from Dmsp-Ols and Spot Vgt Data." *Remote Sensing of Environment* 113 (10): 2205–2209. doi:10.1016/j.rse.2009.06.001.

Chen, X., and W. D. Nordhaus. 2011. "Using Luminosity Data as a Proxy for Economic Statistics." *Proceedings of the National Academy of Sciences* 108 (21): 8589–8594. doi:10.1073/pnas.1017031108.

Croft, T. A. 1978. "Nighttime Images of the Earth from Space." *Scientific American* 239: 86–98. doi:10.1038/scientificamerican0778-86.

Davies, T. W., J. Bennie, R. Inger, N. H. De Ibarra, and K. J. Gaston. 2013. "Artificial Light Pollution: Are Shifting Spectral Signatures Changing the Balance of Species Interactions?" *Global Change Biology* 19 (5): 1417–1423. doi:10.1111/gcb.12166.

Doll, C. N. H., and S. Pachauri. 2010. "Estimating Rural Populations without Access to Electricity in Developing Countries through Night-Time Light Satellite Imagery." *Energy Policy* 38 (10): 5661–5670. doi:10.1016/j.enpol.2010.05.014.

Doll, C. N. H., ed. 2008. *Ciesin Thematic Guide to Night-Time Light Remote Sensing and Its Applications*. Palisades, NY: Center for International Earth Science Information Network of Columbia University.

Elvidge, C. D., E. H. Erwin, K. E. Baugh, D. Ziskin, B. T. Tuttle, T. Ghosh, and P. C. Sutton. "Overview of Dmsp Nightime Lights and Future Possibilities." Paper presented at the 2009 Joint Urban Remote Sensing Event, 20-22 May 2009 2009a.

Elvidge, C., K. Baugh, V. Hobson, E. Kihn, H. Kroehl, E. Davis, and D. Cocero. 1997a. "Satellite Inventory of Human Settlements Using Nocturnal Radiation Emissions: A Contribution for the Global Toolchest." *Global Change Biology* 3 (5): 387–395. doi:10.1046/j.1365-2486.1997.00115.x.

Elvidge, C., D. Ziskin, K. Baugh, B. Tuttle, T. Ghosh, D. Pack, E. Erwin, and M. Zhizhin. 2009b. "A Fifteen Year Record of Global Natural Gas Flaring Derived from Satellite Data." *Energies* 2 (3): 595. doi:10.3390/en20300595.

Elvidge, C. D., K. E. Baugh, J. B. Dietz, T. Bland, P. C. Sutton, and H. W. Kroehl. 1999. "Radiance Calibration of Dmsp-Ols Low-Light Imaging Data of Human Settlements." *Remote Sensing of Environment* 68 (1): 77–88. doi:10.1016/S0034-4257(98)00098-4.

Elvidge, C. D., K. E. Baugh, E. A. Kihn, H. W. Kroehl, and E. R. Davis. 1997b. "Mapping City Lights with Nighttime Data from the Dmsp Operational Linescan System." *Photogrammetric Engineering and Remote Sensing* 63 (6): 727–734.

Elvidge, C. D., K. E. Baugh, M. Zhizhin, and F.-C. Hsu. 2013. "Why Viirs Data are Superior to Dmsp for Mapping Nighttime Lights." *Proceedings of the Asia-Pacific Advanced Network* 35: 62–69. doi:10.7125/APAN.35.7.

Elvidge, C. D., F.-C. Hsu, K. E. Baugh, and T. Ghosh. 2014. "National Trends in Satellite-Observed Lighting." *Global Urban Monitoring and Assessment through Earth Observation* 97–118.

Elvidge, C. D., M. L. Imhoff, K. E. Baugh, V. R. Hobson, I. Nelson, J. Safran, J. B. Dietz, and B. T. Tuttle. 2001. "Night-Time Lights of the World: 1994–1995." *ISPRS Journal of Photogrammetry and Remote Sensing* 56 (2): 81–99. doi:10.1016/S0924-2716(01)00040-5.

Elvidge, C. D., B. T. Tuttle, P. C. Sutton, K. E. Baugh, A. T. Howard, C. Milesi, B. Bhaduri, and R. Nemani. 2007. "Global Distribution and Density of Constructed Impervious Surfaces." *Sensors* 7 (9): 1962–1979. doi:10.3390/s7091962.

Falchi, F., P. Cinzano, D. Duriscoe, C. C. M. Kyba, C. D. Elvidge, K. Baugh, B. A. Portnov, N. A. Rybnikova, and R. Furgoni. 2016. "The New World Atlas of Artificial Night Sky Brightness." *Science Advances* 2: 6. doi:10.1126/sciadv.1600377.

Gallo, K. P., C. D. Elvidge, L. Yang, and B. C. Reed. 2004. "Trends in Night-Time City Lights and Vegetation Indices Associated with Urbanization within the Conterminous USA." *International Journal of Remote Sensing* 25 (10): 2003–2007. doi:10.1080/01431160310001640964.

Gong, P., J. Wang, L. Yu, Y. Zhao, Y. Zhao, L. Liang, Z. Niu, et al. 2013. "Finer Resolution Observation and Monitoring of Global Land Cover: First Mapping Results with Landsat Tm and Etm+ Data." *International Journal of Remote Sensing* 34 (7): 2607–2654. doi:10.1080/01431161.2012.748992.

Guo, W., L. Dengsheng, W. Yanlan, and J. Zhang. 2015. "Mapping Impervious Surface Distribution with Integration of Snnp Viirs-Dnb and Modis Ndvi Data." *Remote Sensing* 7 (9): 12459–12477. doi:10.3390/rs70912459.

He, C., Z. Liu, J. Tian, and M. Qun. 2014. "Urban Expansion Dynamics and Natural Habitat Loss in China: A Multiscale Landscape Perspective." *Global Change Biology* 20 (9): 2886–2902. doi:10.1111/gcb.12553.

He, C., P. Shi, L. Jinggang, J. Chen, Y. Pan, L. Jing, L. Zhuo, and T. Ichinose. 2006. "Restoring Urbanization Process in China in the 1990s by Using Non-Radiance-Calibrated Dmsp/Ols Nighttime Light Imagery and Statistical Data." *Chinese Science Bulletin* 51 (13): 1614–1620. doi:10.1007/s11434-006-2006-3.

Henderson, M., E. T. Yeh, P. Gong, C. Elvidge, and K. Baugh. 2003. "Validation of Urban Boundaries Derived from Global Night-Time Satellite Imagery." *International Journal of Remote Sensing* 24 (3): 595–609. doi:10.1080/01431160304982.

Huang, Q., X. Yang, B. Gao, Y. Yang, and Y. Zhao. 2014. "Application of Dmsp/Ols Nighttime Light Images: A Meta-Analysis and A Systematic Literature Review." *Remote Sensing* 6 (8): 6844. doi:10.3390/rs6086844.

Huang, X., A. Schneider, and M. A. Friedl. 2016. "Mapping Sub-Pixel Urban Expansion in China Using Modis and Dmsp/Ols Nighttime Lights." *Remote Sensing of Environment* 175: 92–108. doi:10.1016/j.rse.2015.12.042.

Imhoff, M. L., W. T. Lawrence, D. C. Stutzer, and C. D. Elvidge. 1997. "A Technique for Using Composite Dmsp/Ols "City Lights" Satellite Data to Map Urban Area." *Remote Sensing of Environment* 61 (3): 361–370. doi:10.1016/S0034-4257(97)00046-1.

Letu, H., M. Hara, G. Tana, and F. Nishio. 2012. "A Saturated Light Correction Method for Dmsp/Ols Nighttime Satellite Imagery." *IEEE Transactions on Geoscience and Remote Sensing* 50 (2): 389–396. doi:10.1109/TGRS.2011.2178031.

Letu, H., M. Hara, H. Yagi, K. Naoki, G. Tana, F. Nishio, and O. Shuhei. 2010. "Estimating Energy Consumption from Night-Time Dmps/Ols Imagery after Correcting for Saturation Effects." *International Journal of Remote Sensing* 31 (16): 4443–4458. doi:10.1080/01431160903277464.

Levin, N., K. Johansen, J. M. Hacker, and S. Phinn. 2014. "A New Source for High Spatial Resolution Night Time Images — The Eros-B Commercial Satellite." *Remote Sensing of Environment* 149 (0): 1–12. doi:10.1016/j.rse.2014.03.019.

Li, D., X. Zhao, and L. Xi. 2016a. "Remote Sensing of Human Beings – a Perspective from Nighttime Light." *Geo-Spatial Information Science* 19 (1): 69–79. doi:10.1080/10095020.2016.1159389.

Li, Q., L. Linlin, Q. Weng, Y. Xie, and H. Guo. 2016b. "Monitoring Urban Dynamics in the Southeast U.S.A. Using Time-Series Dmsp/Ols Nightlight Imagery." *Remote Sensing* 8 (7): 578. doi:10.3390/rs8070578.

Li, X., X. Chen, Y. Zhao, X. Jia, F. Chen, and L. Hui. 2013. "Automatic Intercalibration of Night-Time Light Imagery Using Robust Regression." *Remote Sensing Letters* 4 (1): 45–54. doi:10.1080/2150704X.2012.687471.

Li, X., and L. Deren. 2014. "Can Night-Time Light Images Play a Role in Evaluating the Syrian Crisis?" *International Journal of Remote Sensing* 35 (18): 6648–6661. doi:10.1080/01431161.2014.971469.

Li, X., and P. Gong. 2016a. "An "Exclusion-Inclusion" Framework for Extracting Human Settlements in Rapidly Developing Regions of China from Landsat Images." *Remote Sensing of Environment* 186: 286–296. doi:10.1016/j.rse.2016.08.029.

Li, X., and P. Gong. 2016b. "Urban Growth Models: Progress and Perspective." *Science Bulletin* 61 (21): 1637–1650. doi:10.1007/s11434-016-1111-1.

Li, X., P. Gong, and L. Liang. 2015. "A 30-Year (1984–2013) Record of Annual Urban Dynamics of Beijing City Derived from Landsat Data." *Remote Sensing of Environment* 166 (0): 78–90. doi:10.1016/j.rse.2015.06.007.

Liu, L., and Y. Leung. 2015. "A Study of Urban Expansion of Prefectural-Level Cities in South China Using Night-Time Light Images." *International Journal of Remote Sensing* 36 (22): 5557–5575. doi:10.1080/01431161.2015.1101650.

Liu, X., H. Guohua, A. Bin, L. Xia, and Q. Shi. 2015a. "A Normalized Urban Areas Composite Index (Nuaci) Based on Combination of Dmsp-Ols and Modis for Mapping Impervious Surface Area." *Remote Sensing* 7 (12): 17168–17189. doi:10.3390/rs71215863.

Liu, Y., Y. Wang, J. Peng, D. Yueyue, X. Liu, L. Shuangshuang, and D. Zhang. 2015b. "Correlations between Urbanization and Vegetation Degradation across the World's Metropolises Using Dmsp/Ols Nighttime Light Data." *Remote Sensing* 7 (2): 2067. doi:10.3390/rs70202067.

Liu, Z., H. Chunyang, Q. Zhang, Q. Huang, and Y. Yang. 2012. "Extracting the Dynamics of Urban Expansion in China Using Dmsp-Ols Nighttime Light Data from 1992 to 2008." *Landscape and Urban Planning* 106 (1): 62–72. doi:10.1016/j.landurbplan.2012.02.013.

Liu, Z., H. Chunyang, Y. Zhou, and W. Jianguo. 2014. "How Much of the World's Land Has Been Urbanized, Really? A Hierarchical Framework for Avoiding Confusion." *Landscape Ecology* 29 (5): 763–771. doi:10.1007/s10980-014-0034-y.

Lo, C. P. 2002. "Urban Indicators of China from Radiance-Calibrated Digital Dmsp-Ols Nighttime Images." *Annals of the Association of American Geographers* 92 (2): 225–240. doi:10.1111/1467-8306.00288.

Lu, D., H. Tian, G. Zhou, and G. Hongli. 2008. "Regional Mapping of Human Settlements in Southeastern China with Multisensor Remotely Sensed Data." *Remote Sensing of Environment* 112 (9): 3668–3679. doi:10.1016/j.rse.2008.05.009.

Lu, D., and Q. Weng. 2006. "Use of Impervious Surface in Urban Land-Use Classification." *Remote Sensing of Environment* 102 (1–2): 146–160. doi:10.1016/j.rse.2006.02.010.

Ma, Q., H. Chunyang, W. Jianguo, Z. Liu, Q. Zhang, and Z. Sun. 2014. "Quantifying Spatiotemporal Patterns of Urban Impervious Surfaces in China: An Improved Assessment Using Nighttime Light Data." *Landscape and Urban Planning* 130: 36–49. doi:10.1016/j.landurbplan.2014.06.009.

Ma, T., C. Zhou, T. Pei, S. Haynie, and J. Fan. 2012. "Quantitative Estimation of Urbanization Dynamics Using Time Series of Dmsp/Ols Nighttime Light Data: A Comparative Case Study from China's Cities." *Remote Sensing of Environment* 124: 99–107. doi:10.1016/j.rse.2012.04.018.

Martinuzzi, S., W. A. Gould, and O. M. Ramos González. 2007. "Land Development, Land Use, and Urban Sprawl in Puerto Rico Integrating Remote Sensing and Population Census Data." *Landscape and Urban Planning* 79 (3–4): 288–297. doi:10.1016/j.landurbplan.2006.02.014.

Mertes, C. M., A. Schneider, D. Sulla-Menashe, A. J. Tatem, and B. Tan. 2015. "Detecting Change in Urban Areas at Continental Scales with Modis Data." *Remote Sensing of Environment* 158 (0): 331–347. doi:10.1016/j.rse.2014.09.023.

Milesi, C., C. D. Elvidge, R. R. Nemani, and S. W. Running. 2003. "Assessing the Impact of Urban Land Development on Net Primary Productivity in the Southeastern United States." *Remote Sensing of Environment* 86 (3): 401–410. doi:10.1016/S0034-4257(03)00081-6.

Miller, S. D., S. P. Mills, C. D. Elvidge, D. T. Lindsey, T. F. Lee, and J. D. Hawkins. 2012. "Suomi Satellite Brings to Light a Unique Frontier of Nighttime Environmental Sensing Capabilities." *Proceedings of the National Academy of Sciences* 109 (39): 15706–15711. doi:10.1073/pnas.1207034109.

Pandey, B. P., K. Joshi, and K. C. Seto. 2013. "Monitoring Urbanization Dynamics in India Using Dmsp/Ols Night Time Lights and Spot-Vgt Data." *International Journal of Applied Earth Observation and Geoinformation* 23: 49–61. doi:10.1016/j.jag.2012.11.005.

Schneider, A., M. A. Friedl, and D. Potere. 2010. "Mapping Global Urban Areas Using Modis 500-M Data: New Methods and Datasets Based on 'Urban Ecoregions'." *Remote Sensing of Environment* 114 (8): 1733–1746. doi:10.1016/j.rse.2010.03.003.

Shao, Z., and C. Liu. 2014. "The Integrated Use of Dmsp-Ols Nighttime Light and Modis Data for Monitoring Large-Scale Impervious Surface Dynamics: A Case Study in the Yangtze River Delta." *Remote Sensing* 6 (10): 9359. doi:10.3390/rs6109359.

Sharma, R. C., R. Tateishi, K. Hara, S. Gharechelou, and K. Iizuka. 2016. "Global Mapping of Urban Built-Up Areas of Year 2014 by Combining Modis Multispectral Data with Viirs Nighttime Light Data." *International Journal of Digital Earth* 9: 1–17.

Shi, K., C. Huang, Y. Bailang, B. Yin, Y. Huang, and W. Jianping. 2014. "Evaluation of Npp-Viirs Night-Time Light Composite Data for Extracting Built-Up Urban Areas." *Remote Sensing Letters* 5 (4): 358–366. doi:10.1080/2150704X.2014.905728.

Small, C., C. D. Elvidge, and K. Baugh. "Mapping Urban Structure and Spatial Connectivity with Viirs and Ols Night Light Imagery." Paper presented at the Urban Remote Sensing Event (JURSE), 2013 Joint, 2013.

Small, C., F. Pozzi, and C. D. Elvidge. 2005. "Spatial Analysis of Global Urban Extent from Dmsp-Ols Night Lights." *Remote Sensing of Environment* 96 (3–4): 277–291. doi:10.1016/j.rse.2005.02.002.

Su, Y., X. Chen, C. Wang, H. Zhang, J. Liao, Y. Yuyao, and C. Wang. 2015. "A New Method for Extracting Built-Up Urban Areas Using Dmsp-Ols Nighttime Stable Lights: A Case Study in the Pearl River Delta, Southern China." *Giscience & Remote Sensing* 52 (2): 218–238. doi:10.1080/15481603.2015.1007778.

Sutton, P., D. Roberts, C. Elvidge, and K. Baugh. 2001. "Census from Heaven: An Estimate of the Global Human Population Using Night-Time Satellite Imagery." *International Journal of Remote Sensing* 22 (16): 3061–3076. doi:10.1080/01431160010007015.

Sutton, P., D. Roberts, C. Elvidge, and H. Meij. 1997. "A Comparison of Nighttime Satellite Imagery and Population Density for the Continental United States." *Photogrammetric Engineering and Remote Sensing* 63 (11): 1303–1313.

Sutton, P. C., C. Elvidge, and T. Obremski. 2003. "Building and Evaluating Models to Estimate Ambient Population Density." *Photogrammetric Engineering & Remote Sensing* 69 (5): 545–553. doi:10.14358/PERS.69.5.545.

Tan, M. 2016. "Use of an inside Buffer Method to Extract the Extent of Urban Areas from Dmsp/Ols Nighttime Light Data in North China." *Giscience & Remote Sensing* 53 (4): 444–458. doi:10.1080/15481603.2016.1148832.

Townsend, A. C., and D. A. Bruce. 2010. "The Use of Night-Time Lights Satellite Imagery as a Measure of Australia's Regional Electricity Consumption and Population Distribution." *International Journal of Remote Sensing* 31 (16): 4459–4480. doi:10.1080/01431160903261005.

United Nations. 2015. *World Urbanization Prospects: The 2014 Revision.* UN: Department of Economic and Social Affairs, Population Division.

Wu, J., H. Shengbin, J. Peng, L. Weifeng, and X. Zhong. 2013. "Intercalibration of Dmsp-Ols Night-Time Light Data by the Invariant Region Method." *International Journal of Remote Sensing* 34 (20): 7356–7368. doi:10.1080/01431161.2013.820365.

Xiao, P., X. Wang, X. Feng, X. Zhang, and Y. Yang. 2014. "Detecting China's Urban Expansion over the past Three Decades Using Nighttime Light Data." *IEEE Journal of Selected Topics in Applied Earth Observations and Remote Sensing* 7 (10): 4095–4106. doi:10.1109/JSTARS.2014.2302855.

Yi, K., H. Tani, L. Qiang, J. Zhang, M. Guo, Y. Bao, X. Wang, and L. Jing. 2014. "Mapping and Evaluating the Urbanization Process in Northeast China Using Dmsp/Ols Nighttime Light Data." *Sensors* 14 (2): 3207. doi:10.3390/s140203207.

Yu, B., S. Shu, H. Liu, W. Song, W. Jianping, L. Wang, and Z. Chen. 2014. "Object-Based Spatial Cluster Analysis of Urban Landscape Pattern Using Nighttime Light Satellite Images: A Case

Study of China." *International Journal of Geographical Information Science* 28 (11): 2328–2355. doi:10.1080/13658816.2014.922186.

Zhang, Q., and K. Seto. 2013. "Can Night-Time Light Data Identify Typologies of Urbanization? A Global Assessment of Successes and Failures." *Remote Sensing* 5 (7): 3476. doi:10.3390/rs5073476.

Zhang, Q., B. Pandey, and K. C. Seto. 2016. "A Robust Method to Generate a Consistent Time Series From DMSP/OLS Nighttime Light Data." *IEEE Transactions on Geoscience and Remote Sensing* 54 (10): 5821–5831. doi:10.1109/TGRS.2016.2572724.

Zhang, Q., C. Schaaf, and K. C. Seto. 2013. "The Vegetation Adjusted Ntl Urban Index: A New Approach to Reduce Saturation and Increase Variation in Nighttime Luminosity." *Remote Sensing of Environment* 129: 32–41. doi:10.1016/j.rse.2012.10.022.

Zhang, Q., and K. C. Seto. 2011. "Mapping Urbanization Dynamics at Regional and Global Scales Using Multi-Temporal Dmsp/Ols Nighttime Light Data." *Remote Sensing of Environment* 115 (9): 2320–2329. doi:10.1016/j.rse.2011.04.032.

Zhao, N., and E. L. Samson. 2012. "Estimation of Virtual Water Contained in International Trade Products Using Nighttime Imagery." *International Journal of Applied Earth Observation and Geoinformation* 18: 243–250. doi:10.1016/j.jag.2012.02.002.

Zhao, N., Y. Zhou, and E. L. Samson. 2015. "Correcting Incompatible Dn Values and Geometric Errors in Nighttime Lights Time-Series Images." *Geoscience and Remote Sensing, IEEE Transactions On* 53 (4): 2039–2049. doi:10.1109/TGRS.2014.2352598.

Zhou, Y., S. J. Smith, C. D. Elvidge, K. Zhao, A. Thomson, and M. Imhoff. 2014. "A Cluster-Based Method to Map Urban Area from Dmsp/Ols Nightlights." *Remote Sensing of Environment* 147: 173–185. doi:10.1016/j.rse.2014.03.004.

Zhou, Y., S. J. Smith, K. Zhao, M. Imhoff, A. Thomson, B. Bond-Lamberty, G. R. Asrar et al. 2015. "A Global Map of Urban Extent from Nightlights." *Environmental Research Letters* 10 (5): 1–11. DOI:10.1088/1748-9326/10/5/054011.

Zhou, Y., and Y. Q. Wang. 2008. "Extraction of Impervious Surface Areas from High Spatial Resolution Imagery by Multiple Agent Segmentation and Classification." *Photogrammetric Engineering & Remote Sensing* 74 (7): 857–868. doi:10.14358/PERS.74.7.857.

Zhuo, L., T. Ichinose, J. Zheng, J. Chen, P. J. Shi, and X. Li. 2009. "Modelling the Population Density of China at the Pixel Level Based on Dmsp/Ols Non-Radiance-Calibrated Night-Time Light Images." *International Journal of Remote Sensing* 30 (4): 1003–1018. doi:10.1080/01431160802430693.

Analysis of urbanization dynamics in mainland China using pixel-based night-time light trajectories from 1992 to 2013

Yang Ju ⓘ, Iryna Dronova, Qin Ma ⓘ and Xiang Zhang

ABSTRACT

Understanding urbanization dynamics, or how intensity of urbanization changes over time, is an important basis for urban planning and management, which has been investigated using various data-driven approaches. Considering the advantages and constraints of different data sources, we use pixel-based, time-series night-time light (NTL) trajectories to characterize urbanization dynamics in mainland China where massive urban development has been occurring in recent decades. After pre-processing the data, we extracted time-series NTL trajectories for each 1 km × 1 km pixel between 1992 and 2013 and used the unsupervised k-means classification to identify the major typologies of these trajectories as urbanization dynamics based on their main statistical parameters. The classification identified five urbanization dynamics, namely, stable urban activity, high-level steady growth, acceleration, low-level steady growth, and fluctuation. Their distributions and spatial patterns were further summarized and compared among different Chinese administrative divisions. We specifically analysed the acceleration trajectories that showed rapid transitions from rural to urban, as we considered these trajectories as potential indicators for aggressive urbanization. We found several clusters at prefecture city and county levels with high proportion of the acceleration, and referred to the underlying socioeconomic characteristics and developmental history to understand how these clusters could had been formed. Through this study, we revealed the dominant tendencies of urbanization in China over space and time, and developed an analysis framework that could be extended to other regions.

1. Introduction

Urbanization, a process where population and economy transform from rural to urban and land cover changes from natural to predominantly built-up (Buhaug and Urdal 2013; Zhang and Seto 2013), continues to be an important component of global environmental change. The United Nations' 2014 projection showed that 30% of the world's population was urban in 1950, and this number would increase to 66% by 2050 (United Nations 2014). Over the next several decades, developing countries and countries in Asian and Africa are expected to be the hotspots for population growth and urbanization (United Nations 2014; Jiang and O'Neill 2015). With its significant land-cover change and high concentration of population and development, urbanization could cause unfavourable impacts, including pollution (Doygun and Alphan 2006; Schetke and Haase 2008), increased flooding (Hollis 1975; Kalnay and Cai 2003), changes in climate (Kalnay and Cai 2003), and degraded ecosystem function (Wu 2008; Li et al. 2010). Well-informed planning and management should be applied to urban areas to mitigate these consequences. We argue that to inform the decisions for the future, an important step is to understand different forms of urbanization dynamics, or how intensity of urbanization changes over time (Zhang and Seto 2011), in the recent history. Planners and decision-makers can correlate the dynamics with socioeconomic factors to find urbanization drivers and areas undergoing non-typical transitions, and make subsequent planning and management interventions.

China is playing an important role in global urbanization due to its rapid urban development and large urban population. The country has seen a massive transition from rural to urban since its economic reform in 1978. At the national level, China's fraction of urban population increased from 17.9% to 52.6% between 1978 and 2012 (Bai, Shi, and Liu 2014), and its urban built-up area expanded from 7438 to 45566 km^2 between 1981 and 2012 (Chen, Liu, and Lu 2015). Even in its present, advanced urbanization stage, China was still expected to add 292 million in urban population between 2014 and 2050, making the country one of the three major sources for the world's urban population growth (United Nations 2014). At the local level, urbanization rate varies spatially and shows a gradient from the east to the west, where the east, particularly the coastal regions, are more urbanized (Bai, Shi, and Liu 2014). Because of such variation and different underlying socioeconomic drivers, China is expected to manifest different types of urbanization dynamics across the country.

A particular phenomenon, ghost city, has attracted the attention from researchers, decision makers, and the public (Chen, Liu, and Lu 2015; Woodworth 2015). Characterized by fast development beyond the actual population needs and business speculation, ghost city potentially leads to several issues including financial risk and social injustice (Chen, Liu, and Lu 2015; Wang et al. 2015). To our knowledge, there is no study to date focusing on measuring the urbanization dynamics during the formation of a ghost city. However, since such a city tends to be under fast development within a limited period, we assume it is likely to be dominated by urbanization dynamics showing rural to urban transformations with a rapid and accelerating speed. Ghost city, along with many other urbanization phenomena, need to be better understood by researchers and decision makers to make any necessary interventions to the current urbanization process. Studying urbanization dynamics provides a basis to understand those phenomena and to inform decision making.

To study urbanization dynamics and urban systems, researchers have relied on different types of information, particularly (1) socioeconomic data on population and economy, (2) remote-sensing-derived data such as land cover and night-time lights (NTLs), and (3) user generated, location-based social network data such as Twitter and Flickr Photos. Socioeconomic data directly reflect urbanization as population and economic change, and are often available in time-series forms. However, the main limitation is that such data are aggregated into coarse administrative divisions, such as city, county, and province, which eliminates the finer-scale heterogeneity within the divisions. Urban land-cover data interpreted from remote-sensing images preserve spatial heterogeneity in the form of urbanized pixels at finer spatial resolutions, such as 30 m for a global coverage (Gong et al. 2013). With such data from multiple years, it is possible to analyse how individual cities expand spatially (Weng 2002; Xiao et al. 2006; Jat, Garg, and Khare 2008; Liu et al. 2012) and how urbanization unfolds at the national level (Wang et al. 2012; Sleeter et al. 2013). However, as binary categories of 'urban' and 'non-urban', land cover does not reflect how population and economy change within urbanized areas. Finally, location-based social network data are generated by human activities and thus are often used to locate people's position (Chi et al. 2015) and activity centres (Hu et al. 2015), and to delineate urban boundaries (Jiang and Miao 2015). However, such data can be biased as they only represent certain population groups using the social media (Chi et al. 2015; Hu et al. 2015). Furthermore, social network data have a fine temporal resolution due to their near real-time collection, but only cover a limited time span due to their short history, reducing the possibility to establish a long-term history (e.g. decadal) of urbanization dynamics.

DMSP-OLS NTL (National Geophysical Data Center 2015), collected by the US Air Force Defence Meteorological Satellite Programs (DMSP) Operational Linescan System (OLS), offer the research community a new perspective to understand long-term urbanization dynamics. These data have a relatively long time span, an annual temporal resolution, a nearly global coverage, a moderate spatial resolution, and a capability to reflect the intensity of human activities, as demonstrated by various studies (Chen et al. 2015; Elvidge et al. 2001; Ghosh et al. 2010; Ma et al. 2012; Shi et al. 2014b). In mapping urbanization, researchers have mainly applied thresholds on NTL intensity to identify urban extent (Liu et al. 2012; Ma et al. 2014; Shi et al. 2014a; Tan 2016; Yu et al. 2014; Zhou et al. 2014, 2015). In addition, researchers have recently started to extract time-series NTL trajectories to characterize urbanization dynamics at administrative division level and pixel level. At administrative division level, Stathakis, Tselios, and Faraslis (2015) calculated the sum of lights (SoL) index for European regions, and used linear regression to calculate overall and decadal slope of the SoL trend to represent urbanization process in each region. At pixel level, Qingling Zhang and Seto (2011) proposed five archetypes of urbanization dynamics, extracted pixel-based, time-series NTL trajectories, and classified those trajectories based on the archetypes. Jiansheng Wu et al. (2014) classified the time-series NTL trajectories in China to calculate the composition of the classes for each administrative division, grouped the divisions based the compositions, and finally compared the average composition of each group with a baseline composition to evaluate China's urbanization.

Although these studies (Wu et al. 2014; Zhang and Seto 2011) established an important foundation for NTL-based urbanization analysis, they have not yet extensively discussed the detected trajectories nor explicitly related them to socioeconomic data and underlying urbanization context. Furthermore, the number of classes in the unsupervised classifications was determined mainly through a subjective process, which might not produce an optimal setting to maximize inter-class differences and improve intra-class similarities. Therefore, our study had two primary objectives. First, we aimed to use the dynamics to investigate urbanization in mainland China. We mapped the areas under different urbanization dynamics, calculated the proportion of each type of dynamics in different administrative divisions, and quantified spatial patterns of these proportions to reveal the underlying drivers. In particular, we focused on the dynamics showing accelerated growth in NTL between 1992 and 2013, as it might be an indicator for rapid and even aggressive urbanization, which could lead to ghost cities in some cases. Second, we aimed to develop a framework to classify urbanization dynamics at pixel level using the time-series NTL trajectories. Based on previous methods by Qingling Zhang and Seto (2011) and Jiansheng Wu et al. (2014), our classification extracted a set of eight trajectory parameters to conduct a k-means unsupervised classification, and used a silhouette analysis (Rousseeuw 1987) as the basis to determine the number of classes. In all, we hypothesized that administrative divisions with high proportion of accelerated growth tended to be clustered, and the formation of the clusters could be explained by their unique socioeconomic characteristics and developmental history.

2. Materials and methods

2.1. *NTL data and data pre-processing*

We used the stable light composite from the version 4 DMSP-OLS NTL time-series in this study. Version 4 DMSP-OLS NTL time-series provides three annual products: cloud-free composite, average visible light composite, and stable light composite, from 1992 to 2013 (National Geophysical Data Center 2015). DMSP-OLS sensors collect a set of NTL images of the Earth with 30" (approximately 1 km) spatial resolution twice a day (Elvidge et al. 2009), and the images together cover −180° to 180° longitude and −65° to 75° latitude (National Geophysical Data Center 2015). Each pixel has a digital number (DN) for NTL intensity. An annual composite is made from the selected, highest quality images that exclude sunlit data, glare, moonlit data, observations with clouds, and lighting features from the aurora (National Geophysical Data Center 2015). We used the stable light composite, as it provided persistent lights from cities and towns, and excluded ephemeral events, such as fire and explosion. In this composite, the background noise is replaced with 0, and the DNs range from 1 to 63 (National Geophysical Data Center 2015). This composite provides 34 annual images from six satellite missions (i.e. F10, F12, F14, F15, F16, and F18). Some years may have two images as two satellite missions were operating at the same times. For such years, we used the average of the two images to represent the NTL.

There are three major issues with NTL data, namely, the 'overglow' effect, the saturation in urban centres, and the inconsistency between satellite missions (Li and Zhou 2017). The 'overglow' effect results from the radiation of non-coherent light in all directions from its source (Townsend and Bruce 2010), and causes an overestimation of urban extent (Ma et al. 2012). Both the 'overglow' removal model (ORM) by Townsend and Bruce (2010) and the thresholding method (Amaral et al. 2006; Imhoff et al. 1997; Wu et al. 2014) can reduce this effect. We applied the thresholding method mainly for its simplicity, and set the threshold as DN value of 12, following Jiansheng Wu et al. (2014). Pixels with an average DN less than 12 during the study period were considered as non-urban and were excluded from the analysis.

The saturation occurs due to NTL sensor's limited detection range between 0 and 63, which is insufficient to measure urban centres where the actual NTL intensity might exceed the upper limitation. However, the proportion of saturated pixels in China is less than 1.2% for each NTL image in the study time period (Liu and Li 2014), making saturation a minor and acceptable issue. In those saturated areas, we acknowledge that the dataset is not able to show further variations in NTL intensity.

The third and particularly important issue is the inconsistency between NTL satellite missions. Due to the differences between the sensors and a lack of on-board calibration, NTL annual composites of the same years from different satellite missions may have different DNs for the same locations (Wu et al. 2013). As Figure 1(a) shows, before the calibration, China's sum of DNs in the overlapping years were different between the missions. Such inconsistency is an obstacle for a time-series analysis; therefore, we calibrated the annual images to restore the consistency before further analysis using a quadratic regression model (Equation (1)) by Elvidge et al. (2009). We obtained the model parameters a_t, b_t, and c_t for each year t from a study by Liu and Li (2014), which used 2007's NTL image from mission F16 as the baseline and conducted the quadratic regressions over the invariant areas that should have little change of actual DNs over time. In addition, we excluded the pixels with 0 pre-calibration value from the calibration, as 0 indicated no light in the original data. Figure 1(b) shows the sum of DNs in our study area from different missions after the calibration, where the inconsistency at certain years, such as 2000 and 2010, were bridged, providing a continuous change of DNs over time and indicating the calibrated data can be used for the subsequent analysis:

$$L'_t = a_t \times L_t^2 + b_t \times L_t + c_t, \tag{1}$$

where L'_t is a pixel's DN after calibration in year t, L_t is the pixel's original DN in that year, and a_t, b_t, and c_t are the corresponding model parameters.

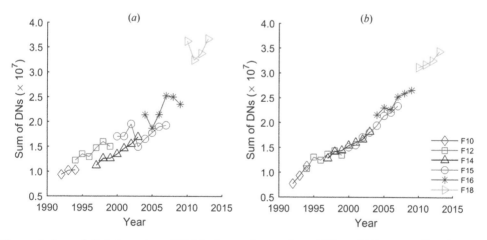

Figure 1. Sum of digital numbers (DNs) in the study area from 1993 to 2013 across all night-time light (NTL) satellite missions, (a) before calibration; (b) after calibration. F10, F12, F14, F15, F16, and F18 are the six NTL satellite missions included in the dataset.

2.2. *Trajectory extraction, noise removal, and principal component analysis on the parameters*

We extracted and smoothed pixel-based, time-series NTL trajectories from the calibrated NTL images. First, for any year with two NTL images, we used the average of the two for that year. Second, we extracted time-series NTL trajectories for each 1 km × 1 km pixel. Finally, we applied the Savitzky–Golay filter (Savitzky and Golay 1964) to remove the noise and to smooth each trajectory, as we assumed the change of DNs in each pixel should be a continuous process.

We then extracted two sets of four parameters to represent each trajectory. The first set utilized the DNs of each trajectory, which included the maximum, minimum, mean, and standard deviation. The second set used the slope of each trajectory indicating how much the DNs changed between two adjacent years (i.e. t to $t + 1$) (Equation (2)). We included the maximum, minimum, and standard deviation of the slopes, with an additional mean slope calculated from a linear regression (Equation (3)) of each trajectory's DNs. Figure 2(a,b) show a conceptual diagram of the main parameters:

$$S_{t+1} = L'_{t+1} - L'_t, \tag{2}$$

$$L' = \beta_0 + \beta_1 T + \varepsilon, \tag{3}$$

where $S_{t+1} >$ is the slope at year $t + 1 >$ in a pixel, L'_{t+1} and L'_t are the calibrated DNs. L' is a vector of the calibrated DNs for the years studied in a pixel, β_0 is the intercept, β_1 is the slope of a linear regression, or the mean slope in our study, T is a vector of the years, ε is the random error.

Finally, we conducted the principal component analysis (PCA) (Pohl and Van Genderen 1998) over the eight parameters to reduce dimensionality of the data and computational intensity of the subsequent analysis. PCA is a widely-adopted dimension reduction technique that linearly combines the input data to get a set of principal components (PCs) that are more independent of each other. Each PC explains certain percentage of variation, or certain

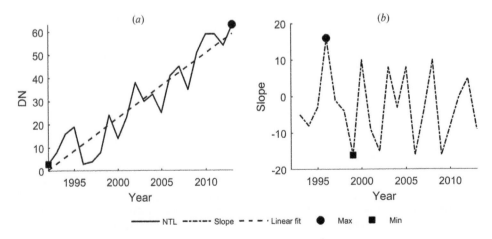

Figure 2. Conceptual diagram of main trajectory parameters, (a) parameters based on DN, including maximum and minimum DN, and a mean slope (β_1) from a linear fit defined by Equation (3); (b) parameters based on slope, including maximum and minimum slope. Slope is defined by Equation (2).

amount of information, in the original data. We used the first three PCs that together explained 97.97% of the variation, in the following classification.

2.3. K-means unsupervised classification and determining the optimal number of classes

We applied the k-means unsupervised classification (Lloyd 1982) to classify the NTL trajectories into a limited number of classes, with the first three PCs as the inputs. Each class was expected to represent a type of urbanization dynamics. K-means iteratively partitions n samples in to k classes, and minimizes the within-class sum of square distance from each member to the mean centre (Lloyd 1982). The algorithm starts with allocating a set of random initial mean centres, and the result is dependent on the centres' locations. Therefore, it's possible to have locally optimal results, rather than globally optimal ones (Steinley 2003). To reduce the dependence of results on the initial mean centre allocation, we iterated the classification 10 times to generate initial mean centres, and used the iteration with the optimal results in our analysis.

Another important consideration in this classification was the number of classes. We used the silhouette analysis (Rousseeuw 1987) as a basis to determine this number, which was different from previous studies that were more relied on subjective decisions (Zhang and Seto 2011; Wu et al. 2014). Given a classification with a certain number of classes, the silhouette analysis measures the closeness of each member in one class to members in the neighbouring classes (Scikit-learn 2015), and provides an average silhouette coefficient that ranges from −1 to 1 for all classes. 1 indicates a good classification where one class is far away from the neighbouring classes, and −1 indicates a poor classification where one class is very close to the neighbouring classes. After testing with a series numbers of classes, we chose the classifications with higher silhouette coefficients as the candidates. In addition, we also examined whether a candidate can properly represent urbanization dynamics in the study area, to find the optimal classification.

We chose five classes in the k-means unsupervised classification. The silhouette analysis showed that a two-class classification had the highest coefficient (Figure 3). However, we considered that having only two classes would be insufficient to differentiate urbanization dynamics. Thus we instead used a five-class classification with the

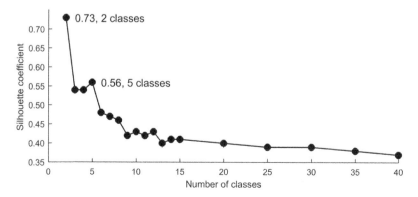

Figure 3. Change of silhouette coefficient value with different numbers of classes.

second highest coefficient where the number of classes could sufficiently represent various urbanization dynamics.

Finally, we labelled each class with a unique type of urbanization dynamics by examining the shape of each class's average trajectory, the range of the members, and the shapes of 1% randomly sampled members in this class. We also used the average value of the eight trajectory parameters in each class to help the labelling.

2.4. *Validation*

Validating the assignment of the unsupervised classes to specific urbanization dynamics was challenging, as the classification results were at a finer spatial resolution compared with candidate validation datasets, such as socioeconomic data typically reported at coarse administrative divisions. Several studies verified NTL-derived urbanization dynamics at pixel scale with fine resolution Google Earth images from different years (Ma et al. 2014; Zhang and Seto 2013). In particular, using Google Earth images, Qian Zhang and Seto (2013) validated their urbanization dynamics assignment, such as rapid urbanization and slow urbanization, for their unsupervised classes of NTL trajectories.

To validate the urbanization dynamics assignment, we examined whether changes in NTL and impervious-surface-derived urbanization ratio within each class had similar patterns in a set of selected cities. We used 30 m resolution Landsat remote-sensing images for every two years between 1992 and 2013 (i.e. 1992, 1994, ..., 2010, 2011, 2013) to estimate the extent of impervious surface as described below, and calculated urbanization ratio for each unsupervised class i, or the fraction of impervious surface within the spatial extent of i, in a city j in year t (Equation (4)). Impervious surface was used as an indicator for urbanization in many studies (Jat, Garg, and Khare 2008; Schneider and Woodcock 2008; Bhatta 2009; Bhatta, Saraswati, and Bandyopadhyay 2010). If changes in the urbanization ratio and NTL were similar, we assumed that we had a reasonable assignment of the unsupervised classes to urbanization dynamics. We further measured the similarity as the correlation between a class's average NTL trajectory and the class's average urbanization ratio in the set of cities (Equation (5)):

$$U_{i,j,t} = \frac{D_{i,j,t}}{C_{i,j}}, \tag{4}$$

$$r_i = \frac{\sum \left(\overline{\mathbf{L}'}_{i,t} - \overline{\overline{\mathbf{L}'}}_i \right) \left(\overline{\mathbf{U}}_{i,t} - \overline{\overline{\mathbf{U}}}_i \right)}{\sqrt{\sum \left(\overline{\mathbf{L}'}_{i,t} - \overline{\overline{\mathbf{L}'}}_i \right)^2} \sqrt{\sum \left(\overline{\mathbf{U}}_{i,t} - \overline{\overline{\mathbf{U}}}_i \right)^2}} \tag{5}$$

where $U_{i,j,t}$ is the urbanization ratio for class i in a city j and year t (i.e. 1992, 1994, ..., 2010, 2011, 2013), $D_{i,j,t}$ is the impervious surface area within class i's spatial extent in city j and year t, $C_{i,j}$ is the area of class i's spatial extent in city j. r_i is the correlation between class i's average NTL trajectory ($\overline{\mathbf{L}'}_{i,t}$) over the class members, and average urbanization ratio trajectory ($\overline{\mathbf{U}}_{i,t} >$) over the selected cities, $\overline{\overline{\mathbf{L}'}}_i$ and $\overline{\overline{\mathbf{U}}}_i$ are average value for $\overline{\mathbf{L}'}_{i,t}$ and $\overline{\mathbf{U}}_{i,t}$ over the years.

We obtained the impervious surface by a normalized difference spectral vector (NDSV) method (Patel et al. 2015) on Landsat remote-sensing images for the selected years in five cities including Harbin in the north, Nanning in the south, Hangzhou and Nanchang in the east, and Hanzhong in the west. The cities were selected mainly based on representativeness for geographic region and urbanization dynamics, and availability of high quality Landsat images with limited cloud cover. Two cities in the east were included as most Chinese cities were in this region. NDSV contains all the possible two-band normalized spectral indices from an image, such as the normalized difference vegetation index (NDVI), the normalized difference water index (NDWI), and the normalized built-up index (NDBI), which can be used to differentiate various land-cover types (Patel et al. 2015). We constructed a NDSV containing 15 indices from Landsat images, and conducted supervised classification with Support Vector Machine (Huang, Davis, and Townshend 2002) to identify vegetation, water, and impervious surface in the five cities. We queried orthorectified Landsat Top of Atmosphere (TOA) images with cloud cover less than 10% in each year as the input. Depending on cloud cover, we typically found a collection of several images from different times in a year, and we reduced such a collection to one cloud-free image using the median of DNs from the different times at each pixel. Median helped to remove clouds that had a high DN, and shadows that had a low DN (Bian et al. 2016). Since a pixel was not always under cloud or shadow, this process would yield a cloud-free image. Training samples were collected where land cover was consistent across all the selected years. We then classified the images using the Support Vector Machine algorithm (Huang, Davis, and Townshend 2002). Finally, we visually inspected the classification results and found them satisfactory. We accessed the images, collected the samples, and conducted the classification using the Google Earth Engine, an online environmental data monitoring platform that analysed publicly available remote-sensing images such as Landsat (Gorelick 2013; Patel et al. 2015), to reduce processing time and enhance feasibility.

In this study, the impervious surface-derived urbanization ratio had to be used with caution, for two reasons. First, the impervious surface used in the calculation may contain some non-urban areas that were excluded in the analysis due to low NTL intensity. Second, changes in the impervious surface cover do not necessarily capture every dimension of urbanization. Urbanization is a combination of land-cover change, population migration, and economic development. Only looking at the change of impervious surface or the derived urbanization ratio may neglect the other two dimensions of urbanization. However, NTL may capture all the three dimensions, as NTL is essentially an indicator for human activity. Given these two reasons, the change of impervious surface does not fully equal to the change of NTL. However, we still used impervious surface from Landsat images for validation as the surface was available in a finer resolution compared with the NTL-derived unsupervised classes. Other advantages of Landsat images included a broad temporal and spatial coverage, and being publicly accessible, making it possible to apply this validation method in other areas of interest.

2.5. *Urbanization pattern recognition*

Following the classification and validation, we calculated the proportion of each type of urbanization dynamics in different Chinese administrative divisions including province (primary administrative division), prefecture city (secondary administrative division), and county (tertiary administrative division). Such proportions showed how different urbanization dynamics contributed to the overall urbanization between 1992 and 2013 in the administrative divisions.

To measure the spatial pattern, we conducted both global and local Moran's I analysis to quantify the spatial autocorrelations in those proportions. Moran's I is a commonly used measure for spatial autocorrelation (Su et al. 2011), which indicates how a variable correlates with itself over space. Calculated by Equation (6) (ESRI 2016), global Moran's I measures the overall spatial autocorrelation across all the administrative divisions in the study area. Its value ranges from −1 to 1, where 1 means positive autocorrelation or similar values neighbouring with each other, 0 means spatial randomness, and −1 means negative autocorrelation or dissimilar values neighbouring with each other (Su et al. 2011). Local Moran's I (Anselin 1995), calculated by Equations (7–8) (ESRI 2016), measures spatial autocorrelation for each administrative division i, and further finds clusters and outliers in the divisions. Clusters are administrative divisions with similar values, which include high-high clusters where a high value is surrounded by high values, and low-low clusters where a low value is surrounded by low values. Outliers are administrative divisions with contrasting values, which further include high-low outliers where a high value is surrounded by low values, and low-high outliers where a low value is surrounded by high values.

Given the same set of data, different spatial weight (w_{ij}) may lead to different results in both global and local Moran's I. In this study, we performed both global and local Moran's I using ArcMap 10.4 (ESRI 2016), and chose 'contiguity edges corners' to generate the spatial weight. In 'contiguity edges corners', administrative divisions that share an edge and/or a corner with the target administrative division have weights of 1 to be included in the Moran's I computation, and the rest have weights of 0 to be excluded from the computation:

$$I = \frac{n\sum_j\sum_k w_{j,k}\left(x_j - \bar{\mathbf{x}}\right)\left(x_k - \bar{\mathbf{x}}\right)}{\left(\sum_j\sum_k w_{j,k}\right)\left(\sum_j\left(x_j - \bar{\mathbf{x}}\right)^2\right)}, \tag{6}$$

$$I_j = \frac{\left(x_j - \bar{\mathbf{x}}\right)}{S_j^2}\sum_k w_{j,k}(x_k - \bar{\mathbf{x}}), \tag{7}$$

$$S_j^2 = \frac{\sum_k (x_k - \bar{\mathbf{x}})^2}{n - 1}, \tag{8}$$

where I is the global Moran's I, I_j is the local Moran's I for administrative division j, x_j is the value of administrative division j, x_k is the value of j's neighbour k ($k{\neq}j$), $\bar{\mathbf{x}}$ is the mean value of all administrative divisions, $w_{j,k}$ is a spatial weight between j and k, n is the number of administrative divisions.

We computed global Moran's I for the proportion of each type of urbanization dynamics at provincial, prefecture city, and county levels. For dynamics showing accelerated urbanization, we further employed local Moran's I to identify local clusters and outliers at prefecture city and county levels. The spatial patterns of the proportions, particularly the clusters, help to inform the underlying urbanization process. For example, positive global spatial autocorrelation of a certain type of urbanization dynamics indicates the clustering of administrative divisions with similar proportions of the dynamics, which may further suggest similarity in other characteristics, such as policy, development history, and natural resources that form those clusters. Local clusters identify which administrative divisions belong to the same cluster, which allow synthesizing the commonalities within a cluster and differences between clusters. Those commonalities and differences may explain the observed pattern of the proportions.

3. Results

3.1. *Five types of urbanization dynamics in China*

We labelled the five classes as stable urban activity, high-level steady growth, acceleration, low-level steady growth, and fluctuation. Figure 4(a–f) show each class's average trajectory, the 90% range (i.e. 5 percentile to 95 percentile), and 1% randomly sampled trajectories in the class. Table 1 shows average value of the eight trajectory parameters for each class. Class 1 represented stable urban activity, as its trajectories had relatively constant DNs that were close to the maximum value (Figure 4(a)) and the smallest mean slope indicating the least changes between 1992 and 2013 (Table 1). Both class 2 and class 4 represented steady growth as their average trajectories were close to straight

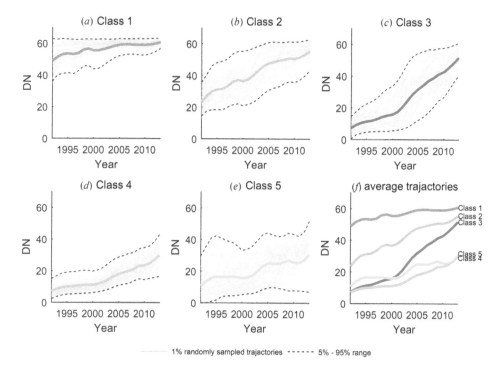

Figure 4. (a–f) Average trajectories for each class from the *k*-means unsupervised classification with *k* = 5. Trajectories from 1% randomly selected samples from each class and 90% range of all members are also plotted.

Table 1. Average value of the eight trajectory parameters in each class.

	Class 1	Class 2	Class 3	Class 4	Class 5
Number of trajectories	23503	54542	127887	150504	28413
Average maximum DN	61.69	56.51	52.52	31.44	40.80
Average minimum DN	47.76	22.03	7.16	6.53	7.46
Average mean DN	56.88	41.71	26.23	16.16	20.87
Average standard deviation DN	3.86	10.49	15.27	7.44	10.00
Average maximum slope	4.69	7.97	9.16	5.96	10.89
Average minimum slope	−2.48	−3.64	−2.47	−2.64	−8.05
Average mean slope	0.47	1.44	2.19	1.00	0.81
Average standard deviation slope	1.72	2.95	3.00	2.13	4.42

Class 1: stable urban activity; class 2: high-level steady growth; class 3: Acceleration; class 4: low-level steady growth; class 5: fluctuation.

lines (Figure 4(b,d)), and trajectories in both classes had smaller mean slopes with less variation (Table 1). However, class 2 on average had higher mean DN compared with class 4, therefore we considered class 2 as high-level steady growth and class 4 as low-level steady growth. Class 3 represented acceleration as its average trajectory showed a concave shape where the increase of DN started to accelerate between 2000 and 2005 (Figure 4(c)). Trajectories in this class also had the highest mean slopes, indicating the most rapid growth (Table 1). Finally, class 5 represented fluctuation as its sampled trajectories had a wide range, and were not stable compared with other classes (Figure 4(e)). Trajectories in this class on average also had the largest average standard deviations for slope (Table 1), indicating high variation in annual changes of NTL intensity.

3.2. *Validation*

Validation results were consistent with our interpretation of the unsupervised classes as the types of urbanization dynamics. The comparison between the NTL trajectories and the urbanization ratio trajectories showed certain level of agreement between the two: for each mean NTL trajectory, the corresponding mean urbanization ratio trajectory had a similar change pattern with correlation coefficient between 0.80 and 0.99 (Figure 5(a, b)). We also noticed that from class 2 to class 5, the NTL trajectories tended to achieve relatively higher levels than the urbanization ratio trajectories did in years. Such differences were partially due to urbanization ratio only capturing changes in impervious surface while NTL capturing other dimensions of urbanization such as increased human and economic activities. Human and economic activities could continue to intensify while land transformation to impervious surface stopped. In addition, the 'overglow' effect could cause overestimation of NTL-derived urbanized areas (i.e.$Class_{i,j}$ in Equation (4)), which in turn underestimated the urbanization ratio. In general, the high correlation between the two types of trajectories and their relatively similar growth pattern suggested that the interpreted NTL dynamics might be consistent with the temporal change of urbanization ratio.

In addition, spatial distribution of the classes in individual cities was consistent with our understanding about Chinese cities. For most cities, we found a concentric gradient of urbanization dynamics (Figure 6). From the urban core to the periphery, there was a transition from stable urban behaviour (class 1) to high-level steady growth (class 2) to

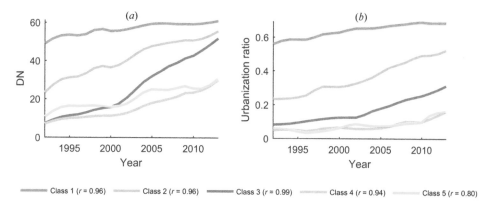

Figure 5. Comparison between the trajectories from NTL and urbanization ratio, (a) mean NTL trajectories, (b) mean urbanization ratio trajectories.

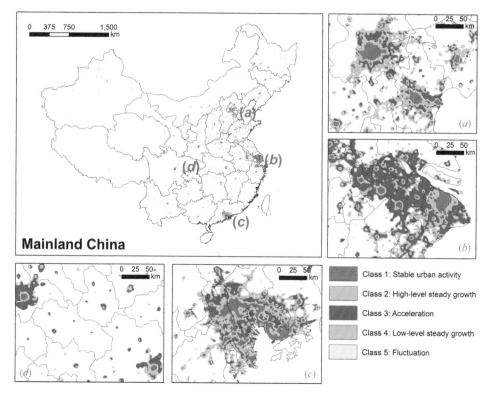

Figure 6. Map of different urbanization types from the unsupervised classification with five classes in four major urban agglomerations in China, (a) Beijing–Tianjin, (b) The Yangtze River Delta, (c) The Pearl River Delta, (d) Chendu–Chongqing.

acceleration (class 3) to low-level stead growth (class 4) to fluctuation (class 5). This transition could be explained by the following: (1) Chinese urban cores (class 1 and class 2) were more stable and developed due to longer development and higher concentrations of population, jobs, amenities, and other resources, (2) the middle zone (class 3) tended to be the most active area for its abundant land resource, low land price, and close distance to urban cores and existing resources, and (3) the periphery areas were too far away from the cores and resources to have enough incentives for urbanization (class 4), and might become unstable (class 5). Both validation using the urbanization ratio and spatial distribution of classes indicated that our labels we were likely to match the dominant urbanization dynamics reasonably well.

3.3. Composition of the urbanization dynamics in Chinese administrative divisions

We summarized the proportion and the global Moran's I value for each type of urbanization dynamics at provincial level in Table 2, and plotted the composition of the urbanization dynamics in each province in Figure 7. The stable urban activity contributed to an average of 5.63% of urban area in a province. There was a positive but not significant spatial autocorrelation of stable urban activity's proportion at provincial level. Several provinces in the north, and Guangdong and Guangxi Province in the south tended to more stable urban activity (Figure 8(a)). Shanghai, Beijing, Guangdong, Tianjing, and Liaoning were the five administrative divisions with the highest proportion. An average of 13.35% of a province's urban area was under the high-level steady growth. Guangdong, Shanghai, Beijing, Tianjing, and Heilongjiang had the highest proportion of this type of dynamics. The proportion of high-level steady growth also showed a positive but not significant spatial auto-correlation between the provinces. The acceleration urbanization dynamics on aver-age occupied 33.90% of a province's urban area. Jiangsu, Xizang, Zhejiang, Ningxia, and Chongqing had the highest proportions of such dynamics. The proportion of acceleration had a significant ($p = 0.030$), positive spatial autocorrelation. The low-level steady growth urbanization dynamics on average accounted for 39.27% of a province's urban area, and its proportion had a significant ($p = 0.010$), positive spatial autocorrelation. Provinces with the highest proportion of low-level steady growth

Table 2. Summary statistics and global Moran's I of the composition of the five types of urbanization dynamics in provinces, prefecture cities and counties.

Administrative division	Urbanization dynamics pattern	Proportion in each administrative division				Global Moran's I
		Mean (%)	Minimum (%)	Maximum (%)	Standard deviation	
Province ($n = 31$)	Stable urban activity	5.63	1.82	17.82	4.18	0.1
	High-level steady growth	13.35	7.62	22.86	4.36	0.15
	Acceleration	33.90	13.49	56.17	10.60	0.21**
	Low-level steady growth	39.27	19.31	55.90	8.06	0.25**
	Fluctuation	7.84	0.24	27.70	6.35	0.23***
Prefecture city ($n = 361$)	Stable urban activity	3.55	0.00	49.93	5.17	0.21***
	High-level steady growth	12.13	0.00	43.15	7.15	0.35***
	Acceleration	30.87	0.00	85.71	16.72	0.33***
	Low-level steady growth	45.23	5.37	100.00	16.47	0.28***
	Fluctuation	8.22	0.00	68.28	9.03	0.34***
County ($n = 2081$)	Stable	1.45	0.00	50.94	4.51	0.06***
	High-level steady growth	8.17	0.00	46.51	9.69	0.28***
	Acceleration	27.21	0.00	100.00	22.20	0.34***
	Low-level steady growth	54.04	0.00	100.00	24.75	0.31***
	Fluctuation	9.13	0.00	100.00	13.94	0.27***

: Significant at 95% confidence interval ($p \leq 0.05$); *: significant at 99% confidence interval ($p \leq 0.01$).

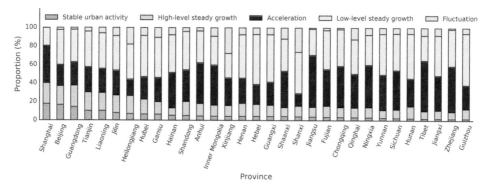

Figure 7. Composition of the five types of urbanization dynamics in Chinese provinces.

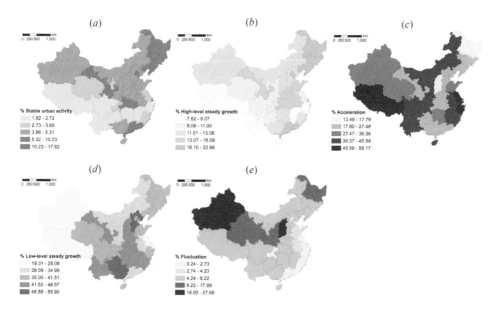

Figure 8. Proportion of the five types of urbanization dynamics at provincial level. (a) stable urban activity, (b) high-level steady growth, (c) acceleration, (d) low-level steady growth, (e) fluctuation.

included Guizhou, Hebei, Guangxi, Hunan and Henan. Compare with the spatial patterns for stable urban activity and fluctuation (Figure 8(a,e)), those of the two steady growth and the acceleration (Figure 8(b–d)) were more ambiguous: we found less obvious local clusters or geographic boundaries between high and low proportions for the latter three (i.e. high-level steady growth, low-level steady growth, and acceleration). An average of 7.84% of a province's urban area was under the fluctuation. The proportion of this type of dynamics also had a significant ($p = 0.010$), positive spatial autocorrelation. Provinces with lower proportion of fluctuation were concentrated along the eastern coast of China, and the ones with higher proportion were mainly clustered in the northwest (Figure 8(e)). The provinces with the highest proportion of fluctuation were Xinjiang, Shanxi, Heilongjiang, Qinghai, and Shanxi.

We further summarized composition of the five urbanization dynamics for prefecture cities and counties in China (Table 2). On average, 3.55% of the urbanized area in a prefecture city was stable, 12.13% was under the high-level steady growth, 30.87% was under the acceleration, 45.23% was under the low-level steady growth, and 8.22% was under the fluctuation. In an average county, 1.45% of the urbanized area was stable, 8.17% was under the high-level steady growth, 27.21% was under the acceleration, 54.04% was under the low-level steady growth, and 9.13% was under the fluctuation. Global Moran's I showed that those proportions were all significantly and spatially auto-correlated.

The acceleration was used as the dynamics of interest to investigate its proportion in prefecture cities and counties, and the corresponding local clusters and outliers (Figures 9(a,b) and 10(a,b)). We chose this type of dynamics as it highlighted areas transforming from rural with low NTL intensity to urban with high NTL intensity at an accelerating speed (Figure 4(c)), which might indicate an aggressive urbanization process and potentially lead to ghost cities. Acceleration also appeared to be a ubiquitous phenomenon in China, as on average 33.90% of a province's urban area, 30.87% of a prefecture city's urban area, 27.21% of a county's urban area were composed of such dynamics. In addition, the ambiguous spatial pattern of this dynamics at provincial level required further examination at the other two levels. Local Moran's I analysis showed that at both county and prefecture city levels, the

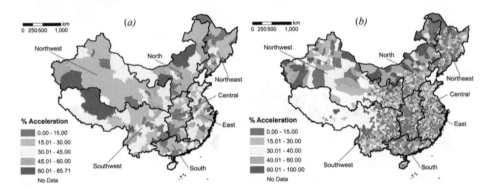

Figure 9. Proportion of class 3 (acceleration) for China's administrative divisions, (a) the proportion for prefecture cities, (b) the proportion for counties.

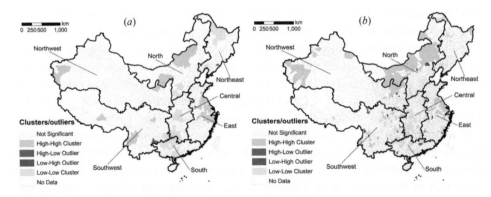

Figure 10. Clusters and outliers for proportion of class 3 (acceleration) for China's administrative divisions, (a) prefecture cities, (b) counties.

clusters with high proportion of the accelerated growth were mainly located in the northern, northwestern, southwestern, and eastern part of the country, while those with the low proportion tended to be in the central and southern part of the country. There were also several outliers that had dissimilar values with their neighbours in the country (Figure 10(a,b)).

4. Discussion

4.1. *Understanding urbanization through the new lens*

Understanding different urbanization dynamics and their spatial patterns is an important prerequisite to inform planning in the face of growing urban population and increasing environmental footprint of expanding urban regions (Deng et al. 2009; Jiang and O'Neill 2015; Stathakis, Tselios, and Faraslis 2015; Wu et al. 2014; Zhang and Seto 2011). With our innovative approach, we identified several important trajectories describing urbanization dynamics in mainland China. The five types of urbanization dynamics provided a spatially explicit reconstruction of urban development as manifested in the intensity of version 4 DMSP-OLS NTL signal. When combined with other socioeconomic datasets and knowledge about local and national developmental history and policy, the urbanization dynamics may help to reveal the large-scale urbanization processes and confirm the underlying mechanisms, which provide more insights into the on-going urbanization in China.

We found the acceleration urbanization dynamics to be an especially useful indicator of fast and even aggressive urbanization, as it showed an accelerating transformation from rural to urban between 1992 and 2013. Spatial pattern of these dynamics revealed urbanization characteristics at different spatial scales. At the finer scale (i.e. 1 km × 1 km pixel), the acceleration outlined areas under rapid urbanization, which provided a base map for more formal and labour-intensive verification methods such as *in situ* land use survey. At the aggregated scales, administrative divisions with high proportion of such dynamics formed several clusters in different parts of China (Figure 10(a,b)).

The underlying processes that formed these clusters varied among different geographic regions. The clusters in the eastern part of China largely overlapped with the Yangtze River Delta (YRD), where urbanization was partially caused by development of non-agricultural industries through rural collective accumulation (Gu et al. 2011). Since 1991, particularly in the northern YRD, such industries have started to concentrate, and stimulated urbanization and construction of small towns (Gu et al. 2011). These industries also attracted significant amount of labour and caused population growth. Other clusters likely resulted from different processes, such as aggressive urbanization due to financial speculation (Sorace and Hurst 2016) and land-centred urbanization policy (Chen, Liu, and Lu 2015; Sorace and Hurst 2016) that encouraged construction of urban areas beyond population needs. A typical example of this aggressive process is Ordos in the Inner Mongolia Autonomous Region. The city was the poorest one in its province before 2004 and then rapidly became one of China's wealthiest cities due to the discovery and production of significant amount of natural resources such as rare earth minerals and natural gas (Woodworth 2015; Sorace and Hurst 2016). Such a great shift in economy stimulated an unsustainable urbanization, which was characterized by massive land development and an inflated real estate market, while lacked a broader industrial base other than the extraction of natural resources (Woodworth 2015). When the real estate market collapsed in 2011 and China's economy slowed down, the resulting vacant residential and commercial projects made the city known as a 'ghost town' in many major media outlets (Woodworth 2015). Data from China city statistical survey (National Bureau of Statistics of the People's Republic of China 2016) showed that urbanization in Ordos was more likely driven by its economy rather than population when compared with YRD's typical city, Suzhou. With 1% increase in urban area between 2001 and 2013, Suzhou had 0.2% increase in population while Ordos only had 0.06% increase, suggesting that Ordos' urbanization was beyond the actual population growth needs (Table 3). Qianwen Zhang and Su (2016) also found that population were the main urban expansion driver for Nanjing and Hangzhou in the YRD, while economy became important for Hohhot in Inner Mongolia, suggesting different urbanization forces between the two regions. In both Suzhou and Ordos, our analysis was able to detect the areas under rapid urbanization (Figure 11).

For cases like Ordos, policy interventions are required to control rapid urbanization and to ensure that urban development is compatible with population and economic growth. As a result, China's central government announced the 'National New-type Urbanization Plan (2014–2020)' in 2014, which regulated aggressive urban development as the plan required urban construction land to be less than 100 m^2 per capita in 2020 (Wang et al. 2015). However, local governments may still find other ways to get around this regulation (Wang et al. 2015). A potential solution to enforce the plan is to develop a monitoring framework that includes large scale mapping tools, such as the approach of our study, to feasibly achieve a global understanding about urbanization and to find hidden issues under the overall pattern. The framework should also incorporate fine scale methods, such as in situ land use survey, to accurately identify places under aggressive urbanization.

Urbanization dynamics found in this study indicated growth in general, and we were not able to identify shrinking areas. Urban shrinkage is a world-wide phenomenon that cities experience population loss in large areas and economic transformations indicating a structural crisis (Wiechmann and Pallagst 2012). In the Chinese context, Long and Wu (2016) argued that urban shrinkage was due to migrant flow from rural areas and small cities to large and developed cities, and discovered 180 shrinking cities in China using Chinese population census in 2000 and 2010. Due to the reduction in active urban land use, population, and economic activities, we assumed that shrinking cities would tend to show declining NTL trajectories.

Table 3. Percentage increase in built-up area, population, and GDP between 2001 and 2013 in Suzhou and Ordos.

	Percentage increase (%) between 2001 and 2013 – original value (normalized value by increase in built-up area)	
Parameter	Suzhou	Ordos
Built-up area	3.02 (1.00)	6.38 (1.00)
Population	0.59 (0.20)	0.36 (0.06)
GDP	9.70 (3.21)	21.19 (3.32)

Original percentages are listed, and normalized percentages by built-up area are listed in the brackets.

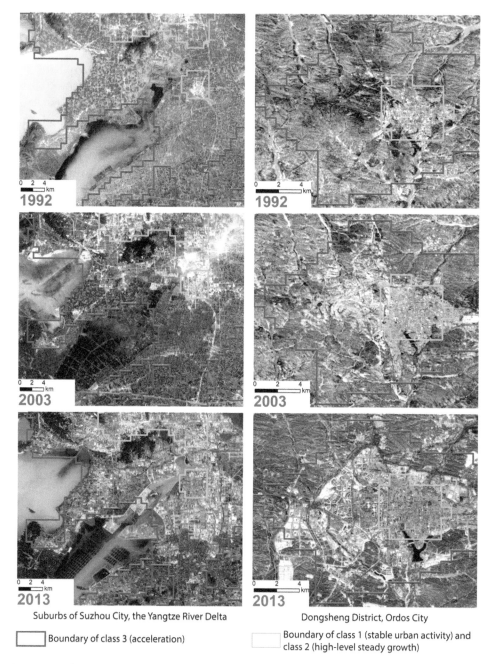

1992

1992

2003

2003

2013

2013

Suburbs of Suzhou City, the Yangtze River Delta

Dongsheng District, Ordos City

☐ Boundary of class 3 (acceleration)

☐ Boundary of class 1 (stable urban activity) and class 2 (high-level steady growth)

Figure 11. Urbanization outlined by class 3 (acceleration) in suburbs of Suzhou City, the Yangtze River Delta, and in Dongsheng District, Ordos City.

Contrary to the expectations, we were not able to detect declining NTL trajectories, or urban shrinkage, in this study. Such inability was likely due to the following reasons. First, most Chinese urban areas were still developing between 1992 and 2013, which led to generally increasing trends in NTL. We observed decreases in NTL in several time periods, such as years around the 2008 world financial crisis. However, since our method did not segment the trajectories between 1992 and 2013, we were not able to isolate the segments with decreased NTL, and thus failed to find areas that were declining in specific time periods. Second, there could be time lags between urban shrinkage and NTL change: urban areas might be abandoned, but electricity was still used and delayed the pronounced decline in NTL. Since urban shrinkage is an emerging phenomenon in China, our study period (i.e. 1992–2013) might be too short to incorporate such time lags. Finally, as the country is still urbanizing, the shrinkage can be limited and occurred at spatial scales finer than the 1 km × 1 km NTL pixel. The scattered spatial distribution and fine scale may also prevent the shrinking areas from being detected by NTL trajectories.

4.2. *An innovative approach to monitor urbanization dynamics and its challenges*

Urbanization is an important global phenomenon that has both environmental and social implications. With the increased availability of data and analysis methods, there has been a tendency to study urban and regional issues through the data-driven, geospatial approach. Some examples include characterizing post-disaster population displacement pattern through 1.9 million mobile phone users in Haiti (Lu, Bengtsson, and Holme 2012), and identifying 'Ghost Cities' using larger volume positioning data and points of interests (Chi et al. 2015). In general, those studies tried to use the pattern within the datasets to inform the underlying process. We consider our study as another example of such type of methods.

Remote sensing allows monitoring large scale urbanization over relatively long time periods. However, previous remote-sensing studies focused more on mapping the changing urban extent (Gao et al. 2016; Jat, Garg, and Khare 2008; Wang et al. 2012), which were insufficient to capture other dimensions of urbanization including changes in population and economic activities (Zhang and Seto 2011). We provided an innovative and low-cost approach that used pixel-based, time-series NTL trajectories to monitor urbanization dynamics. Compare to other data, NTL provides a moderate spatial resolution, a long temporal coverage, and a low cost in computation. NTL's ability to characterize intensity of human activities also captures multiple dimensions in urbanization including urban extent, population and economic activities. Compare with previous studies that also employed time-series NTL trajectories (Zhang and Seto 2011; Wu et al. 2014), we provided an alternative approach that addressed potential challenges in earlier studies. First, we applied the silhouette analysis to more objectively inform the number of classes based on the pattern of the data. Second, we validated our labelling of the unsupervised classes with time-series urbanization ratio estimated from impervious surface. With this approach, we could yield a reasonable set of urbanization dynamics for mainland China between 1992 and 2013. In addition, through visual inspection we found that the spatial distribution of our results was consistent with the one from (Zhang and Seto 2011) for the Pearl River Delta (Figure 6(c)) and the Yangtze River Delta (Figure 6(b)), and with the one from Wu et al. (2014) for Beijing and Tianjin Area (Figure 6(a)). Finally, we reduced the amount of input data for the classifier. To our knowledge, we were the first to use secondary descriptors for the trajectories, the eight trajectory parameters, in the classification. Furthermore, we could accelerate the classification by reducing the eight parameters to three input variables using the dimension reduction approach.

4.3. *Limitations and future research directions*

Data-driven methods to understand urbanization generally provide a broad picture about this phenomenon and may reveal some unexpected findings that cannot be uncovered with traditional methods such as survey, interview, and observation. However, data-driven methods have their own uncertainties, which mainly result from the properties of the data itself and the analysis method. For example, our study's outcome was affected by the NTL data quality and the *k*-means clustering method, even though we performed our analysis with the best of our knowledge. With an improved NTL dataset, such as the multi-month VIIRS (http://ngdc.noaa.gov/eog/viirs/download_monthly.html), and different sets of parameters for the *k*-means clustering, we might expect some potential differences in the analysis outcomes. Therefore, such data-driven studies should not only rely on a single dataset and a single method. Instead, multiple datasets (Liu et al. 2016) and methods should be implemented in the future work to cross validate their results. When results from such data-driven studies are applied, they should be used as complimentary to those from the traditional methods (Liu et al. 2016).

Several specific challenges should be considered in interpreting our study results. First, identifying the number of classes for the unsupervised classification may have important effects on the derived urbanization dynamics, because too few classes may oversimplify the urbanization dynamics, while too many classes may be too difficult for interpretation. In addition, different number of classes may lead to different conclusions. The underlying problem is the difficulty to draw distinct decision boundaries for the NTL trajectories, as they showed a broad and continuous variation of shapes and levels where the differences could be subtle (shown by the 1% randomly sampled trajectories in Figure 4(a–f)). To overcome the problem, future studies should invite local planners to verify if the number of classes can properly identify different typologies of urbanization in the planners' communities and cities. Researchers could also select a few representative cities, and verify the urbanization dynamics from NTL with land-cover change from high resolution remote-sensing images (Zhang and Seto 2013). Furthermore, a sensitivity analysis should be carried out to understand how different numbers of classes may affect the analysis.

Second, the differences in timing of the NTL changes during the study period should be incorporated to characterize more specific urbanization dynamics and to reduce variation in each type of dynamics. The eight trajectory parameters we used did not show the timing of the changes, therefore our urbanization dynamics were relatively general with broad range in certain classes (e.g. class 2, 3, and 5 in Figure 4(a–f)), and required us to interpret the classes with caution when referring the temporal aspect of them. Such general urbanization dynamics were particularly insufficient to identify urban shrinkage as it might occur during specific time periods in mainland China. We suggest future researchers apply temporal segmentation methods that identify the timing of major changes to in this type of analysis, which would help to understand different phases in urbanization.

Finally, a more systematic validation of the NTL-derived urbanization dynamics in more cities and with multiple data sources is necessary. While our study only used five cities as validation sites, future researchers should include more representative cities in terms of urbanization type and level, spatial pattern of land use, economy, and geography. After stratifying the five validation cities into eastern cities including Hangzhou and Nanchang, and non-eastern cities including Harbin, Nanning, and Hanzhong, we found that non-eastern cities on average reached lower urbanization ratio in each NTL urbanization dynamics outlined zones (i.e. $Class_{ij}$ in Equation (4)). We hypothesize that the lower urbanization ratios of the non-eastern cities were due to their stronger 'overglow' effect. The non-eastern cities are characterized by lower density development (Zhang et al. 2005), which could produce more scattered urban areas and light sources, and therefore stronger 'overglow' effect and overestimation of possible urbanizing areas from NTL-derived urbanization dynamics. Qian Zhang and Seto (2013) also found that NTL had high accuracy of characterizing urbanization dynamics in places where urbanization actually occurred, and lower accuracy in places that were not urbanized but influenced by 'overglow'. If possible, we recommend that future researchers identify NTL-derived urbanization dynamics within urbanized areas determined by land cover to improve accuracy. In addition, finer resolution socioeconomic data such as population and economy should be used together with land cover to verify if they show similar changes with NTL. With such more systematic validation and reduction in 'overglow' effect, NTL-derived dynamics can more accurately reflect the actual urbanization process.

5. Conclusions

Urbanization becomes one of the central drivers of global environmental change. Monitoring and understanding urbanization process becomes increasingly critical for planning and designing towards sustainable cities. With the increased availability of data and analytical power, we can utilize data-driven approaches along with traditional approaches to study urbanization, identify the patterns, and reveal the underlying process. By understanding the pattern and the process, planners, designers, and decision makers can apply proper strategies to enhance or intervene the ongoing urbanization. Our study, which used pixel-based, time-series NTL trajectories to identify different urbanization dynamics, demonstrates the power of such data-driven approach.

Building on existing knowledge and research approaches, we proposed an alternative method that better determined the types of urbanization dynamics and reduced the input information. Taking mainland China as a case study example, we could identify five major types of urbanization dynamics between 1992 and 2013. With a focus on accelerated growth, we found clusters with high proportion of this urbanization dynamics, and interpreted the formation of the clusters by relating them to existing knowledge on policy and developmental history. This approach provides a low-cost, timely, and large-scale understanding about how urbanization happens over space and time, and should be used in combination with socioeconomic data, land-cover mapping, and *in situ* land use survey to provide a solid basis for better urban planning, design, monitoring, and management.

Acknowledgements

We would like to thank the two anonymous reviewers and editors for their insightful comments on this manuscript.

Disclosure statement

No potential conflict of interest was reported by the authors.

ORCID

Yang Ju ⓘ http://orcid.org/0000-0002-1947-7533
Qin Ma ⓘ http://orcid.org/0000-0002-6995-6663

References

Amaral, S., A. M. V. Monteiro, G. Camara, and J. A. Quintanilha. 2006. "DMSP/OLS Night-Time Light Imagery for Urban Population Estimates in the Brazilian Amazon." *International Journal of Remote Sensing* 27 (5): 855–870. doi:10.1080/01431160500181861.

Anselin, L. 1995. "Local Indicators of Spatial Association—LISA." *Geographical Analysis* 27 (2): 93–115. doi:10.1111/j.1538-4632.1995.tb00338.x.

Bai, X., P. Shi, and Y. Liu. 2014. "Society: Realizing China's Urban Dream." *Nature* 509 (7499): 158–160. doi:10.1038/509158a.

Bhatta, B. 2009. "Analysis of Urban Growth Pattern Using Remote Sensing and GIS: A Case Study of Kolkata, India." *International Journal of Remote Sensing* 30 (18): 4733–4746. doi:10.1080/01431160802651967.

Bhatta, B., S. Saraswati, and D. Bandyopadhyay. 2010. "Urban Sprawl Measurement from Remote Sensing Data." *Applied Geography, Climate Change and Applied Geography – Place, Policy, and Practice* 30 (4): 731–740. doi:10.1016/j.apgeog.2010.02.002.

Bian, J., A. Li, Q. Liu, and C. Huang. 2016. "Cloud and Snow Discrimination for CCD Images of HJ-1A/B Constellation Based on Spectral Signature and Spatio-Temporal Context." *Remote Sensing* 8 (1): 31. doi:10.3390/rs8010031.

Buhaug, H., and H. Urdal. 2013. "An Urbanization Bomb? Population Growth and Social Disorder in Cities." *Global Environmental Change* 23 (1): 1–10. doi:10.1016/j.gloenvcha.2012.10.016.

Chen, M., W. Liu, and D. Lu. 2015. "Challenges and the Way Forward in China's New-Type Urbanization." *Land Use Policy*. doi:10.1016/j.landusepol.2015.07.025.

Chen, Z., B. Yu, Y. Hu, C. Huang, K. Shi, and J. Wu. 2015. "Estimating House Vacancy Rate in Metropolitan Areas Using NPP-VIIRS Nighttime Light Composite Data." *IEEE Journal of Selected Topics in Applied Earth Observations and Remote Sensing* 8 (5): 2188–2197. doi:10.1109/JSTARS.2015.2418201.

Chi, G., Y. Liu, Z. Wu, and H. Wu. 2015. "Ghost Cities Analysis Based on Positioning Data in China." *arXiv:1510.08505 [Cs]*, October http://arxiv.org/abs/1510.08505

Deng, J. S., K. Wang, Y. Hong, and J. G. Qi. 2009. "Spatio-Temporal Dynamics and Evolution of Land Use Change and Landscape Pattern in Response to Rapid Urbanization." *Landscape and Urban Planning* 92 (3–4): 187–198. doi:10.1016/j.landurbplan.2009.05.001.

Doygun, H., and H. Alphan. 2006. "Monitoring Urbanization of Iskenderun, Turkey, and Its Negative Implications." *Environmental Monitoring and Assessment* 114 (1–3): 145–155. doi:10.1007/s10661-006-2524-0.

Elvidge, C. D., M. L. Imhoff, K. E. Baugh, V. R. Hobson, I. Nelson, J. Safran, J. B. Dietz, and B. T. Tuttle. 2001. "Night-Time Lights of the World: 1994–1995." *ISPRS Journal of Photogrammetry and Remote Sensing* 56 (2): 81–99. doi:10.1016/S0924-2716(01)00040-5.

Elvidge, C. D., D. Ziskin, K. E. Baugh, B. T. Tuttle, T. Ghosh, D. W. Pack, E. H. Erwin, and M. Zhizhin. 2009. "A Fifteen Year Record of Global Natural Gas Flaring Derived from Satellite Data." *Energies* 2 (3): 595–622. doi:10.3390/en20300595.

ESRI. 2016. "Arcgis Desktop (Version 10.4)." ArcMap 10.4: ESRI.

Gao, B., Q. Huang, C. He, Z. Sun, and D. Zhang. 2016. "How Does Sprawl Differ across Cities in China? A Multi-Scale Investigation Using Nighttime Light and Census Data." *Landscape and Urban Planning* 148: 89–98. doi:10.1016/j.landurbplan.2015.12.006.

Ghosh, T., R. L. Powell, C. D. Elvidge, K. E. Baugh, P. C. Sutton, and S. Anderson. 2010. "Shedding Light on the Global Distribution of Economic Activity." *The Open Geography Journal* 3: 1.

Gong, P., J. Wang, L. Yu, Y. Zhao, Y. Zhao, L. Liang, Z. Niu, et al. 2013. "Finer Resolution Observation and Monitoring of Global Land Cover: First Mapping Results with Landsat TM and ETM+ Data." *International Journal of Remote Sensing* 34 (7): 2607–2654. doi:10.1080/01431161.2012.748992.

Gorelick, N. 2013. "Google Earth Engine." *EGU General Assembly Conference Abstracts*. 15:11997. http://adsabs.harvard.edu/abs/2013EGUGA..1511997G

Gu, C., L. Hu, X. Zhang, X. Wang, and J. Guo. 2011. "Climate Change and Urbanization in the Yangtze River Delta." *Habitat International* 35 (4): 544–552. doi:10.1016/j.habitatint.2011.03.002.

Hollis, G. E. 1975. "The Effect of Urbanization on Floods of Different Recurrence Interval." *Water Resources Research* 11 (3): 431–435. doi:10.1029/WR011i003p00431.

Hu, Y., S. Gao, K. Janowicz, B. Yu, W. Li, and S. Prasad. 2015. "Extracting and Understanding Urban Areas of Interest Using Geotagged Photos." *Computers, Environment and Urban Systems* 54: 240–254. doi:10.1016/j.compenvurbsys.2015.09.001.

Huang, C., L. S. Davis, and J. R. G. Townshend. 2002. "An Assessment of Support Vector Machines for Land Cover Classification." *International Journal of Remote Sensing* 23 (4): 725–749. doi:10.1080/01431160110040323.

Imhoff, M. L., W. T. Lawrence, D. C. Stutzer, and C. D. Elvidge. 1997. "A Technique for Using Composite DMSP/OLS City Lights" Satellite Data to Map Urban Area." *Remote Sensing of Environment* 61 (3): 361–370. doi:10.1016/S0034-4257(97)00046-1.

Jat, M. K., P. K. Garg, and D. Khare. 2008. "Monitoring and Modelling of Urban Sprawl Using Remote Sensing and GIS Techniques." *International Journal of Applied Earth Observation and Geoinformation* 10 (1): 26–43. doi:10.1016/j.jag.2007.04.002.

Jiang, B., and Y. Miao. 2015. "The Evolution of Natural Cities from the Perspective of Location-Based Social Media." *The Professional Geographer* 67 (2): 295–306. doi:10.1080/00330124.2014.968886.

Jiang, L., and B. C. O'Neill. 2015. "Global Urbanization Projections for the Shared Socioeconomic Pathways." *Global Environmental Change*. doi:10.1016/j.gloenvcha.2015.03.008.

Kalnay, E., and M. Cai. 2003. "Impact of Urbanization and Land-Use Change on Climate." *Nature* 423 (6939): 528–531. doi:10.1038/nature01675.

Li, X., and Y. Zhou. 2017. "Urban Mapping Using DMSP/OLS Stable Night-Time Light: A Review." *International Journal of Remote Sensing* 0 (0): 1–17. doi:10.1080/01431161.2016.1274451.

Li, Y., X. Zhu, X. Sun, and F. Wang. 2010. "Landscape Effects of Environmental Impact on Bay-Area Wetlands under Rapid Urban Expansion and Development Policy: A Case Study of Lianyungang, China." *Landscape and Urban Planning* 94 (3–4): 218–227. doi:10.1016/j.landurbplan.2009.10.006.

Liu, J., J. Li, W. Li, and J. Wu. 2016. "Rethinking Big Data: A Review on the Data Quality and Usage Issues." *ISPRS Journal of Photogrammetry and Remote Sensing, Theme Issue State-Of-The-Art in Photogrammetry, Remote Sensing and Spatial Information Science,"* 115: 134–142. doi:10.1016/j.isprsjprs.2015.11.006.

Liu, J., and W. Li. 2014. "A Nighttime Light Imagery Estimation of Ethnic Disparity in Economic Well-Being in Mainland China and Taiwan (2001–2013)." *Eurasia Geography and Economics* 55 (6): 691–714. doi:10.1080/15387216.2015.1041147.

Liu, Z., C. He, Q. Zhang, Q. Huang, and Y. Yang. 2012. "Extracting the Dynamics of Urban Expansion in China Using DMSP-OLS Nighttime Light Data from 1992 to 2008." *Landscape and Urban Planning* 106 (1): 62–72. doi:10.1016/j.landurbplan.2012.02.013.

Lloyd, S. P. 1982. "Least Squares Quantization in PCM." *Information Theory, IEEE Transactions On* 28 (2): 129–137. doi:10.1109/TIT.1982.1056489.

Long, Y., and K. Wu. 2016. "Shrinking Cities in a Rapidly Urbanizing China." *Environment and Planning A* 48 (2): 220–222. doi:10.1177/0308518X15621631.

Lu, X., L. Bengtsson, and P. Holme. 2012. "Predictability of Population Displacement after the 2010 Haiti Earthquake." *Proceedings of the National Academy of Sciences* 109 (29): 11576–11581. doi:10.1073/pnas.1203882109.

Ma, Q., C. He, J. Wu, Z. Liu, Q. Zhang, and Z. Sun. 2014. "Quantifying Spatiotemporal Patterns of Urban Impervious Surfaces in China: An Improved Assessment Using Nighttime Light Data." *Landscape and Urban Planning* 130: 36–49. doi:10.1016/j.landurbplan.2014.06.009.

Ma, T., C. Zhou, T. Pei, S. Haynie, and J. Fan. 2012. "Quantitative Estimation of Urbanization Dynamics Using Time Series of DMSP/OLS Nighttime Light Data: A Comparative Case Study from China's Cities." *Remote Sensing of Environment* 124: 99–107. doi:10.1016/j. rse.2012.04.018.

National Bureau of Statistics of the People's Republic of China. 2016. *China City Statistical Yearbook*. Beijing: China Statistics Press. http://tongji.cnki.net/kns55/navi/HomePage.aspx?id= N2014050073&name=YZGCA&floor=1

National Geophysical Data Center. 2015. "Version 4 DMSP-OLS Nighttime Lights Time Series." http://ngdc.noaa.gov/eog/gcv4_readme.txt

Patel, N. N., E. Angiuli, P. Gamba, A. Gaughan, G. Lisini, F. R. Stevens, A. J. Tatem, and G. Trianni. 2015. "Multitemporal Settlement and Population Mapping from Landsat Using Google Earth Engine." *International Journal of Applied Earth Observation and Geoinformation* 35: 199–208. doi:10.1016/j.jag.2014.09.005.

Pohl, C., and J. L. Van Genderen. 1998. "Review Article Multisensor Image Fusion in Remote Sensing: Concepts, Methods and Applications." *International Journal of Remote Sensing* 19 (5): 823–854. doi:10.1080/014311698215748.

Rousseeuw, P. J. 1987. "Silhouettes: A Graphical Aid to the Interpretation and Validation of Cluster Analysis." *Journal of Computational and Applied Mathematics* 20: 53–65. doi:10.1016/0377-0427 (87)90125-7.

Savitzky, A., and M. J. E. Golay. 1964. "Smoothing and Differentiation of Data by Simplified Least Squares Procedures." *Analytical Chemistry* 36 (8): 1627–1639. doi:10.1021/ac60214a047.

Schetke, S., and D. Haase. 2008. "Multi-Criteria Assessment of Socio-Environmental Aspects in Shrinking Cities. Experiences from Eastern Germany." *Environmental Impact Assessment Review* 28 (7): 483–503. doi:10.1016/j.eiar.2007.09.004.

Schneider, A., and C. E. Woodcock. 2008. "Compact, Dispersed, Fragmented, Extensive? A Comparison of Urban Growth in Twenty-Five Global Cities Using Remotely Sensed Data, Pattern Metrics and Census Information." *Urban Studies* 45 (3): 659–692. doi:10.1177/ 0042098007087340.

Scikit-learn. 2015. "Selecting the Number of Clusters with Silhouette Analysis on Kmeans Clustering — Scikit-Learn 0.17 Documentation." http://scikit-learn.org/stable/auto_examples/ cluster/plot_kmeans_silhouette_analysis.html

Shi, K., C. Huang, B. Yu, B. Yin, Y. Huang, and J. Wu. 2014a. "Evaluation of NPP-VIIRS Night-Time Light Composite Data for Extracting Built-Up Urban Areas." *Remote Sensing Letters* 5 (4): 358– 366. doi:10.1080/2150704X.2014.905728.

Shi, K., B. Yu, Y. Huang, Y. Hu, B. Yin, Z. Chen, L. Chen, and J. Wu. 2014b. "Evaluating the Ability of NPP-VIIRS Nighttime Light Data to Estimate the Gross Domestic Product and the Electric Power Consumption of China at Multiple Scales: A Comparison with DMSP-OLS Data." *Remote Sensing* 6 (2): 1705–1724. doi:10.3390/rs6021705.

Sleeter, B. M., T. L. Sohl, T. R. Loveland, R. F. Auch, W. Acevedo, M. A. Drummond, K. L. Sayler, and S. V. Stehman. 2013. "Land-Cover Change in the Conterminous United States from 1973 to 2000." *Global Environmental Change* 23 (4): 733–748. doi:10.1016/j.gloenvcha.2013.03.006.

Sorace, C., and W. Hurst. 2016. "China's Phantom Urbanisation and the Pathology of Ghost Cities." *Journal of Contemporary Asia* 46 (2): 304–322. doi:10.1080/00472336.2015.1115532.

Stathakis, D., V. Tselios, and I. Faraslis. 2015. "Urbanization in European Regions Based on Night Lights." *Remote Sensing Applications: Society and Environment* 2: 26–34. doi:10.1016/j. rsase.2015.10.001.

Steinley, D. 2003. "Local Optima in K-Means Clustering: What You Don't Know May Hurt You." *Psychological Methods* 8 (3): 294–304. doi:10.1037/1082-989X.8.3.294.

Su, S., Z. Jiang, Q. Zhang, and Y. Zhang. 2011. "Transformation of Agricultural Landscapes under Rapid Urbanization: A Threat to Sustainability in Hang-Jia-Hu Region, China." *Applied Geography* 31 (2): 439–449. doi:10.1016/j.apgeog.2010.10.008.

Tan, M. 2016. "Use of an inside Buffer Method to Extract the Extent of Urban Areas from DMSP/OLS Nighttime Light Data in North China." *Giscience & Remote Sensing* 53 (4): 444–458. doi:10.1080/ 15481603.2016.1148832.

Townsend, A. C., and D. A. Bruce. 2010. "The Use of Night-Time Lights Satellite Imagery as a Measure of Australia's Regional Electricity Consumption and Population Distribution." *International Journal of Remote Sensing* 31 (16): 4459–4480. doi:10.1080/01431160903261005.

United Nations. 2014. *World Urbanization Prospects: The 2014 Revision*. New York: United Nations, Department of Economic and Social Affairs (DESA), Population Division.

Wang, L., C. Li, Q. Ying, X. Cheng, X. Wang, X. Li, L. Hu, et al. 2012. "China's Urban Expansion from 1990 to 2010 Determined with Satellite Remote Sensing." *Chinese Science Bulletin* 57 (22): 2802–2812. doi:10.1007/s11434-012-5235-7.

Wang, X.-R., E. C.-M. Hui, C. Choguill, and S.-H. Jia. 2015. "The New Urbanization Policy in China: Which Way Forward?" *Habitat International* 47: 279–284. doi:10.1016/j.habitatint.2015.02.001.

Weng, Q. 2002. "Land Use Change Analysis in the Zhujiang Delta of China Using Satellite Remote Sensing, GIS and Stochastic Modelling." *Journal of Environmental Management* 64 (3): 273–284. doi:10.1006/jema.2001.0509.

Wiechmann, T., and K. M. Pallagst. 2012. "Urban Shrinkage in Germany and the USA: A Comparison of Transformation Patterns and Local Strategies." *International Journal of Urban and Regional Research* 36 (2): 261–280. doi:10.1111/j.1468-2427.2011.01095.x.

Woodworth, M. D. 2015. "Ordos Municipality: A Market-Era Resource Boomtown." *Cities* 43: 115–132. doi:10.1016/j.cities.2014.11.017.

Wu, J. 2008. "Toward a Landscape Ecology of Cities: Beyond Buildings, Trees, and Urban Forests." In *Ecology, Planning, and Management of Urban Forests.* edited by M. M. Carreiro, Y.-C. Song, and J. Wu, 10–28. New York: Springer. http://link.springer.com/chapter/10.1007/978-0-387-71425-7_2

Wu, J., S. He, J. Peng, W. Li, and X. Zhong. 2013. "Intercalibration of DMSP-OLS Night-Time Light Data by the Invariant Region Method." *International Journal of Remote Sensing* 34 (20): 7356–7368. doi:10.1080/01431161.2013.820365.

Wu, J., L. Ma, W. Li, J. Peng, and H. Liu. 2014. "Dynamics of Urban Density in China: Estimations Based on DMSP/OLS Nighttime Light Data." *IEEE Journal of Selected Topics in Applied Earth Observations and Remote Sensing* 7 (10): 4266–4275. doi:10.1109/JSTARS.2014.2367131.

Xiao, J., Y. Shen, J. Ge, R. Tateishi, C. Tang, Y. Liang, and Z. Huang. 2006. "Evaluating Urban Expansion and Land Use Change in Shijiazhuang, China, by Using GIS and Remote Sensing." *Landscape and Urban Planning* 75 (1–2): 69–80. doi:10.1016/j.landurbplan.2004.12.005.

Yu, B., S. Shu, H. Liu, W. Song, J. Wu, L. Wang, and Z. Chen. 2014. "Object-Based Spatial Cluster Analysis of Urban Landscape Pattern Using Nighttime Light Satellite Images: A Case Study of China." *International Journal of Geographical Information Science* 28 (11): 2328–2355. doi:10.1080/13658816.2014.922186.

Zhang, F.-G., J.-M. Hao, G.-H. Jiang, Z.-Y. Ding, X.-B. Li, and T. Li. 2005. "Spatial-Temporal Variance of Urban Land Use Intensity [J]." *China Land Science* 1: 3.

Zhang, Q., and K. C. Seto. 2013. "Can Night-Time Light Data Identify Typologies of Urbanization? A Global Assessment of Successes and Failures." *Remote Sensing* 5 (7): 3476–3494. doi:10.3390/rs5073476.

Zhang, Q., and S. Su. 2016. "Determinants of Urban Expansion and Their Relative Importance: A Comparative Analysis of 30 Major Metropolitans in China." *Habitat International* 58: 89–107. doi:10.1016/j.habitatint.2016.10.003.

Zhang, Q., and K. C. Seto. 2011. "Mapping Urbanization Dynamics at Regional and Global Scales Using Multi-Temporal DMSP/OLS Nighttime Light Data." *Remote Sensing of Environment* 115 (9): 2320–2329. doi:10.1016/j.rse.2011.04.032.

Zhou, Y., S. J. Smith, C. D. Elvidge, K. Zhao, A. Thomson, and M. Imhoff. 2014. "A Cluster-Based Method to Map Urban Area from DMSP/OLS Nightlights." *Remote Sensing of Environment* 147: 173–185. doi:10.1016/j.rse.2014.03.004.

Zhou, Y., S. J. Smith, K. Zhao, M. Imhoff, A. Thomson, B. Bond-Lamberty, G. R. Asrar, X. Zhang, C. He, and C. D. Elvidge. 2015. "A Global Map of Urban Extent from Nightlights." *Environmental Research Letters* 10 (5): 54011. doi:10.1088/1748-9326/10/5/054011.

Assessing urban growth dynamics of major Southeast Asian cities using night-time light data

Shankar Acharya Kamarajugedda ⓘ, Pradeep V. Mandapaka ⓘ and Edmond Y.M Lo

ABSTRACT

This study analysed urban growth patterns for 15 Southeast Asian cities using remotely sensed night-time light data from 1992 to 2012. We extracted three urban categories (countryside, peri-urban, and core-urban) for each city using objectively derived thresholds from the brightness gradient (BG) approach. The peri-urban and core-urban combined categories were generally found to increase over time for all cities whereas countryside urban category decreased implying strong spatial and temporal trends in urbanization. These trends were also found to be sensitive to geographic characteristics of cities. The study showed that the BG approach can be successfully applied to extract and study growth dynamics of different urban categories for Southeast Asian cities having range of demographic and socioeconomic conditions. The BG derived urban categories compared favourably with Landsat derived impervious areas, where the former was found to envelope the high percentage impervious region derived from the latter. The BG-derived urban areas are lastly compared against the population data to explore linkages with population growth.

1. Introduction

It is well recognized that the effect of urbanization on the environment is manifold, including the loss of agricultural land, urban heat island effect and associated increase in near-surface temperature, increases in carbon emissions and air pollution, and the depletion of natural resources (Seto, Sánchez-Rodríguez, and Fragkias 2010; Schneider, Friedl, and Potere 2010). The rapid rate of urbanization also poses numerous challenges including infrastructure design, water resource management, and public health. The last century has witnessed an increasing concentration of people in urban regions. For example, the world-wide urban population in the past 25 years has grown by 73% compared 38% for total population (UNESCAP 2014). Similarly, the urban population percentage increased from 43% in 1990 to 54% in 2015, and is expected to reach 66% by 2050 (UNPD 2014). These urbanization rates depend on various factors such as the geography, socioeconomic conditions, and the political environment. For instance, the urban population percentage in developed and high-income countries is projected to increase at a lower rate compared to less developed and low-income countries.

Currently, the world's fastest growing cities are concentrated in the Global South (i.e. Africa, Latin America) and in developing Asia (UNPD 2014). In particular, Asia and Southeast Asia (SEA) are home to sprawling megacities such as Tokyo, Bangkok, Jakarta, and Manila, with population exceeding 10 million. SEA region is steadily urbanizing with about 245 million people (42% of total population) living in urban areas and it is predicted that the urban population will increase to 47% by 2025 (Aritenang 2014). Further the SEA region possesses a wide range of urbanization levels ranging from economically advanced countries (e.g. Singapore and Malaysia) to less developed countries (e.g. Cambodia and Myanmar). SEA cities also have large geographic variability compared to those in US, China, and India because of the complex land-sea structure of the SEA region. Understanding and modelling the spatiotemporal dynamics of urbanization of these cities as conditioned on multiple factors (e.g. economic, geographic, and political) is essential for developing appropriate planning and management strategies.

Socioeconomic and census data are good data sources in urbanization studies (Henderson et al. 2003; Chen et al. 2014). However in many less-developed and developing nations, such data is frequently unavailable at desired spatial and temporal resolutions for modelling urban growth dynamics (e.g. Masek, Lindsay, and Goward 2000). In such situations, remote sensing can be a dependable alternative, providing data at multiple spatial and temporal scales. Satellite data is more suited to extract urban areas because of the satellites' high visit rates and the availability of archived datasets over a long period. Satellite-based observations with various spatiotemporal resolutions have been applied to investigate urban growth patterns and urbanization impacts on natural systems (Masek, Lindsay, and Goward 2000; Seto et al. 2011; Ma et al. 2012; Ma et al. 2015). Optical data, and especially the Landsat data has been extensively used in many studies to analyse urban growth patterns, and urban land-use and land-cover changes (Masek, Lindsay, and Goward 2000; Seto et al. 2002; Bhatta 2010) and shown to be useful in planning for sustainable urban growth (Seto et al. 2002). Spatial patterns of land cover changes for metropolitan cities have also been assessed using fine and medium resolution remotely sensed data, e.g. the Moderate Resolution Imaging Spectroradiometer (MODIS) with a resolution of 500 m that can be used to produce maps of urban regions at a global scale (Schneider, Friedl, and Potere 2009; 2010).

Another source of remote-sensing data is the Defense Meteorological Satellite Program (DMSP) OLS night-time light (NTL) data. The NTL data has been widely used to monitor urbanization at regional and global scales (e.g. Henderson et al. 2003; Small, Pozzi, and Elvidge 2005; Liu et al. 2012; Ma et al. 2012; Zhang and Seto 2013; Elvidge et al. 2014; Ma et al. 2015). The major advantage of using NTL data is that it captures information related to anthropogenic activity. Thus urban growth patterns extracted from NTL data analysis are shown to be strongly correlated with social and economic variables such as population, GDP and electric power consumption (Sutton et al. 2001; Doll, Muller, and Morley 2006; Chand et al. 2009), making it a good proxy to model urbanization.

Extracting and mapping urban extent from NTL raw imagery is a key task. Hence, researchers developed various techniques such as thresholding (Imhoff et al. 1997; Elvidge et al. 2009; Henderson et al. 2003) and also image classification method (Cao et al. 2009) to generate urban extent maps from NTL imagery at different spatial scales.

Of these methods, thresholding approach has been widely used to extract urban information from NTL data. However, a major issue of underestimation or overestimation arises when using a single threshold, and as such, no single brightness threshold is universally applicable for mapping urban boundaries from NTL imagery (Small, Pozzi, and Elvidge 2005; Small et al. 2011). An alternative cluster based approach was developed by Zhou et al. (2014) to reduce the constraints from general thresholding approaches. In this approach, an optimal threshold for each cluster is estimated using the cluster size and the overall night-light magnitude in each cluster. A global urban map was then generated using this cluster based method. However, such single optimal threshold techniques also have drawbacks; they fail to capture different degrees of anthropogenic activity at a local scales (e.g. within a city). In order to overcome this constraint, Ma et al. (2015) proposed an approach to better reflect the spatiotemporal dynamics of urban process from NTL brightness at a local scale. This approach partitions the NTL imagery into different urban categories which are then linked with anthropogenic activity and urbanization level at local scales.

Most studies related to the dynamics of urbanization are for the regions of US, Europe, China, and India with very few studies for SEA (e.g. Herold, Goldstein, and Clarke 2003; Batty 2007; Luo and Wei 2009; Chand et al. 2009; Seto et al. 2011). Schneider et al. (2015) characterized urban transitions in more than 1000 cities in East Asia and SEA by analysing MODIS data for years 2000 and 2010. Small and Elvidge (2013) studied the decadal changes in anthropogenic night light in Asia, focusing on India and China and indicated that growth trajectory in SEA was different to that of India and China. A comprehensive spatiotemporal characterization of urban growth across Southeast Asian cities is still lacking.

The main objective of this work is to analyse the dynamics of urban growth in 15 major SEA cities using NTL data from 1992 to 2012. We evaluated three recently developed inter-sensor calibration approaches (Wu et al. 2013; Elvidge et al. 2014; Zhang, Pandey, and Seto 2016) for the SEA region as the raw NTL data varies significantly with sensor and over time. Specifically for each city, we categorized NTL data for each year as countryside (CS), peri-urban (PU) and core-urban (CU) categories, and quantified the spatiotemporal trends for each category. The thresholds employed were obtained using a recently proposed brightness gradient (BG) approach (Ma et al. 2015). Section 2 provides a description of the data, the cities chosen, the BG approach, and the three calibration techniques. The results are presented in Section 3 followed by concluding remarks in Section 4.

2. Data and methods

2.1. Data

This study uses stable lights NTL annual composite dataset version 4 for the years 1992 to 2012 as downloaded from the National Centers for Environmental Information (NCEI) website (http://ngdc.noaa.gov/eog/dmsp/downloadV4composites.html). This stable lights dataset captures persistent lights from global cities and sub-urban areas under cloud free conditions. These data products have 30 arc-sec spatial resolution (~1 km near the equator), global coverage ranging between −180° to 180° longitude and −65°

to 75° latitude, with 6-bit radiometric resolution (i.e. 64 levels). These datasets for the years 1992–2012 have been acquired by six different satellite missions namely DMSP F10, F12, F14, F15, F16, and F18.

The study also used data from Landsat – Thematic Mapper (Landsat 5 TM), containing seven spectral bands (six visible and one thermal). The spatial resolution is 30 m for visible bands and 120 m for the thermal band. This dataset is downloaded from the U.S. Geological Survey (USGS) earth explorer website (http://earthexplorer.usgs.gov/).

2.2. *NTL data processing*

It is well known that the non-availability of on-board sensor calibration limits the usage of the raw NTL data for urban growth studies (Elvidge et al. 2009; Zhang and Seto 2011; Elvidge et al. 2014). Sensor degradation, difference in sensor orbits and average annual composites can give different magnitude of lit pixels on the ground even when there is no change detected (Zhang and Seto 2011). Furthermore inter-sensor differences needs to be minimised through inter-sensor calibration (Wu et al. 2013; Elvidge et al. 2009; Elvidge et al. 2014; Zhang and Seto 2013). Bridging the inter-sensor gap in NTL data has received more attention in the past few years with the increasing use of NTL for temporal analysis (e.g. characterizing trends). Three calibration methods have been proposed since 2009 to ensure continuity in NTL data across sensors (Elvidge et al. 2009; Wu et al. 2013; Zhang, Pandey, and Seto 2016). Although, assessment of these approaches has been conducted at global scales, their merits and disadvantages at a regional scale is not fully understood.

The technique proposed by Elvidge et al. (2009) involves the selection of a reference area that has undergone least changes in NTL patterns over time, and fitting a second-order polynomial as follows:

$$D_{cal} = C_0 \times D_{raw}^2 + C_1 \times D_{raw} + C_2,   \tag{1}$$

where D_{cal} denotes calibrated NTL digital number (DN) data; D_{raw} denotes raw NTL digital number (DN) data; and C_0, C_1, and C_2 are coefficients of fitted quadratic equation.

This calibration approach has been shown to reduce the discrepancies between raw DN values captured by two different sensors in a single year. We used coefficient values from Elvidge et al. (2014) for years 1992–2012 here.

Zhang, Pandey, and Seto (2016) proposed a calibration approach which also involves fitting a second-order polynomial, but with coefficients estimated using a technique called ridge line sampling (RSL). The corresponding coefficients for this study were taken from Zhang, Pandey, and Seto (2016).

Wu et al. (2013) proposed an alternative strategy for inter-sensor calibration, which performs radiometric calibration as well as reducing errors caused by pixel saturation and inter-annual variation. This approach uses the following power function to calibrate raw DN values:

$$D_{cal} = a_1 (D_{raw} + 1)^{b_1} - 1   \tag{2}$$

where D_{cal} denotes calibrated NTL digital number (DN) data; D_{raw} denotes raw NTL digital number (DN) data; and a_1 and b_1 are the coefficients of the fitted power law.

In this study, the coefficients a_1 and b_1 for each year from 1992 to 2010 were taken from Wu et al. (2013).

We first extracted the NTL data for the entire SEA region (Figure 1) and applied the three calibration approaches discussed above. It is noted that in Wu et al.'s (2013) approach, calibration coefficients were available over 1992–2010, while coefficients were available over 1992–2012 for other the two approaches. We analysed their sensitivity using the sum of normalized difference index (SNDI) developed by Zhang, Pandey, and Seto (2016) and selected the calibration approach that performs best for the SEA region. The calibrated NTL data is then extracted for 15 major SEA cities based on a bounding box for each city as shown in Figure 1. These 15 cities have been selected based on the criteria that population size exceeds one million according to 2015 year values (UNPD 2014). The bounding box coordinates and the population for each city are listed in Table 1.

For years having annual composites collected by two different satellite sensors, we have selected the latest sensor data in order to reduce the effect of sensor degradation. For compactness, the results from the three calibration approaches will be referred to as WU2013, EL2014, and ZH2016.

2.3. Extraction of urban categories from calibrated NTL data

Capturing the pixel level fluctuations is important in detecting spatial changes in the NTL data. We employed the BG approach, proposed recently by Ma et al. (2015). The BG refers to the spatial gradient in NTL over a pixel w.r.t. its neighbours. By considering spatial gradient instead of absolute NTL value, one can delineate urban areas according

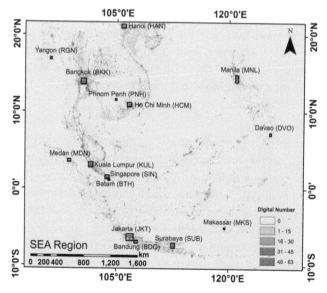

Figure 1. 2012 year night-time light (NTL) image for 15 selected SEA cities with population exceeding 1 million.

Table 1. Coordinates of the bounding box and 2015 population for each city.

City	NW corner	SE corner	Population ($\times 10^3$ people)
Bandung (BDO)	−6.76° 107.41°	−7.11° 107.88°	2544
Bangkok (BKK)	14.26° 100.19°	13.53° 100.89°	9270
Batam (BTH)	1.18° 103.88°	1.01° 104.14°	1391
Davao (DVO)	7.34° 125.46°	7.00° 125.70°	1630
Ho Chi Minh (HCM)	11.16° 106.35°	10.61° 106.96°	7298
Hanoi (HAN)	21.36° 105.53°	20.85° 106.12°	3629
Jakarta (JKT)	−5.92° 106.34°	−6.78° 107.28°	10,323
Kuala Lumpur (KUL)	3.42° 101.25°	2.74° 101.87°	6837
Medan (MDN)	3.81° 98.44°	3.45° 98.88°	2204
Makassar (MKS)	−5.00° 119.38°	−5.20° 119.56°	1489
Manila (MNL)	14.86° 120.84°	13.91° 121.24°	12,946
Phnom Penh (PNH)	11.66° 104.75°	11.45° 104.97°	1731
Singapore (SIN)	1.71° 103.49°	1.23° 104.04°	5619
Surabaya (SUB)	−7.06° 112.36°	−7.72 112.84	2853
Yangon (RGN)	17.02° 96.01°	16.76° 96.28°	4802

to different degrees of urbanization. For example, a flatter spatial gradient implies minimal NTL changes in space, which is expected either within core-urban regions or in relatively countryside/rural regions. Conversely, steeper spatial gradient implies a transition zone between core urban and countryside, i.e. peri-urban areas.

There are multiple ways in which the spatial gradient can be computed. Here we follow Ma et al.'s (2015) approach and used the maximum gradient technique of Burrough, McDonnell, and Lloyd (2015). The relationship between pixel-level DN and corresponding BG is modelled using the quadratic polynomial:

$$B = a_0 \times D_{cal}^2 + a_1 \times D_{cal} + a_2, \qquad (3)$$

where B denotes brightness gradient value; D_{cal} denotes calibrated NTL digital number (DN) data; and a_0, a_1, and a_2 are the coefficients of the fitted polynomial.

The fitted polynomial is a downward parabola, which is used later to derive urban categories. While Ma et al. (2015) used five urban categories (low, medium-low, medium, medium-high, and high), we have chosen three categories as some cities in our study are relatively smaller. The DN thresholds for extracting above three categories were calculated using the fitted polynomial (Equation (3)). First, the brightness gradient value B_1 was calculated as $(B_0 + 2B_{max})/3$, where B_0 and B_{max} are the lowest and highest BG of the rising limb of Equation (3), respectively. Similarly, B_2 was calculated as $(B_{max} + 2B_3)/3$, where B_3 is the lowest BG of the falling limb of Equation (3). The DN values corresponding to B_1 and B_2 were chosen as DN thresholds D_1 and D_2 to extract three urban categories. Specifically, all pixels with DN values ranging from 3 to D_1 were categorized as countryside (CS), pixels with $D_1 < DN < D_2$ as peri-urban (PU), and pixels with $DN > D_2$ as core-urban (CU). The CU category pixels should represent the highly developed urban core with intense anthropogenic activity along with a high percentage of impervious area or built-up land. The PU category is attributed to the periphery of the core urban area with reasonable urban activity. The CS category includes areas with lower human activity such as agricultural sites and small villages. This approach involving DN thresholds which are city-specific avoids setting an arbitrary threshold for heavily urbanized areas that has large DN values. Lastly a minimum threshold of 3 is used because of high interannual variability of very low DN values. In this work, we applied the BG approach on the calibrated NTL data for all 15 cities.

2.4. Comparison of NTL urban extent with Landsat TM-derived impervious area

The three urban categories extracted from BG approach are compared with the impervious area derived from Landsat TM data. The supervised maximum likelihood classification technique is applied on the Landsat TM datasets using ArcMap 10 software to extract the impervious areas or artificial surfaces. Training samples for the supervised classification were selected based on visual inspection (e.g. comparing with Google Earth imagery).

3. Results and discussion

3.1. Evaluation of inter-sensor calibration results

The total light index (TLI) is the sum of all grey values in an NTL image, and is widely used to display and analyse interannual variability and volatility in NTL time series (e.g. Wu et al. 2013). The TLI results in a single value for a specific year for both uncalibrated and calibrated NTL datasets. The TLI values are then used to compute the SNDI (Zhang, Pandey, and Seto (2016)) to evaluate the performance of the calibration approaches. The SNDI measures the level of convergence between the TLI values of a given region over the overlapping years. Figure 2 shows the interannual variability of TLI for the uncalibrated NTL data for the entire SEA region. There is a clear gap in TLI of two sensors during overlapping years in the uncalibrated NTL data. The interannual volatility within each sensor is also evident in Figure 2 (e.g. TLI from F15 takes a sudden dip after 2002), thus illustrating the necessity of inter-sensor calibration before any quantitative temporal analysis of NTL data.

Figure 3 shows the interannual variability of TLI for the NTL data post application of three calibration approaches described in Section 2.2. The gap in TLI for overlapping years has reduced considerably after calibration in all three approaches. Even the interannual volatility for the same sensor has reduced significantly. The overall TLI for each sensor derived from WU2013 are significantly larger compared to the other two approaches. This pattern is partly because of the power law used (Equation (2)) in calibration which unlike the other two approaches involving second order polynomials, can over-correct high raw DN values. This is consistent with Wu et al. (2013) reporting that the higher TLI values might be due to the use of a non-saturated radiance-calibrated image for calibration. The results derived using EL2014 and ZH2016 follow a similar pattern for most of the sensors except F18 sensor, for which TLI for the years 2010 and 2012 have contradictory patterns; this also reported by

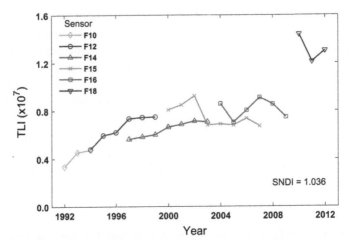

Figure 2. Total light index (TLI) value for Southeast Asia (as shown in Figure 1) for uncalibrated NTL images.

Zhang, Pandey, and Seto (2016) as a sensor bias for these two years. This artefact is also reflected later during SEA urban growth analysis. ZH2016 approach gives the lowest SNDI value (0.225) compared to WU2013 (0.258) and EL2014 (0.379). Hence, we selected ZH2016 calibration approach for the rest of this paper. Further, we have selected data from the latest sensor when more than one sensor is available for a particular year, in order to reduce the discrepancy caused by sensor degradation.

3.2. Spatiotemporal analysis of urban categories

We applied the BG approach to delineate the CS, PU, and CU categories from the calibrated NTL annual composites for all 15 cities. For illustration, the BG-DN scatterplot and corresponding fitted relationships are shown for Bangkok and Singapore for the year 2012 in Figure 4. The fitted polynomials are then used to obtain the DN thresholds, which are then employed to extract three urban categories. Figure 5 shows the inter-annual variability of CS, PU and CU categories for 15 SEA cities. Here the CS, PU and CU values for each city are normalised with the total areas (i.e. sum of CS, PU and CU). The CS (CU) category for all cities showed a decreasing (increasing) trend with time. In addition, the decreasing (increasing) trends in CS (CU) urban categories are stronger compared to the PU category, which is an indication of increasing urbanization, and can lead to a flattening in the PU category (e.g. Bangkok, Kuala Lumpur and Singapore). Generally all these 15 cities have undergone increase in urbanization (CU and PU) over time. Bangkok, Singapore and Kuala Lumpur further have a stronger urbanization trend compared to other cities since the CU category crosses the PU and CS, i.e. forming the largest category. This rapid urbanization trend can associated with the nature of these cities having developed larger metropolitans associated with rapid development when compared to other SEA cities.

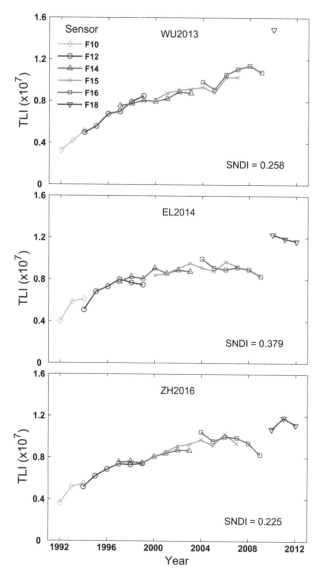

Figure 3. Comparison of inter-calibration results for Southeast Asia (as shown in Figure 1) based on three different calibration approaches.

3.3. Comparison of NTL urban extent with Landsat TM-derived impervious area

We compared the urban categories extracted from the NTL data with the urban land use map generated from Landsat TM data for years 1994 and 2008. Landsat TM data with minimal cloud cover (<5% within the bounding box) were extracted for the 11 cities and the two selected years, and supervised classification applied to extract the urban extent. Landsat TM data for Surabaya and Davao was not available for the year 2008 and for Manila and Medan, had more than 20% cloud cover for 2008. Hence these four cities were not included in this analysis.

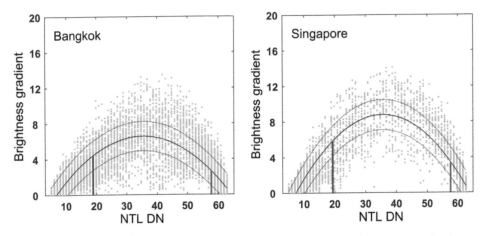

Figure 4. Quadratic polynomial fitted to NTL-DN vs corresponding BG for two cities and for the year 2012. The blue and red vertical lines indicate the DN thresholds used for extracting countryside, peri-urban and core-urban categories.

Figure 6 shows spatiotemporal comparison of Landsat TM derived urban extent with NTL-derived urban categories for Bangkok and Kuala Lumpur for the years 1994 and 2008. The NTL derived CU category is in good agreement with the Landsat TM derived urban land use map with the NTL high urban region consistently enveloping the Landsat TM urban extent. The difference in area between NTL and Landsat TM is expected given the different spatial resolutions (1 km for NTL vs 30 m for Landsat TM) and that the Landsat map shows the impervious surfaces (i.e. built-up areas) as opposed to lit pixels in NTL maps. Nevertheless, both maps clearly show the spatial evolution of the urban extent (NTL CU and PU categories) from 1994 to 2008. The decrease in CS and the increase in CU categories in the NTL maps are particularly evident as is the impervious surface from the Landsat map.

Figure 7 shows similar results as Figure 6 for two smaller cities Singapore and Ho Chi Minh while Figures S-1 and S-2 (supplementary material) show results for Bandung, Batam, Hanoi, Jakarta, Makassar, Phnom Penh and Yangon. As expected, urban extent maps from NTL images are unable to accurately capture inland waterbodies such as lakes and rivers, and land–sea boundaries, especially when the feature sizes are comparable to the spatial resolution. To further examine the effect of land-sea boundaries, Figure 8 provides a closer view between NTL derived urban categories and Landsat TM derived land-use land-cover (LULC) classes for Singapore for the year 1994. Note that the bounding box for Singapore also includes a part of southern Johor, Malaysia. This is because Singapore is small (approx. 700 sq.km), and the southern Johor region is influenced by Singapore's urban growth, with the urban areas of Singapore and southern Johor merging. As such the waterbody, primarily the narrow Johor Straits that separates Singapore from southern Johor seen in the Landsat TM LULC map is not captured in the NTL urban categorization. While spatial resolution difference is one reason (the Johore Straits is 1.3 km at its narrowest point), another reason is the saturation in NTL imagery in the highly developed areas on both sides of the Johor Straits. This indicates that NTL derived urban extents for island cities could be affected with pixel saturation and unresolved waterbodies. A similar phenomena is also observed in Batam which is also an island city where saturation effects are seen along its coastline.

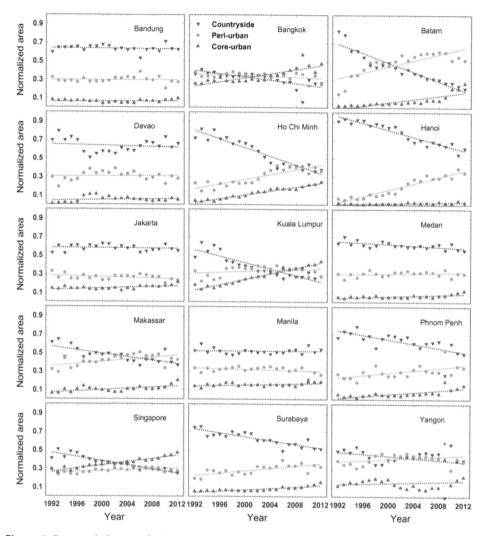

Figure 5. Temporal change of urban categories (CS, PU, and CU) for 15 SEA cities.

Figures 6 and 7 and supplementary Figures S-1 and S-2 readily show that NTL derived core-urban (CU) extent is in overall good agreement with the Landsat derived impervious area. Figure 9 further compares the percentage of Landsat derived impervious area (expressed as a percentage of the NTL CU area) for each city with the NTL CU extent (expressed as a percentage of the total extent CS + PU + CU) for years 1994 and 2008. There is a very good correlation between the Landsat derived impervious area with NTL derived CU area. Cities such as Singapore and Bangkok which have large percentages of CU area also have large percentages in their built-up impervious surfaces. Kuala Lumpur and Ho Chi Minh have also significantly expanded (both in NTL and Landsat) between 1994 and 2008 when compared to other cities, an indication of their rapid development. Phnom Penh, Batam, Bandung, Makassar, Hanoi, and Jakarta (except Yangon) have also expanded though to a lesser degree.

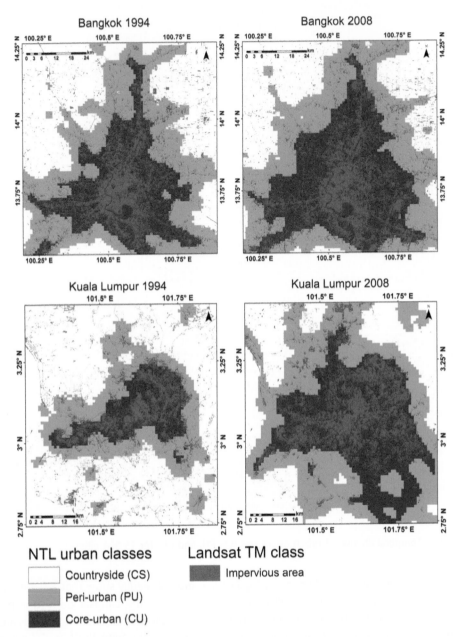

Figure 6. Comparison of Landsat TM derived impervious area with NTL derived urban categories for Bangkok and Kuala Lumpur.

3.4. *Analysis of NTL-derived urban transitions*

Transition maps for each of the 15 SEA cities were constructed to assess the spatiotemporal transitions of NTL urban categories as shown in Figure 10 and in Figures S-4 and S-5. Pixels which changed from CS in 1994 to PU in 2008 (i.e. increasing urbanization) are denoted as 'CS-PU', and similarly for the other five transition classes seen (CS-CU, PU-PU, PU-CU, CU-CU, and CU-PU). There is a low number of pixels (<1% in all cities except for 2

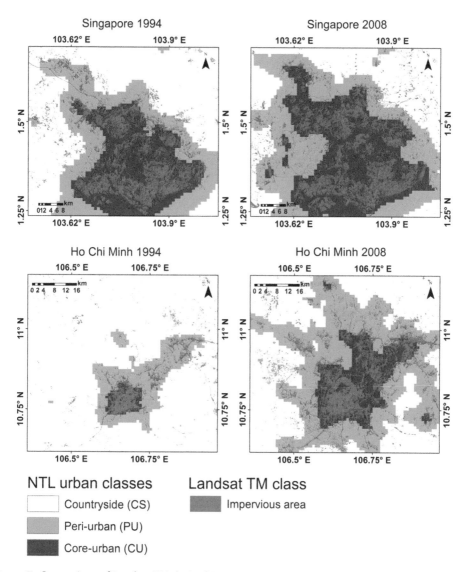

Figure 7. Comparison of Landsat TM derived impervious area with NTL derived urban categories for Singapore and Ho Chi Minh.

(Jakarta and Yangon) which had <5%) which show the reverse trend of going from a higher urbanization category to a lower (specifically CU-PU) and these are neglected below. The NTL TLI data for Yangon in year 2008 also has an unexpected dip and this would be a reason for such a reverse trend as also seen in Figure 5 for Yangon where the CU shows a broad dip.

As expected, cities which already had a larger percentage CU category (e.g. Bangkok and Singapore shown in Figure 10 and Jakarta shown in supplementary Figure S-4) in 1994 show large amounts of CU-CU pixels. Transition percentages calculated by normalizing the areas under a particular transition type with the total NTL derived urban area (i.e. sum of CU, PU, and CU) in 1994 is tabulated in Table 2 for all 15 cities. It is evident

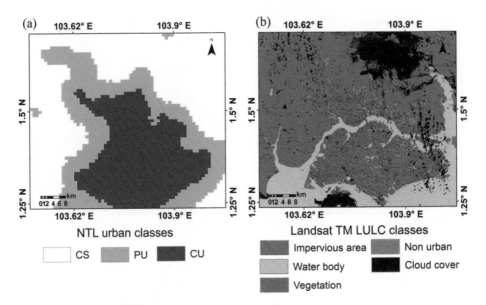

Figure 8. Map of NTL urban categories using calibration of (a) ZH2016 and (b) Landsat TM LULC classes for Singapore for the year 1994.

that all cities have experienced significant CS-PU transitions, with values ranging from 5% for Manila to 50% for Ho Chi Minh, and with Phnom Penh, Kuala Lumpur, Hanoi, Batam and Davao also having >20% transitions. Most cities have negligible CS-CU transition with the exception of Kuala Lumpur having a 9.5% and Ho Chi Minh with 5.9%. The Indonesian cities of Batam, Medan and Makassar have large PU area in 1994 and remained so in 2008. This behaviour can be attributed to them being smaller island/coastal cities having geographical constraints (e.g. sea and/or surrounding high terrain) for expansion. The PU-CU transition was found to be highest for Ho Chi Minh followed by Kuala Lumpur and then Bangkok.

3.5. Comparison with population

We compared the urban extents extracted from NTL with population data to explore linkages with population growth. We obtained population data from UNPD (2014) as available for years of 1995, 2000, 2005, and 2010 and for all 15 cities. Here it is noted that only NTL derived PU and CU categories were considered in this comparison as the CS is expected be sparely populated. Figure 11 shows the comparison between the TLI (computed using PU and CU pixels) and population for all 15 cities in the 4 years of 1995, 2000, 2005, and 2010. It is observed that TLI is proportionately increasing with population except for Yangon, Manila, Kuala Lumpur, and Bangkok. Yangon and Manila have their TLI not increasing proportionately with population while Kuala Lumpur and Bangkok show a stronger increases above linear. Figure 12 further shows the comparison between population and urban extent (PU and CU only) for all 15 cities for the same years. Similar results were observed as in Figure 11. Bangkok and Kuala Lumpur have

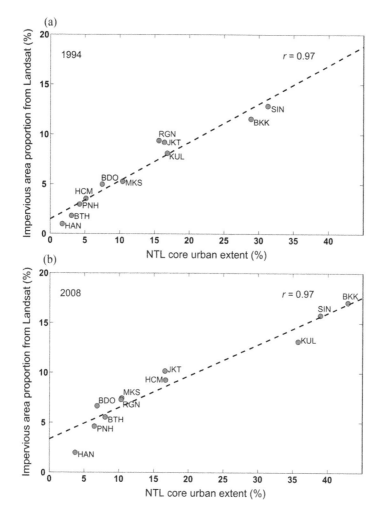

Figure 9. Comparison of Landsat-derived impervious area percentage for (a) 1994 and (b) 2008 with NTL derived core urban extent percentage. Please refer Table 1 for meaning of city abbreviations.

significantly increased in terms of urban extent but the population has not proportionally increased.

We finally show in Figure 13 the rate of change of NTL derived urban extent (PU + CU) versus rate of change of population over 1995–2010, normalized w.r.t. 1995. Specifically, the percentage change in urbanized land was obtained as $100\times ((PU + CU)_{2010} - (PU + CU)_{1995})/(PU + CU)_{1995}$ where $(PU + CU)_{1995}$ and $(PU + CU)_{2010}$ represent the combined peri-urban and core-urban areas for years 1995 and 2010, respectively. The percentage change in urbanized land ranges from 8% to >300%, with lower values for Manila and Yangon and higher values for Hanoi and Ho Chi Minh, and has a correlation r of 0.78 with rate of change of population. Inland cities such as Hanoi, Ho Chi Minh and Phnom Penh have their rapid increase in populations accompanied by a corresponding increase in urbanized land. Conversely, increase in urbanized land is not as strongly linked to increase in population for Kuala Lumpur, Singapore and Davao which are geographically constrained. It is interesting to note that the Vietnamese cities

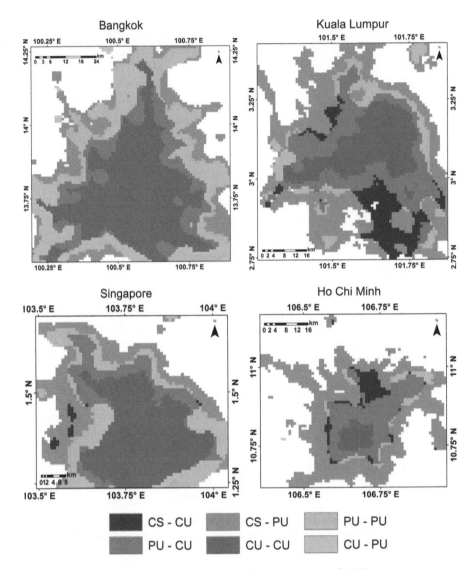

Figure 10. NTL derived urban class transition map between 1994 and 2008.

of Hanoi and Ho Chi Minh have dramatic increases in population and urbanized land over 1995–2010 which could be possibly linked to the extensive developmental activities following the post-Vietnam war era. Batam showed an even larger (416%) increase in population, this arising because it only started growing from 1995 when its population was <200,000, and as such is not shown in Figure 13.

4. Conclusions

SEA has been going through rapid urbanization over the last few decades. In this work, we analysed urban growth patterns for 15 SEA cities using remotely sensed NTL data from 1992 to 2012. We evaluated three recently developed inter-sensor calibration approaches (WU2013, EL2014, and ZH2016) and observed that ZH2016 is best suited

Table 2. Transition percentages obtained using ZH2016 and calculated as the area of a transition region normalized by the NTL derived total area (CS + PU + CU) for 1994 year.

City	CS-PU	CS-CU	PU-PU	PU-CU	CU-CU
			Transition as proportion of total area (%)		
BDO	6.92	0.00	26.57	0.80	7.06
BKK	11.45	0.00	22.37	10.88	28.26
BTH	32.52	0.81	28.62	4.72	2.60
DVO	31.99	0.00	23.02	5.10	2.11
HCM	50.49	5.93	4.51	20.37	5.16
HAN	24.42	0.00	7.32	2.24	1.75
JKT	5.94	0.49	24.97	3.13	15.58
KUL	32.59	9.46	9.46	19.11	16.79
MDN	5.97	0.00	27.14	1.74	6.18
MKS	18.98	0.00	39.04	4.01	9.09
MNL	4.97	0.00	28.36	1.48	13.85
PNH	40.52	1.31	16.01	7.52	4.25
SIN	17.21	0.85	15.82	10.26	30.64
SUB	11.95	0.00	23.34	3.64	6.20
RGN	10.30	0.00	34.37	0.12	11.79

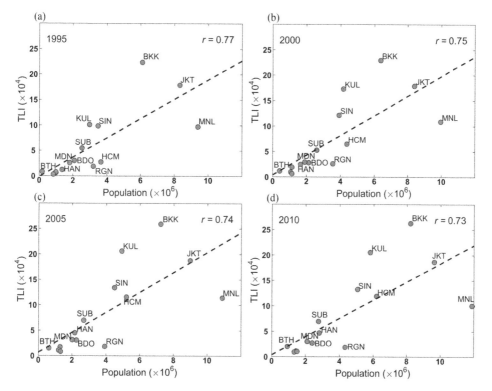

Figure 11. NTL total light index (PU and CU pixels) versus population for (a) 1995, (b) 2000, (c) 2005, and (d) 2010.

for the SEA region. In contrast to most prior studies on urban growth patterns that have a single arbitrary and city-specific NTL threshold, this analysis here extracted three urban categories (countryside (CS), peri-urban (PU), and core-urban (CU)) for all cities following a BG approach.

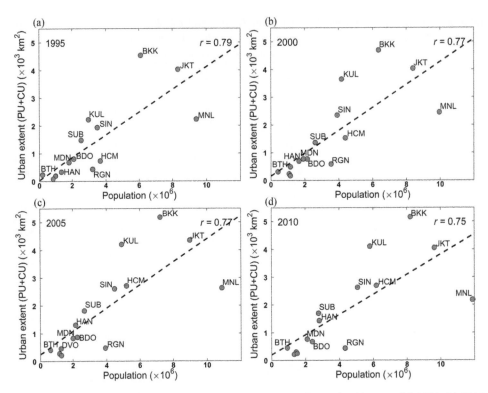

Figure 12. NTL derived urban extent (PU and CU) versus population for (a) 1995, (b) 2000, (c) 2005, and (d) 2010.

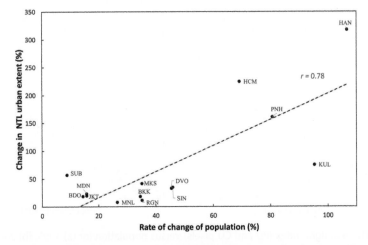

Figure 13. Rate of change in urban extent from 1995 to 2010 (normalized w.r.t. 1995 urban extent) with rate of change in population. (Note: Batam (BTH) has an exceptional 416% increase in population and not included.)

The analysis showed that NTL data together with the BG approach can successfully delineate urban categories for SEA cities. The derived core-urban and peri-urban categories compared favourably with Landsat derived impervious areas with the NTL spatial extents enveloping the impervious cover. The combined PU and CU categories were further found to increase over time for all 15 SEA cities studied. It is evident that all cities have experienced significant CS-PU transitions, with values as high as 50% for Ho Chi Minh. Most cities have negligible CS-CU transition with the exception of Kuala Lumpur and Ho Chi Minh. The Indonesian cities of Batam, Medan and Makassar have large PU area in 1994 and remained so in 2008 which can be attributed to them being smaller island/coastal cities having geographical constraints for expansion. Ho Chi Minh shows the highest PU-CU transition followed by Kuala Lumpur and Bangkok. The complex land-sea structure of SEA region presented accuracy issues in extracting urban categories for smaller island cities such as Singapore which is also affected by narrow waterbodies with size comparable to the NTL 1 km resolution.

The urbanization trends showed strong correlation with the population growth for the SEA cities with inland cities showing an unrestrained spatial growth with population. The SEA cities also expand spatially into the surrounding countryside generally following a linear trend with population increase. Extensions of this study are to analyse the urban categories at higher resolution and assess linkages with other socioeconomic indicators (e.g. GDP, energy consumption) and critical infrastructures (e.g. road network, power grid, and water supply).

Acknowledgements

This study is supported by National Research Foundation Singapore - Future Resilient Systems programme. The first author also acknowledges graduate assistantship support from the Interdisciplinary Graduate School (IGS), Nanyang Technological University. We thank Prof. Tim Warner, the editors of this special issue and the anonymous reviewers for their valuable suggestions and comments which improved the quality of the manuscript.

Disclosure statement

There is no potential conflict of interest by the authors.

Funding

This work was supported by the Future Resilient Systems program of the National Research Foundation, Singapore. The first author also acknowledges support from the Interdisciplinary Graduate School (IGS), Nanyang Technological University.

ORCID

Shankar Acharya Kamarajugedda ⓘ http://orcid.org/0000-0002-9876-3916
Pradeep V Mandapaka ⓘ http://orcid.org/0000-0001-7353-3172

References

Aritenang, A. F. 2014. "Urbanization in Southeast Asia: Issues and Impacts." *Bulletin of Indonesian Economic Studies* 50 (1): 144–144. doi:10.1080/00074918.2014.896312.
Batty, M. 2007. *Cities and Complexity: Understanding Cities with Cellular Automata, Agent-Based Models, and Fractals.* Cambridge, Massachusetts: MIT Press.

Bhatta, B. 2010. *Analysis of Urban Growth and Sprawl from Remote Sensing Data. Advances in Geographic Information Science.* Berlin, Heidelberg: Springer Berlin Heidelberg. doi:10.1007/978-3-642-05299-6.

Burrough, P. A., R. A. McDonnell, and C. D. Lloyd. 2015. *Principles of Geographical Information Systems.* Oxford, UK: Oxford University Press.

Cao, X., J. Chen, H. Imura, and O. Higashi. 2009. "A SVM-Based Method to Extract Urban Areas from DMSP-OLS and SPOT VGT Data." *Remote Sensing of Environment* 113 (10): 2205–2209. doi:10.1016/j.rse.2009.06.001.

Chand, T. R. K., K. V. S. Badarinath, C. D. Elvidge, and B. T. Tuttle. 2009. "Spatial Characterization of Electrical Power Consumption Patterns over India Using Temporal DMSP-OLS Night-Time Satellite Data." *International Journal of Remote Sensing* 30 (3): 647–661. doi:10.1080/01431160802345685.

Chen, M., H. Zhang, W. Liu, and W. Zhang. 2014. ""The Global Pattern of Urbanization and Economic Growth: Evidence from the Last Three Decades." Edited by A. R. H. Montoya." *PLoS ONE* 9 (8): e103799. doi:10.1371/journal.pone.0103799.

Doll, C. N. H., J. Muller, and J. G. Morley. 2006. "Mapping Regional Economic Activity from Night-Time Light Satellite Imagery." *Ecological Economics* 57 (1): 75–92. doi:10.1016/j.ecolecon.2005.03.007.

Elvidge, C. D., F. Hsu, K. E. Baugh, and T. Ghosh. 2014. "National Trends in Satellite-Observed Lighting." *Global Urban Monitoring and Assessment through Earth Observation* 97–118. doi:10.1201/b17012-9

Elvidge, C. D., D. Ziskin, K. E. Baugh, B. T. Tuttle, T. Ghosh, D. W. Pack, E. H. Erwin, and M. Zhizhin. 2009. "A Fifteen Year Record of Global Natural Gas Flaring Derived from Satellite Data." *Energies* 2 (3): 595–622. doi:10.3390/en20300595.

Henderson, M., E. T. Yeh, P. Gong, C. Elvidge, and K. Baugh. 2003. "Validation of Urban Boundaries Derived from Global Night-Time Satellite Imagery." *International Journal of Remote Sensing* 24 (3): 595–609. doi:10.1080/01431160304982.

Herold, M., N. C. Goldstein, and K. C. Clarke. 2003. "The Spatiotemporal Form of Urban Growth: Measurement, Analysis and Modeling." *Remote Sensing of Environment* 86 (3): 286–302. Urban Remote Sensing. doi:10.1016/S0034-4257(03)00075-0.

Imhoff, M. L., W. T. Lawrence, D. C. Stutzer, and C. D. Elvidge. 1997. "A Technique for Using Composite DMSP/OLS 'City Lights' Satellite Data to Map Urban Area." *Remote Sensing of Environment* 61 (3): 361–370. doi:10.1016/S0034-4257(97)00046-1.

Liu, Z., C. He, Q. Zhang, Q. Huang, and Y. Yang. 2012. "Extracting the Dynamics of Urban Expansion in China Using DMSP-OLS Nighttime Light Data from 1992 to 2008." *Landscape and Urban Planning* 106 (1): 62–72. doi:10.1016/j.landurbplan.2012.02.013.

Luo, J., and Y. H. D. Wei. 2009. "Modeling Spatial Variations of Urban Growth Patterns in Chinese Cities: The Case of Nanjing." *Landscape and Urban Planning* 91 (2): 51–64. doi:10.1016/j.landurbplan.2008.11.010.

Ma, T., C. Zhou, T. Pei, S. Haynie, and J. Fan. 2012. "Quantitative Estimation of Urbanization Dynamics Using Time Series of DMSP/OLS Nighttime Light Data: A Comparative Case Study from China's Cities." *Remote Sensing of Environment* 124 (September): 99–107. doi:10.1016/j.rse.2012.04.018.

Ma, T., Y. Zhou, C. Zhou, S. Haynie, T. Pei, and T. Xu. 2015. "Night-Time Light Derived Estimation of Spatio-Temporal Characteristics of Urbanization Dynamics Using DMSP/OLS Satellite Data." *Remote Sensing of Environment* 158 (March): 453–464. doi:10.1016/j.rse.2014.11.022.

Masek, J. G., F. E. Lindsay, and S. N. Goward. 2000. "Dynamics of Urban Growth in the Washington DC Metropolitan Area, 1973-1996, from Landsat Observations." *International Journal of Remote Sensing* 21 (18): 3473–3486. doi:10.1080/014311600750037507.

Schneider, A., M. A. Friedl, and D. Potere. 2009. "A New Map of Global Urban Extent from MODIS Satellite Data." *Environmental Research Letters* 4 (4): 044003. doi:10.1088/1748-9326/4/4/044003.

Schneider, A., M. A. Friedl, and D. Potere. 2010. "Mapping Global Urban Areas Using MODIS 500-M Data: New Methods and Datasets Based on 'Urban Ecoregions.'." *Remote Sensing of Environment* 114 (8): 1733–1746. doi:10.1016/j.rse.2010.03.003.

Schneider, A., C. M. Mertes, A. J. Tatem, B. Tan, D. Sulla-Menashe, S. J. Graves, N. N. Patel, et al. 2015. "A New Urban Landscape in East–Southeast Asia, 2000–2010." *Environmental Research Letters* 10 (3): 034002. doi:10.1088/1748-9326/10/3/034002.

Seto, K. C., M. Fragkias, B. Güneralp, and M. K. Reilly. 2011. "A Meta-Analysis of Global Urban Land Expansion." *PloS One* 6 (8): e23777. doi:10.1371/journal.pone.0023777.

Seto, K. C., R. Sánchez-Rodríguez, and M. Fragkias. 2010. "The New Geography of Contemporary Urbanization and the Environment." *Annual Review of Environment and Resources* 35 (1): 167–194. doi:10.1146/annurev-environ-100809-125336.

Seto, K. C., C. E. Woodcock, C. Song, X. Huang, J. Lu, and R. K. Kaufmann. 2002. "Monitoring Land-Use Change in the Pearl River Delta Using Landsat TM." *International Journal of Remote Sensing* 23 (10): 1985–2004. doi:10.1080/01431160110075532.

Small, C., and C. D. Elvidge. 2013. "Night on Earth: Mapping Decadal Changes of Anthropogenic Night Light in Asia." *International Journal of Applied Earth Observation and Geoinformation* 22 (June): 40–52. doi:10.1016/j.jag.2012.02.009.

Small, C., C. D. Elvidge, D. Balk, and M. Montgomery. 2011. "Spatial Scaling of Stable Night Lights." *Remote Sensing of Environment* 115 (2): 269–280. doi:10.1016/j.rse.2010.08.021.

Small, C., F. Pozzi, and C. D. Elvidge. 2005. "Spatial Analysis of Global Urban Extent from DMSP-OLS Night Lights." *Remote Sensing of Environment* 96 (3–4): 277–291. doi:10.1016/j.rse.2005.02.002.

Sutton, P., D. Roberts, C. Elvidge, and K. Baugh. 2001. "Census from Heaven: An Estimate of the Global Human Population Using Night-Time Satellite Imagery." *International Journal of Remote Sensing* 22 (16): 3061–3076. doi:10.1080/01431160010007015.

UNESCAP. 2014. *Statistical Yearbook for Asia and the Pacific 2014*. Bangkok: UNESCAP.

UNPD. 2014. *United Nations,Department of Economic and Social Affairs, Population Division, World Urbanization Prospects: The 2014 Revision*. New York: United Nations.

Wu, J., S. He, J. Peng, W. Li, and X. Zhong. 2013. "Intercalibration of DMSP-OLS Night-Time Light Data by the Invariant Region Method." *International Journal of Remote Sensing* 34 (20): 7356–7368. doi:10.1080/01431161.2013.820365.

Zhang, Q., B. Pandey, and K. C. Seto. 2016. "A Robust Method to Generate a Consistent Time Series from DMSP/OLS Nighttime Light Data." *IEEE Transactions on Geoscience and Remote Sensing* 54 (10): 5821–5831. doi:10.1109/TGRS.2016.2572724.

Zhang, Q., and K. C. Seto. 2011. "Mapping Urbanization Dynamics at Regional and Global Scales Using Multi-Temporal DMSP/OLS Nighttime Light Data." *Remote Sensing of Environment* 115 (9): 2320–2329. doi:10.1016/j.rse.2011.04.032.

Zhang, Q., and K. C. Seto. 2013. "Can Night-Time Light Data Identify Typologies of Urbanization? A Global Assessment of Successes and Failures." *Remote Sensing* 5 (7): 3476–3494. doi:10.3390/rs5073476.

Zhou, Y., S. J. Smith, C. D. Elvidge, K. Zhao, A. Thomson, and M. Imhoff. 2014. "A Cluster-Based Method to Map Urban Area from DMSP/OLS Nightlights." *Remote Sensing of Environment* 147 (May): 173–185. doi:10.1016/j.rse.2014.03.004.

A novel method for urban area extraction from VIIRS DNB and MODIS NDVI data: a case study of Chinese cities

Qiao Zhang, Ping Wang, Hui Chen, Qinglun Huang, Hongbing Jiang, Zijian Zhang, Yanmei Zhang, Xiang Luo and Shujuan Sun

ABSTRACT

Mapping urban areas at the regional and global scales has been used in ecology, environment, sociology, and other subjects. Recently, it has become increasingly popular to extract urban areas from night-time light remote-sensing data. In this article, we reported an alternative method to extract information of urban areas from VIIRS Day/Night Band (DNB) and MODIS normalized differential vegetation index (NDVI) data based on the adaptive mutation particle swarm optimization (AMPSO) algorithm and the Support Vector Machine (SVM) classification algorithm. This method was validated using the urban areas of nine Chinese cities classified from Landsat Enhanced Thematic Mapper (ETM+) images by object-oriented image classification technology. We demonstrated that this new method for urban area extraction had a good classification coherency with the Landsat8 OLI result. In addition, it is more robust than other classification methods, and can be used to characterize the inter-urban texture as well.

1. Introduction

Although urban areas cover only a small fraction of the surface of the Earth, they do have great influence on the environment both regionally and globally, including water quality, air quality, and heat islands (Cao et al. 2009; Foley et al. 2005; Grimm et al. 2008; Lu, Coops, and Hermosilla 2016). Urbanization in China has undergone a rapid expansion process under the Reform and Opening-up policy since 1978. The built-up area in China already exceeded 49,800 km^2 in 2014, compared with 7438 km^2 in 1981 (Fang 2009). By 2015, the urbanization rate in China reached 56.1%, and research on the layout and expansion features of urban area has attracted extensive attention.

Currently, urban areas are mainly obtained by land-usage survey using high-resolution or medium-resolution remote-sensing data. The urban areas in China, which have been mapped using high-resolution aerial and satellite imagery, are approximate to the actual situation presented in the First National Geographic Census starting 2012–2015

(The First National Geographic Census Leading Group Office of the State Council 2013). Pan et al. (2009) detected and analysed the urban area of Chengdu with Landsat-5 TM and CEERS2 data from 1992 to 2006. Hu et al. (2015) extracted urban boundaries in Wuhan from Landsat imagery. Wang et al. (2012) determined the urban areas of China with TM/ETM imagery from 1990 to 2010. Although these high- and medium-resolution remote-sensing data have been widely applied in urbanization mapping, these approaches require a labour-intensive and time-consuming process.

At the regional or even global scales, night-time light data has been regarded as an effective and economic data source for mapping urban areas or human settlements (Cao et al. 2009; Elvidge et al. 2007). A well-known night-time light sensor is the Defense Meteorological Satellite Program's Operational Linescan System (DMSP/OLS), which was first launched in 1965. DMSP/OLS 'City Lights' satellite data have been widely used to map urban areas (Imhoff et al. 1997; Small, Pozzi, and Elvidge 2005; Zhang and Seto 2011; Liu et al. 2012b; Zhou et al. 2014; Yi, Zeng, and Bingfang 2016). More recently, a new night-time light sensor, Visible Infrared Imaging Radiometer Suite (VIIRS), which was equipped on the Suomi NPP satellite and launched on 28 October 2011, was used to extract built-up urban areas, poverty estimation, and big events (Shi et al. 2014; Li et al. 2014; Li and Xi 2015). Compared with the DMSP/OLS data, the Day/Night Band (DNB) data collected by VIIRS has higher spatial resolution, wider radiometric range, on-board calibration, and improved cloud and fire discrimination (Elvidge et al. 2013; Small, Elvidge, and Baugh 2013). It was also found that urban areas extracted from NPP-VIIRS data had higher spatial accuracies than those from DMSP-OLS data (Shi et al. 2014).

Several methods have been utilized to map urban areas using night-time light data. The threshold method has been the most commonly used due to its simplicity (Imhoff et al. 1997; Henderson et al. 2003). However, a single threshold cannot fit various cities at different levels of development. With the recognition of this problem, the multiple threshold method was adopted by many experts (Zhou et al. 2014; Henderson et al. 2003; Fan et al. 2012; Milesi et al. 2003a). Milesi et al. (2003b) selected the optimal thresholds for each state in the southeastern USA by comparing with land-cover data sets and Census tabular data. Liu et al. (2012b) designed a thresholding technique along with 1 km × 1 km land-use/-cover data to extract urban information in China. However, the selection of optimal thresholds is a complex and time-consuming process. Meanwhile, some advanced and automatic approaches have been developed to map urban built-up areas by integrating MODIS data with night-time light data. Jing et al. (Jing et al. 2015) used four machine learning methods to extract urban areas of the eastern part of China by using DMSP-OLS and MODIS normalized differential vegetation index (NDVI) data. Cao et al. (2009) proposed a Support Vector Machine (SVM)-based region-growing algorithm to extract urban areas of Chinese cities from DMSP/OLS and SPOT NDVI data. These classification algorithms could automatically or semi-automatically extract the urban area from night-time light and NDVI data and had comparable performance with the local-optimized threshold method. However, the accuracy of these results was greatly dependent on the training sets and was not robust.

Motivated by the above observation, in this article, we developed an alternative extraction method for urban area by utilizing VIIRS DNB and MODIS NDVI data, based on adaptive mutation particle swarm optimization and SVM theories (hereafter referred to as the 'AMPSO-SVM' method). This new method took advantage of the classification algorithm but also addressed the shortcoming that the accuracy was not robust. To validate the performance of the new method, we applied the new method to map the urban areas of nine Chinese cities with different levels of development. Results were also obtained using the local-optimized threshold method and the SVM-based region-growing method to validate the performance of the new method. The rest of this article is organized as follows. In Sec.2, we provide the data sources and the cities used in this study. In Sec. 3, we present our method in detail. In Sec. 4, we provide our results by comparing our method with the conventional methods. We discuss our result and conclude in Sec. 5 and Sec. 6, respectively.

2. Data and study area

2.1. Data

Five data sources were used in this study. The main specifications of the different data sources are listed in Table 1. The VIIRS DNB data are collected with a constant 742 m x 742 m pixel footprint from nadir out to the edge of scan, and VIIRS has substantially lower detection limits (2E-10 W/cm^2/sr) whereas the OLS detection limit is near 5E-10 W/cm^2/sr (Miller et al. 2012). NOAA National Geophysical Data Center (NDGC) recently reprocessed the VIIRS DNB time series products, which are referred as the Version 1 night-time VIIRS DNB composites (http://ngdc.noaa.gov/eog/viirs/download_monthly.html). The VIIRS DNB data of China's land in October 2015 are shown in Figure 1. The normalized difference vegetation index (NDVI) data were also used to extract urban areas with VIIRS DNB data. The 16 day NDVI products covering the nine cities were

Table 1. Specification of the data used in this study.

Data source	Product description	Spatial resolution (m)	Time of acquisition
NPP-VIIRS DNB	Version 1 monthly products by Earth Observations Group (EOG) at NOAA/NGDC	742	October 2015
MODIS NDVI	16-day MODIS NDVI products	250	September–November 2015
Landsat8 OLI	Bands: Green, Red, NIR, SWIR 1	30	December 2014–October 2015
Planet	Bands: Blue, Green, Red	3–5	2015
ZY-3	MUX images, Bands: Blue, Green, Red	5.8	2015

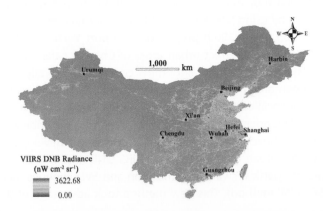

Figure 1. Studied cities and the VIIRS night-time light data of China's land in October 2015.

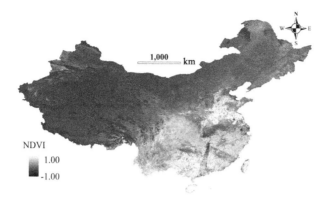

Figure 2. Studied cities and the MODIS NDVI data of China's land in 2015.

Table 2. The GDP and population of the studied cities in 2014.

	Beijing	Chengdu	Guangzhou	Harbin	Hefei	Shanghai	Urumqi	Wuhan	Xi'an
GDP (billion RMB)	2133.08	1005.66	1810.04	533.27	518.06	2356.77	251.00	1006.95	549.26
Population (million persons)	21.516	14.428	13.501	9.873	7.696	24.257	2.669	10.338	8.628

The GDP and population data were obtained from the China Statistical Yearbook 2014.

obtained from the LAADS website (https://ladsweb.nascom.nasa.gov/data/search.html). The MODIS NDVI data of China's land in October 2015 are shown in Figure 2. High-resolution images such as Planet and ZY-3 were also acquired to make cross-validation samples. In addition, Landsat 8 OLI images were acquired to classify urban areas to access the accuracy of urban areas extracted from DNB and NDVI data.

2.2. *Study area*

In total, nine representative Chinese cities were selected for this study: Beijing, Chengdu, Guangzhou, Harbin, Hefei, Shanghai, Urumqi, Wuhan, and Xi'an. These cities are shown in Figure 1 and Figure 2. The nine cities covered the Chinese mainland areas from east to west and from north to south. In addition, the development of these cities was different from each other. As shown in Table 2, the population of these cities varied from approximately 2.7 million to over 24.2 million, and the gross domestic product (GDP) ranged from approximately 251 billion RMB (41 billion USD) to over 2356 billion RMB (385 billion USD).

3. Methods

The urban areas were extracted from DNB and NDVI data by the AMPSO-SVM method as well as two existing and popular methods. These methods are introduced next.

3.1. *Local-optimized threshold method*

The local-optimized threshold method has been the most widely used by several experts (He et al. 2006; Henderson et al. 2003; Milesi et al. 2003b; Liu et al. 2012b). In our study, we set the optimal DNB threshold for each city by comparison with the Landsat OLI-classified urban areas. The OLI urban areas were extracted by the eCognition software, which is well-known for its object-oriented image classification technology. By applying predefined and structured rules to image analysis tasks, eCognition offers a streamlined workflow to quickly extract target information from raw image data (http://www.ecogni tion.com/suite). First, the eCognition software segmented the OLI images by setting certain parameters of homogeneity and heterogeneity in colour and shape. Second, the pure urban and nonurban objects were selected as training samples. Then we used the object feature to train the classifier and classified the urban and non-urban areas based on the segmented image objects. The Landsat results, which were assessed by compar-ison with the results derived from artificial visual interpretation, had accuracies of 90% or higher in the study area. In the local-optimized threshold method, we first obtained the area of urban land in each city from OLI images and then adjusted the threshold to match the area of urban land extracted from the DNB data as closely as that derived from the OLI data. The optimal threshold was obtained when the area of urban land from the DNB data most closely matched the area derived from the OLI data.

3.2. SVM-based region-growing method

An SVM-based region-growing method, which was developed by Cao et al. (2009) to extract urban area from the DMSP/OLS and SPOT VGT data, was applied to the new night-time light data of the VIIRS DNB data in this study. SVM has great potential in generating urban mapping products with a limited number of training samples (Mountrakis, Jungho, and Ogole 2011). The basic idea of SVM is to classify the input vectors into two classes using a hyperplane with the maximal margin, derived by solving the following constrained quadratic programming problem:

$$\max_{a} \left(\sum_{i=1}^{n} ai - \frac{1}{2} \sum_{i,j=1}^{n} aiajyiyjK(xi, xj) \right); i, j = 1, ..., n,$$

$$\text{subject to} : ai \geq 0, \text{and} \sum_{i=1}^{n} aiyi = 0,$$

(1)

where x_i are the training sample vectors, y_i are the corresponding class label, $K(x_i, x)$ is the kernel function, a_i are the Lagrange multipliers, and n is the number of training samples. In this study, the radial basis function (RBF) (Equation (2)) was selected as the kernel function:

$$K(xi, x) = \exp(-y\|xi - x\|^2), i = 1, ..., n,$$

(2)

where y was set to 1.0.

First, two training sets of urban and non-urban samples were selected from VIIRS DNB and MODIS NDVI data. The non-urban samples included water-covered regions and vegetation-covered regions, with the NDVI values greater than 0.4 or smaller than 0.0, and the DNB values smaller than 15. To select urban samples, the complete images were divided into a series of patches with 9 × 9 pixels. The pixels that had the maximum DNB values in the potential urban patches and had a value greater than 15 were selected as urban samples. Then, an iterative classification and training procedure was adopted to identify the urban pixels via region growing, outputting the urban area result. The disadvantage of this method is that the performance applied to VIIRS DNB and MODIS data is greatly influenced by the training sets.

3.3. AMPSO-SVM method

In this study, we designed a new method based on the SVM-based region-growing method, but it performed more accurately and robustly than the previous method. The main principle of this method was to obtain the best training sets by optimizing a couple of DNB thresholds (P_1 and P_2) with an AMPSO-SVM method. P_1 and P_2 were utilized to select non-urban and urban samples, respectively. The non-urban samples included water-covered regions and vegetation-covered regions with NDVI values greater than 0.4 or smaller than 0.0, and DNB values smaller than P_1. Meanwhile, pixels where the DNB had maximum values in the potential urban patches and had values greater than P_2 were selected as the initial urban samples. Each patch is a neighbour-hood containing 9 × 9 pixels or less at the edges of the images.

The AMPSO-SVM method employed an AMPSO algorithm to optimize the parameters (P_1, P_2). Standard particle swarm optimization (PSO) is an evolutionary algorithm dis-covered through the simulation of a simplified social model (Kennedy and Eberhart 1995). In PSO, the potential solutions, called particles, fly through the problem space by following current optimum particles. In this study, the problem space is a three-dimen-sional space as shown in Figure 3, where F denotes the urban area extracting perfor-mance, which is affected by P_1 and P_2. The parameters P_1 and P_2 were used to select the training samples from VIIRS DNB and MODIS NDVI data. Each particle (individual) representing a coupling of P_1 and P_2 was initialized with a group of random values. Each particle owned its position and velocity features:

$$Xi(P1, P2), Vi(P1, P2), i = 1, ..., n, \tag{3}$$

where X_i and V_i are the position and velocity of the particle, respectively, i refers to the i-th particle, and N is the number of particles.

Each particle calculates its own velocity and updates its position at the end of each iteration.

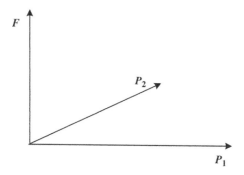

Figure 3. Problem space for optimizing the parameters P_1 and P_2 in this study.

$$X_{i,d}^{t+1} = X_{i,d}^{t} + V_{i,d}^{t+1}, d = 1, 2. \tag{4}$$

$$V_{i,d}^{t+1} = w^{t}V_{i,d}^{t} + c_{1}r_{1}(P_{i,d}^{t} - X_{i,d}^{t}) + c_{2}r_{2}(P_{G}^{t} - X_{i,d}^{t}). \tag{5}$$

In Equation (4), $X_{i,d}^{t}$ and $X_{i,d}^{t+1}$ denote the previous and current positions of the i-th particle in the d-th dimension, respectively, and t represents the t-th iterative. In Equation (5), c_1 and c_2 are the personal and social learning factors, respectively; they were assigned a value of 1.5 in this study. r_1 and r_2 are the random variables within the range [0,1]. $P_{i,d}^{t}$ denotes the best previous position encountered by the i-th particle and P_{G}^{t} denotes the global best previous position so far. w^{t} is the inertia weight that decreases along with the iteration number, defined as

$$w^{t} = w0 \exp(-0.5t^{2}), \tag{6}$$

where w_0 is the initial value of the inertia weight.

The standard PSO has been successfully applied in several research or applicable areas in the past several years. However, PSO always fell into the local optimal solution but not the global optimal solution. It was necessary to provide a mechanism to make the algorithm jump into other areas of the solution space to continue the search until the global optimum was found. The AMPSO method provides this mechanism.

To determine whether the algorithm fell into the local optimal solution, we introduced a stopping criterion to determine the premature convergence. The stopping criterion contained the following two conditions:

$$\begin{aligned} \sigma^2 &= 0, \\ f_G &\geq f_T, \end{aligned} \tag{7}$$

where σ^2 and f_G represent the variance of the fitness of a population and the best fitness of the population, respectively. The first condition could be satisfied by both local optimal solution and global optimal solution, so the second condition is necessary to distinguish the global optimal solution from local optimal solutions. f_T is the theoretical global optimal fitness that was set beforehand. Since we expected to obtain a higher accuracy compared with the other urban area extraction methods, f_T should be set according to the accuracy of the local-optimized threshold method and the SVM-based region-growing method. In this study, f_T was set to be 90%, which was a little higher than the accuracy of the local-optimized threshold method and the SVM-based region-growing method.

If the current PSO algorithm had premature convergence rather than global convergence, the direction of the particles could be changed by an adaptive mutation operation so that particles could continue searching into other areas. The mutation operation was a random operator by which the previous global best position (P_{G}^{t}) was changed. A probability of mutation operation R_m was set by the following formula:

$$R_m = \begin{cases} M, \sigma^2 < \sigma_T^2 \text{ and } f_G < f_T \\ 0, \text{others} \end{cases}, \tag{8}$$

where M is a variable within the range [0.1, 0.3] and σ_T^2 was set as a small value. Theoretically σ^2 should be zero when the particle swarm fell into the local optimal

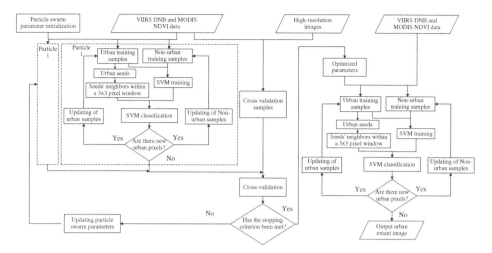

Figure 4. Flow chart of the AMPSO-SVM method.

solution. However, in order to speed up the convergence we supposed when σ^2 was less than a small value (σ_T^2), the particle swarm would fall into a local optimal solution. The value of σ_T^2 should be close to and greater than zero. In this study, σ_T^2 was set 0.05 empirically.

If a random variable r valued in the range [0, 1] was smaller than R_m, the mutation operation occurred and the global best previous position (P_G^t) was updated by the following equation (Liu et al. 2012a):

$$P_G^t = P_G^t(1 + 0.5\eta), \tag{9}$$

where η is a random variable subjected to a Gauss ([0,1]) distribution.

The flow chart of the 'AMPSO-SVM' method is exhibited in Figure 4. The new method includes two major parts: parameter optimization and urban area mapping. First, we obtained the best combination of parameters for each studied city using the AMPSO algorithm. Second, we mapped urban area with the SVM-based region-growing algorithm.

In Figure 4, we collected VIIRS DNB, MODIS NDVI, Landsat8 OLI, Planet, and ZY-3 images covering the study area first. The VIIRS DNB and MODIS NDVI images were co-registered, resampled to the same resolution, and then clipped by the geometrical urban boundary of the nine studied cities. In our new method, we utilized a group of samples for each city to cross-validate the SVM models. A cross-validation sample is shown as follows:

$$\langle S_L \rangle \langle 1 \rangle : \langle S_1 \rangle \langle 2 \rangle : \langle S_2 \rangle, \tag{10}$$

where S_L equals 0 or 1, 0 denotes non-urban pixel, and 1 denotes urban pixel. The value of S_L is visually interpreted from Planet or ZY-3 images. S_1 is the VIIRS DNB value of this sample. S_2 is the MODIS NDVI value of this sample. For obtaining a fine classification accuracy, about 100 cross-validation samples were acquired for each city.

In Figure 4, we initialized 30 particles with random real numbers. The initial positions of particles were produced as positive a value and their associated velocities could be a positive or negative value. These initialized particles were then utilized to select the training samples from VIRRS DNB and MODIS NDVI data.

In the step of SVM-based region-growing in Figure 4, we first completed the SVM model training by the initial training sets. The SVM-based classifier was then applied to classify the unknown pixels according to the DNB and NDVI values. We did not classify all unknown pixels at one time, but addressed them with the 3 × 3 neighbourhood near each urban seed pixel step by step. In each loop, the new urban pixels classified by the SVM classifier were selected as new seeds. Then we trained the new SVM model by updating the non-urban and urban samples again. The previous process was repeated until there was no new urban pixel.

In the step of cross validation in Figure 4, each particle was validated by a set of cross-validation samples using the following fitness function:

$$f = \frac{k_c}{k_c + k_u},$$

(11)

where f is the fitness function, and k_c and k_u are the number of true and false classifications, respectively.

With the global optimal solution achieved by the AMPSO algorithm, the best group of optimized parameters (P_1, P_2) will be obtained. Finally, we used an SVM-based region-growing algorithm to extract the urban areas of the studied cities. Using the optimized values of P_1 and P_2, we selected the non-urban and urban training sets from VIIRS DNB and MODIS NDVI data. Then, we trained an SVM model with the non-urban and urban training sets based on the selected training samples. The unknown pixels were determined by the region-growing algorithm, as shown in Section 3.2. After all unknown pixels are classified, urban areas of the studied cities can be achieved.

4. Results

To compare the AMPSO-SVM algorithm with the common methods, we extracted the urban area of nine cities with the local-optimized threshold method using VIIRS DNB data, the SVM-based region-growing method using VIIRS DNB and MODIS NDVI data, the AMPSO-SVM algorithm using VIIRS DNB and MODIS NDVI data, and the threshold method using only MODSI NDVI data. Urban areas were evaluated by urban classification results from Landsat8 OLI images. It is feasible and accepted to assess the accuracy of urban area extraction from night-time light data based on Landsat data (Liu et al. 2012b; Cao et al. 2009; Small, Pozzi, and Elvidge 2005; Zhou et al. 2014) because the spatial resolution of Landsat data (30 m) is much finer than that of NPP-VIIRS DNB data (742 m).

Figure 5 shown the urban areas of nine cities extracted from multi-sensored data. Figure 5(a,b) show the Landsat 8 OLI images and VIIRS DNB images of the selected cities, respectively. Images in Figure 5(a) were pseudo-colour pictures composed of OLI bands 3, 4, and 5. The OLI results shown in Figure 5(c) were extracted by eCognition software. Urban areas extracted from VIIRS DNB data by the local-optimized threshold method, the SVM-based region-growing method, and the AMPSO-SVM method are presented in Figure 5(d,e,f), respectively. The urban areas shown in Figure 5(g) were classified only using MODIS NDVI data by the threshold method. The pixels where the NDVI values were greater than 0.4 or smaller than 0.0 were classified as non-urban areas, and other

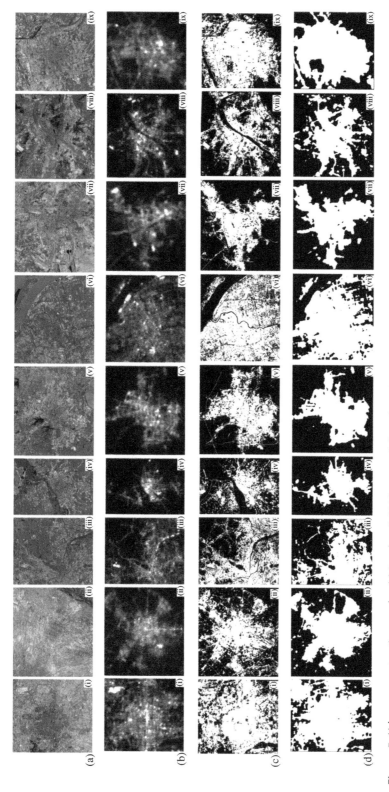

Figure 5. Urban area extraction results: (a) Landsat 8 OLI images of the selected cities, (b) VIIRS DNB images, (c) the urban area from OLI classification, (d) the result of the local-optimized threshold method using VIIRS DNB data, (e) the result of the SVM-based region-growing method using VIIRS DNB and MODIS data, (f) the result of the AMPSO-SVM method using VIIRS DNB and MODIS data, and (g) the results only using MODIS data.

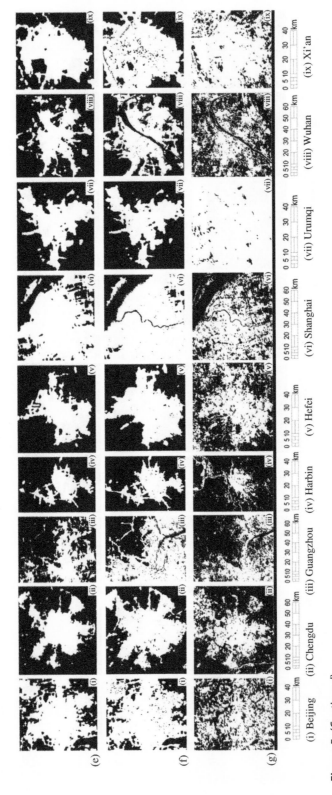

Figure 5. (Continued).

pixels were classified as urban areas. The results extracted from DNB data by the three methods had similar outlines with the urban area derived from OLI classification. Compared with the results from OLI classification, Beijing, Harbin, Hefei, and Urumqi were extracted well by all methods. For Chengdu, Guangzhou, Shanghai, Wuhan, and Xi'an, the AMPSO-SVM method presented the best performance, followed by the local-optimized threshold method. The new method characterized the inter-urban texture well by excluding 'blooming' over vegetated and water pixels with large VIIRS DNB values. Especially for Guangzhou, Wuhan, and Xi'an, the AMPSO-SVM method characterized more inter-urban variability than the other two methods. In general, the AMPSO-SVM method performed better than the other two methods.

The quantitative assessments for the performance of the three methods are listed in Table 3. The accuracy was validated using the overall accuracy (OA) and kappa index based on an error matrix (Congalton 1991). Unlike most researchers who calculate OA and the kappa index based on a set of random samples, we computed the confusion matrix from all pixels in images of each city and then achieved the OA and kappa index. With this approach, the assessment could characterize the 'real' accuracy of the overall images, although the values of OA and the kappa index seem to be lower than the accuracy values reported in other articles.

As is shown in Table 3, the kappa coefficients of the AMPSO-SVM method were the largest in all cities, besides Hefei, compared with the other methods, and the OA coefficients of the AMPSO-SVM method were the largest in all cities compared with those of the other methods. The OA and kappa coefficients of the AMPSO-SVM method were greater than that of the SVM-based region-growing method in all cities. According to the average OA and kappa, the AMPSO-SVM method had the best coherency with OLI results (82.06%, 0.602), followed by the local-optimized threshold method (80.47%, 0.576) and then the SVM-based region-growing method (80.70%, 0.573). The results using only MODIS data had poor performance (76.26%, 0.417). In general, the AMPSO-SVM method had the best performance in all studied cities, and the accuracy of the SVM-based region-growing method was comparable to that of the local-optimized

Table 3. Accuracy assessment of urban area extracted by four methods.

City	Threshold method using MODIS data — Overall accuracy (%)	Kappa	Local-optimized threshold method using VIIRS DNB data — Local-optimized threshold DNB	Overall accuracy (%)	Kappa	SVM-based region-growing method using VIIRS DNB and MODIS NDVI data — Overall accuracy (%)	Kappa	AMPSO-SVM method using VIIRS DNB and MODIS NDVI data — Overall accuracy (%)	Kappa
Beijing	70.81	0.375	12	75.54	0.428	73.97	0.430	**77.41**	**0.448**
Chengdu	81.29	0.611	10	79.10	0.569	82.14	0.623	**84.01**	**0.681**
Guangzhou	75.58	0.510	11	76.43	0.528	73.75	0.470	**77.27**	**0.550**
Harbin	75.51	0.348	5	81.39	0.558	83.00	0.547	**83.12**	**0.573**
Hefei	75.65	0.514	12	**87.09**	**0.716**	86.81	0.700	86.75	0.722
Shanghai	79.84	0.507	17	77.24	0.492	78.73	0.501	**80.31**	**0.510**
Urumqi	34.79	0.017	16	87.28	0.713	87.72	0.719	**87.83**	**0.725**
Wuhan	75.74	0.505	8	81.44	0.604	81.78	0.595	**81.87**	**0.627**
Xi'an	72.13	0.363	10	78.72	0.573	78.44	0.570	**79.99**	**0.579**
Average	71.26	0.417	11.2	80.47	0.576	80.70	0.573	**82.06**	**0.602**

The numbers in bold indicate the maximum values of OA and kappa in each city.

Table 4. The optimized parameters used in the AMPSO-SVM method.

Optimized parameter	Beijing	Chengdu	Guangzhou	Harbin	Hefei	Shanghai	Urumqi	Wuhan	Xi'an	Average
P_1	22.90	24.53	24.59	25.76	20.82	20.80	22.18	20.07	27.51	23.24
P_2	11.75	16.50	15.76	14.54	15.92	16.85	16.00	15.97	16.00	15.48

threshold method. In addition, it can be found that the local-optimized threshold DNB values were very different, ranging from 5 to 17 in all studied cities.

5. Discussion

Unlike daytime MODIS data, night-time light data could provide a unique perspective on human social activities. The urban areas extracted using MODIS NDVI data were much different from the results extracted from VIIRS DNB. For example, many pixels in sub-urban areas were set as urban areas using only MODIS data, whereas they were classified as non-urban areas using VIIRS DNB and MODIS NDVI data or using only VIIRS DNB data.

Compared with the local-optimized threshold method that has a complex and time-consuming process of threshold selection, the SVM-based region-growing method and the AMPSO-SVM method extracted urban areas by automatic classification using night-time light and NDVI data. The AMPSO-SVM method had better performance than the other two methods in the studied cities.

In the SVM-based region-growing method, initial training sets were selected with the same DNB values. However, in the new method, we introduced the AMPSO algorithm to find the global optimum combination of parameters (P_1, P_2) for training sample sets. These optimized parameters ensured the best solution for urban area extraction by the SVM-based region-growing algorithm. The optimized parameters for each of the studied cities are listed in Table 4.

It is notable that the value of P_1 was much greater than the value of P_2 for each of the studied cities. The average values of P_1 and P_2 were 23.24 and 15.48, respectively. This is very different from the SVM-based region-growing method. Simply by this improvement, the new method had a better performance than the SVM-based region-growing method. In our method, the pixels with high DNB but also large NDVI values, such as city parks, vegetation near roads, and golf courses, could be distinguished from urban areas. Thus, the inter-urban texture could be characterized well by excluding blooming over vegetation and water pixels.

6. Conclusions

In this study, we extracted urban areas from new night-time light data (VIIRS DNB). We developed an AMPSO-SVM method to acquire urban areas from VIIRS DNB, MODIS NDVI, and high-resolution images. The new method consisted of two main steps: (1) parameter optimization using the AMPSO algorithm and (2) SVM-based region-growing urban area mapping. Unlike the SVM-based region-growing method with the same parameter, two different parameters are optimized and used to select non-urban samples and initial urban samples in the new method. Nine cities of different development levels were selected to test the performance of the new method, in addition to the local-optimized threshold method and the SVM-based region-growing method. The results showed that the AMPSO-SVM method had the best coherency with the OLI classification results (average OA = 82.06%, average kappa = 0.602). The performance of the new method was better than the local-optimized threshold method (average OA = 80.47%, average kappa = 0.578), the SVM-based region-growing method (average OA = 80.70%, average kappa = 0.573), and the results only using MODIS NDVI data (average OA = 71.26%, average kappa = 0.417). This new method could obtain higher accuracy than the SVM-based region-growing method by selecting the best training sets automatically and intelligently. In addition, the new method characterizes the inter-urban texture well by excluding blooming over vegetation and water pixels. As more night-time light remote-sensing data are becoming available, this new method has great potential applicability for global and regional urbanization and dynamic mapping. Furthermore, accurate urban area results could provide essential data for scientific research, sociology, and economics.

Disclosure statement

No potential conflict of interest was reported by the authors.

Funding

This work was supported by the National basic surveying and mapping science and technology program [2016KJ0300]; National Key Technology Support Program [2012BAH28B04].

References

Cao, X., J. Chen, H. Imura, and O. Higashi. 2009. "A SVM-Based Method to Extract Urban Areas from DMSP-OLS and SPOT VGT Data." *Remote Sensing of Environment* 113 (10): 2205–2209. doi:10.1016/j.rse.2009.06.001.
Congalton, R. G. 1991. "A Review of Assessing the Accuracy of Classifications of Remotely Sensed Data." *Remote Sensing of Environment* 37: 35–46. doi:10.1016/0034-4257(91)90048-B.
Elvidge, C. D., P. Cinzano, D. R. Pettit, J. Arvesen, P. Sutton, C. Small, R. Nemani et al. 2007. "The Nightsat Mission Concept." *International Journal of Remote Sensing* 28 (12): 2645–2670. doi:10.1080/01431160600981525.
Elvidge, C. D., K. Baugh, M. Zhizhin, and F. C. Hsu. 2013. "Why VIIRS Data are Superior to DMSP for Mapping Nighttime Lights." Paper presented at the the the Asia-Pacific Advanced Network. doi:10.7125/APAN.35.7.
Fan, J., M. Ting, C. Zhou, and Y. Zhou. 2012. "A New Approach to the Application of DMSP/OLS Nighttime Light Data to Urbanization Assessment." *Igarss* 6983–6986.
Fang, C. 2009. "The Urbanization and Urban Development in China after the Reform and Opening-Up." *Economic Geography* 29 (1): 19–25. [in chinese].
Foley, J. A., G. P. Ruth Defries, C. B. Asner, G. Bonan, and S. R. Carpenter. 2005. "Global Consequences of Land Use." *Science* 309 (5734): 570–574. doi:10.1126/science.1111772.
Grimm, N. B., S. H. Faeth, N. E. Golubiewski, C. L. Redman, C. Wianguo, X. Bai, and J. M. Briggs. 2008. "Global Change and the Ecology of Cities." *Science* 319 (5864): 756–760. doi:10.1126/science.1150195.
He, C., P. Shi, L. Jinggang, J. Chen, Y. Pan, L. I. Jing, L. Zhou, and T. Ichinose. 2006. "Restoring Urbanization Process in China in the 1990s by Using Non-Radiance-Calibrated DMSP/OLS

Nighttime Light Imagery and Statistical Data." *Chinese Science Bulletin* 51 (13): 1614–1620. doi:10.1007/s11434-006-2006-3.

Henderson, M., E. T. Yeh, P. Gong, C. Elvidge, and K. Baugh. 2003. "Validation of Urban Boundaries Derived from Global Night-Time Satellite Imagery." *International Journal of Remote Sensing* 24 (3): 595–609. doi:10.1080/01431160304982.

Hu, S., L. Tong, A. E. Frazier, and Y. Liu. 2015. "Urban Boundary Extraction and Sprawl Analysis Using Landsat Images: A Case Study in Wuhan, China." *Habitat International* 47 (6): 183–195. doi:10.1016/j.habitatint.2015.01.017.

Imhoff, M. L., W. T. Lawrence, D. C. Stutzer, and C. D. Elvidge. 1997. "A Technique for Using Composite DMSP/OLS "City Lights" Satellite Data to Map Urban Area." *Remote Sensing of Environment* 61 (3): 361–370. doi:10.1016/S0034-4257(97)00046-1.

Jing, W., Y. Yang, X. Yue, and X. Zhao. 2015. "Mapping Urban Areas with Integration of DMSP/OLS Nighttime Light and MODIS Data Using Machine Learning Techniques." *Remote Sensing* 7: 12419–12439. doi:10.3390/rs70912419.

Kennedy, J., and R. Eberhart. 1995. "Particle Swarm Optimization." *IEEE International Conference on Neural Networks* 4 (8): 1942–1948.

Li, D., and L. Xi. 2015. "An Overview on Data Mining of Nighttime Light Remote Sensing." *Acto Geodaetica Et Cartographica Sinica* 44 (6): 591–601. [in chinese].

Li, X., R. Zhang, C. Huang, and L. Deren. 2014. "Detecting 2014 Northern Iraq Insurgency Using Night-Time Light Imagery." *International Journal of Remote Sensing* 36 (13): 3446–3458. doi:10.1080/01431161.2015.1059968.

Liu, Y., B. Zhang, L. Huang, and L. Wang. 2012a. "A Novel Optimization Parameters of Support Vector Machines Model for the Land Use/Cover Classification." *Journal of Food, Agriculture & Environment* 10 (2): 1098–1104.

Liu, Z., C. He, Q. Zhang, Q. Huang, and Y. Yang. 2012b. "Extracting the Dynamics of Urban Expansion in China Using DMSP-OLS Nighttime Light Data from 1992 to 2008." *Landscape and Urban Planning* 106: 62–72. doi:10.1016/j.landurbplan.2012.02.013.

Lu, Y., N. C. Coops, and T. Hermosilla. 2016. "Regional Assessment of Pan-Pacific Urban Environments over 25 Years Using Annual Gap Free Landsat Data." *International Journal of Applied Earth Observation and Geoinformation* 50 (3): 198–210. doi:10.1016/j.jag.2016.03.013.

Milesi, C., C. D. Elvidge, R. R. Nemani, and S. W. Running. 2003a. "Assessing the Environmental Impacts of Human Settlements Using Satellite Data." *Management of Environmental Quality* 14: 99–107. doi:10.1108/14777830310460414.

Milesi, C., C. D. Elvidge, R. R. Nemani, and S. W. Running. 2003b. "Assessing the Impact of Urban Land Development on Net Primary Productivity in the Southeastern United States." *Remote Sensing of Environment* 86 (3): 401–410. doi:10.1016/S0034-4257(03)00081-6.

Miller, S. D., S. P. Mills, C. D. Elvidge, and J. D. Hawkins. 2012. "Suomi Satellite Brings to Light a Unique Frontier of Nighttime Environmental Sensing Capabilities." Paper presented at the Proceedings of the National Academy of Sciences. doi:10.1073/pnas.1207034109

Mountrakis, G., I. Jungho, and C. Ogole. 2011. "Support Vector Machines in Remote Sensing: A Review." *ISPRS Journal of Photogrammetry and Remote Sensing* 66: 247–259. doi:10.1016/j.isprsjprs.2010.11.001.

Pan, H., W. Peng, H. Wei, G. Jiang, J. Zhou, and W. Zhou. 2009. "Dynamical Detection on Urban Sprawl Based on EO Data a Case Study of Chengdu City." In *International Forum on Information Technology and Applications*, IEEE Computer Society, 412–416. doi:10.1109/IFITA.2009.72

Shi, K., C. Huang, B. Yua, B. Yin, Y. Huang, and W. Jianping. 2014. "Evaluation of NPP-VIIRS Night-Time Light Composite Data for Extracting Built-Up Urban Areas." *Remote Sensing Letters* 5 (4): 358–366. doi:10.1080/2150704X.2014.905728.

Small, C., C. D. Elvidge, and K. Baugh. 2013. "Mapping Urban Structure and Spatial Connectivity with VIIRS and OLS Night Light Imagery." Paper presented at the Urban Remote Sensing Event (JURSE), IEEE.

Small, C., F. Pozzi, and C. D. Elvidge. 2005. "Spatial Analysis of Global Urban Extent from DMSP-OLS Night Light." *Remote Sensing of Environment* 96: 277–291. doi:10.1016/j.rse.2005.02.002.

The First National Geographic Census Leading Group Office of the State Council. 2013. *The Contents and Standards for National Geographic Census*. Beijing: Survering and Mapping Press. [in chinese].

Wang, L., L. Congcong, Q. Ying, X. Chen, and Y. Wang. 2012. "China's Urban Expansion from 1990 to 2010 Determined with Satellite Remote Sensing." *Chinese Science Bulletin* 57 (16): 1388–1403. [in chinese].

Yi, K., Y. Zeng, and W. Bingfang. 2016. "Mapping and Evaluation the Process, Pattern and Potential of Urban Growth in China." *Applied Geography* 71 (6): 44–55. doi:10.1016/j.apgeog.2016.04.011.

Zhang, Q., and K. C. Seto. 2011. "Mapping Urbanization Dynamics at Regional and Global Scales Using Multi-Temporal DMSP/OLS Nighttime Light Data." *Remote Sensing of Environment* 115: 2320–2329. doi:10.1016/j.rse.2011.04.032.

Zhou, Y., S. J. Smith, C. D. Elvidge, and M. L. Imhoff. 2014. "A Cluster-Based Method to Map Urban Area from DMSP/OLS Nightlights." *Remote Sensing of Evironment* 147: 173–185. doi:10.1016/j. rse.2014.03.004.

Monitoring urban expansion using time series of night-time light data: a case study in Wuhan, China

Xin Xin, Bin Liu, Kaichang Di, Zhe Zhu, Zhongyuan Zhao, Jia Liu, Zongyu Yue and Guo Zhang

ABSTRACT

Real-time urban expansion information is important for understanding the socio-economic activities and construction policies in urban areas. Night-time light data, which are available at annual and monthly temporal resolutions, can facilitate better analysis of socio-economic activities. In this study, we proposed a novel calibration method for Defense Meteorological Satellite Program's Operational Line-scan System (DMSP/OLS) data based on Rational Function Model (RFM). Stable lit pixels were employed to validate the effectiveness of the proposed model. The deference of mean square error shows that RFM method is better than traditional quadratic polynomials method for 76% of the data sets. Urban areas from 1992 to 2013 were extracted based on the calibrated data. A correlation analysis and multiple linear regression analysis between socio-economic factors and DMSP/OLS data were performed for Wuhan, China. The results of correlation factors showed that the correlation coefficient between night-time lights and socio-economic factors was higher than 0.85. Population produced the highest correlation coefficient among all the socio-economic factors. Multiple linear regression analyses were also performed, and the results showed that population and urban area could enhance the R^2 in Wuhan, and population density could enhance the R^2 in a comparative city Ordos. The development driving forces of the city could be reflected in multiple linear regression analysis of the night-time light data and socio-economic factors. Moreover, we investigated the relationship between construction policy and urban expansion using the time-series night-time light data, and found that the night-time light data could also reflect the construction policy and monitor the urban expansion effectively.

1. Introduction

China has undergone rapid economic development and urbanization in recent decades. Urban area in China increased from 7438 km^2 in 1981 to 32,520.7 km^2 in 2005 (Liu et al.

2012). Additionally, the population in China increased from 602 million in 1953 (First Population Census) to 1,370 million in 2010 (Sixth Population Census), an increase of 1.28 times. However, the censuses that have been carried out every 10 years, have limited the capability for updating urban expansion in real-time, which can be problematic for cities in China that are changing every day (Zhu et al. 2016).

High temporal frequency night-time light satellite data provide a possible solution. Night-time light data are indicator of human activity and urban lighting intensity (Amaral et al. 2005; Liu et al. 2012; Shi et al. 2014b; Sutton et al. 2001; Zhao, Zhou, and Samson 2015). Night-time light satellites can detect weak optical signals, and these signals are not affected by sunlight, shadows, vegetation, or other features on the Earth's surface. Global night-time light composites are available monthly or annually. They are useful for analysing urbanization and human activities in rapidly developing areas. There are two main sources of night-time light data: the Defense Meteorological Satellite Program's Operational Line-scan System (DMSP/OLS) (Elvidge et al. 1995, 1997; Johnson, Flament, and Bernstein 1994) and the Day/Night Band (DNB) on the National Polar-orbiting Partnership Visible Infrared Imaging Radiometer Suite (NPP/VIIRS) (Lee et al. 2006). The NPP/VIIRS data, which have a higher resolution and a wider spectrum detection range, are the upgraded version of DMSP/OLS data. The two satellites are both working at present. DMSP/OLS started working in 1973, and NPP/VIIRS was launched in 2009 (Elvidge et al. 2007). However, the earliest DMSP/OLS night-time light data that can be obtained are from 1992, and the earliest NPP/VIIRS data are available since April 2012 on the National Oceanic and Atmospheric Administration/National Geophysical Data Center (NOAA/NGDC) website (http://ngdc.noaa.gov/eog/dmsp/downloadV4composites.html). These night-time light data have been widely used in many areas, such as fire monitoring (Elvidge et al. 1995), gas flaring (Elvidge et al. 2009), urban area extraction (Imhoff et al. 1997; Shi et al. 2014a; Small, Pozzi, and Elvidge 2005; Zhang and Seto 2011), urban expansion monitoring (Liu et al. 2012; Zhou et al. 2014), socio-economic analysis (Amaral et al. 2005; Sutton et al. 2001), crisis monitoring (Li and Li 2014; Li et al. 2015), CO_2 emissions (Ou et al. 2015), and cloud detection (Xia et al. 2014). Asian cities have also gained much attention through the use of night-time light data (Ma et al. 2012; Shi et al. 2014b). Previous research has shown that DMSP/OLS data can be used to detect long time series of annual dynamic changes at a macro scale, and NPP/VIIRS night-time light data, which feature a higher spatial resolution, can provide better details of urban structure. There is a strong relation between urban development and urban construction policy in some cities of China (Li 2008), but few researchers have studied the correlation between them using night-time light data.

The long time series of night-time light data provide a great opportunity for monitoring urban expansion. However, due to the degradation of each DMSP/OLS satellite and the lack of on-board calibration, the satellites do not provide DMSP/OLS night-time light data with good comparability (Elvidge et al. 2009). Therefore, it is critical to calibrate the DMSP/OLS data before they can be used in long time-series analysis. Several researchers have used relative radiance calibration for DMSP/OLS data, and the most common method is calculating the correction factor using the quadratic regression model (Elvidge et al. 2009; Liu et al. 2012).

Calibrated night-time light data can be used to analyse the socio-economic activities. Correlation analysis is commonly used for this purpose (Amaral et al. 2005; Liu et al.

2012; Ma et al. 2012; Shi et al. 2014b), but this method is only effective for analysing the relation between a single factor and the night-time light data. There have been few studies using the multiple regression method to analyse night-time light data and many socio-economic factors altogether (Levin and Duke 2012). It is necessary to use the multiple regression method for a more detailed analysis between the socio-economic statistical data and the night-time light data, e.g. to reveal the relation between the night-time light data and different combinations of the socio-economic factors.

In this research, we performed a case study of urban expansion monitoring in Wuhan, China, using multi-scale time-series night-time light data in the following four steps. First, we developed a novel DMSP/OLS data calibration method based on the Rational Function Model (RFM), which had a better performance than the traditionally used quadratic polynomial model. We also proposed a new method to verify the calibration accuracy using stable lit pixels, which have a stable mapping relationship between two data sets. Second, we extracted the urban areas using DMSP/OLS data and analysed the expansion dynamics. Third, we investigated the correlations between the urban areas and the socio-economic data through correlation analysis and multiple linear regression analysis. Finally, we analysed the impacts of construction policy and human activity on urban expansion using both DMSP/OLS and NPP/VIIRS time-series data, respectively.

2. Data

Our study area is in Wuhan (30°33′N, 114°19′E, the capital city of Hubei Province), China (Figure 1), which is in the middle-lower Yangtze Plain. The Yangtze River flows across Wuhan and divides Wuhan into three towns: Wuchang, Hanyang, and Hankou. Wuhan is the core city of economic development in Wuhan city circle. Wuhan city circle is one of the city circles that are under national major urban planning and will become the fourth region in China after the Yangtze River Delta, Pearl River Delta, and Bohai Sea Economic Zone. It contains nine cities (Wuhan, Huangshi, Ezhou, Huanggang, Xiaogan, Xianning, Tianmen, Qianjiang, and Xiantao) with different levels of urban development.

Rapid growth of urban area, population, and gross domestic product (GDP) has occurred since the twentieth century. Until 2013, the urban area in Wuhan had reached 534.28 km^2, and the population had reached approximately 8 million, an increase of 20.1% from the population in 1992. Wuhan's GDP rose multiple times from 1992 to 2013 (the GDP in 1992 was 255.42 trillion Chinese Yuan and the GDP in 2013 was 10,069.48 trillion Chinese Yuan).

In this study, the cities in Wuhan city circle, which is an important part of the Yangtze Economic Zone in China, were selected for a comparative study.

Three categories of data, including night-time light data (DMSP/OLS data and DNB of NPP/VIIRS data), statistical data (GDP, urban area, electricity consumption, population, and population density), and administrative boundaries of the study area, were used in this study. The DMSP/OLS data were obtained from the NOAA/NGDC website. We chose the annual stable DMSP/OLS data in the Version 4 data set from 1992 to 2013, which were acquired by six DMSP satellites, including F10, F12, F14, F15, F16, and F18. A detailed list of the DMSP/OLS data we used is shown in Table 1. They all have 3000 km swath with a resolution of 30 arc-second geographic grids. The digital number (DN) values (ranging from 0 to 63 with 0 representing background value) represent the

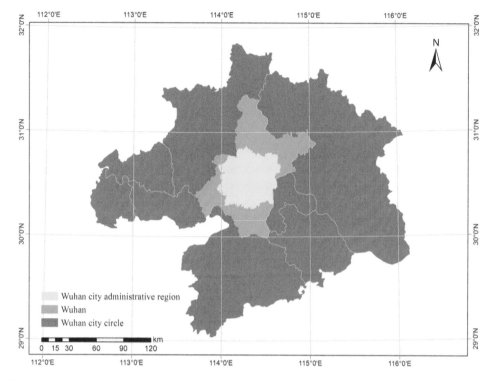

Figure 1. Study area.

Table 1. List of annual DMSP/OLS data.

Satellite Year	F10	F12	F14	F15	F16	F18
1992	F101992					
1993	F101993					
1994	F101994	F121994				
1995		F121995				
1996		F121996				
1997		F121997	F141997			
1998		F121998	F141998			
1999		F121999	F141999			
2000			F142000	F152000		
2001			F142001	F152001		
2002			F142002	F152002		
2003			F142003	F152003		
2004				F152004	F162004	
2005				F152005	F162005	
2006				F152006	F162006	
2007				F152007	F162007	
2008					F162008	
2009					F162009	
2010						F182010
2011						F182011
2012						F182012
2013						F182013

Table 2. The socio-economic statistics in Wuhan.

Year	GDP (×10⁹ Chinese Yuan)	Urban area (km²)	Electricity consumption (×10⁶ kWh)	Population (×10⁴)	Population density (people km⁻²)
1992	255.42	–	845669	684.46	–
1993	357.23	–	888289	691.69	–
1994	485.76	–	967882	700.01	827
1995	606.91	200.00	1011015	710.01	839
1996	782.13	202.00	1049970	715.94	846
1997	912.33	202.00	1086824	723.90	855
1998	1001.89	204.00	1079014	731.79	864
1999	1085.68	208.00	1024472	740.20	874
2000	1206.84	210.00	1186642	749.19	882
2001	1335.40	212.00	1524125	758.23	893
2002	1467.80	214.22	1587842	768.10	904
2003	1622.18	216.22	1788280	781.19	920
2004	1882.24	218.22	1855802	785.90	925
2005	2261.17	220.22	2108743	801.36	943
2006	2679.33	–	2308126	818.84	964
2007	3209.47	–	2586728	828.21	975
2008	4115.51	460.77	2864338	833.24	981
2009	4620.86	484.00	3102749	835.55	984
2010	5565.93	500.00	3536311	836.73	985
2011	6762.20	507.04	3836469	827.24	1180
2012	8003.82	520.30	4032605	821.71	1191
2013	9051.27	534.28	4372338	822.05	1203

average DN value of stable lights without filtering. The data were extracted according to the administrative boundaries of the study area and projected using the Mollweide projection (Snyder 1987), and then resampled to 1 km resolution.

The DNB data of NPP-VIIRS were also downloaded from the NOAA/NGDC website (http://ngdc.noaa.gov/eog/viirs/download_monthly.html). The monthly average radiance cloud-free composite images (January 2014, August 2014, December 2014, July 2015, and August 2015), which have undergone the stray-light correction procedure (Mills, Weiss, and Liang 2013), were chosen for this study. All the original data were in 15 arc-second geographic grids, and the data were resampled to the resolution of 500 m to facilitate analysis. The DN value of each pixel represents the radiation value of the pixel.

The socio-economic statistics used in this study were obtained from the Statistic Yearbook of Wuhan (Wuhan Bureau of Statistics 1992-2013). Table 2 shows the statistical data in details.

3. Methods

In our method, two types of night-time light data are used to monitor urban expansion and analyse correlations among night-time lights, socio-economic activities, and construction policy. Before the analysis, DMSP/OLS data need to be calibrated to improve their comparability. The relative radiance calibration of the DMSP/OLS data from 1992 to 2013 was based on RFM. Urban areas were extracted from the night-time light data after the calibration. Specially, urban areas were extracted from DMSP/OLS data based on the weighted light area, which is defined as the summing lit pixels multiplied by the normalized DN value (Ma et al. 2012). Areas were also extracted from NPP/VIIRS data through a thresholding method (Milesi et al. 2003).

As noted above, the DMSP/OLS data from the Version 4 data set cannot represent the real radiance value and cannot be used directly in long time-series urban analysis because of the lack of on-board calibration. DN values under the same light condition from different satellites are different. To address this problem, we developed a novel calibration method for DMSP/OLS data based on RFM. The new method involves two main steps: 1) inter-calibration of the annual data based on RFM, and 2) intra-annual composition and series correction of the inter-calibrated data. We also extracted stable lit pixels to verify the calibration method.

3.1. Inter-calibration of the annual data

The total number of lit pixels (Figure 2(a) and the average DN values (Figure 2(b)) were counted in each satellite year in China. It is obvious that the data are not directly comparable. There are large differences between different satellites in the same year. To improve the comparability of the time-series data, the data need to be inter-calibrated first. In this study, the reference satellite year was chosen as F162005, which has the minimum variance of the total lit pixel number and the average DN values between different satellites in the same year in Wuhan. In addition, by analysing the socio-economic characteristics of different cities in China, Jixi (a city in Heilongjiang Province), which has a wide spread of DN values and a stable state of socio-economic development, was selected as the reference calibration area (Liu et al. 2012). Inter-calibration was based on the aforementioned reference satellite year and reference calibration area. The most widely used inter-calibration method was developed by Elvidge et al. (2009) using a quadratic regression model. However, due to the characteristic of the quadratic regression curve, the points near the minimum/maximum DN values cannot fit well in some reference areas. RFM offers a good solution to this problem in our study. RFM, which can model functions with poles, is superior to polynomials models. An RFM is the quotient of two polynomials (Press et al. 2007). In this study, quadratic polynomial is used as the numerator and denominator of the RFM, which is given by the following equation:

$$D_C = \frac{p_1 \times D^2 + p_2 \times D + p_3}{D^2 + q_1 \times D + q_2},$$
(1)

where
D_C is the inter-calibrated DN value of the data;
D is the original DN value of the source satellite year data; and
p_1, p_2, p_3, q_1 and q_2 are the coefficients of the second-order RFM;

The objective of this step is to inter-calibrate the lit pixels in each satellite year so that they are more comparable in time-series analysis. It is worth noting that the dark pixels, whose DN values are 0, are not involved in the inter-calibration.

3.2. Intra-annual composition and series correction of the inter-calibration data

The purpose of intra-annual composition is to make full use of the information derived from the data acquired in the same year from different satellites and remove the unstable lit pixels (Liu et al. 2012). Series correction is based on the assumption that

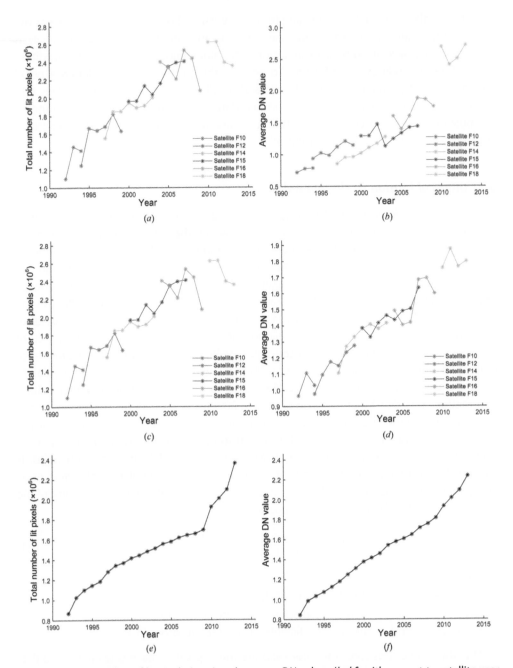

Figure 2. Total number of lit pixels (*a,c,e*) and average DN values (*b,d,f*) with respect to satellite year in China. (*a*) and (*b*) show the data before calibration. (*c*) and (*d*) show the data after inter-calibration. (*e*) and (*f*) show the data after both inter-calibration and intra-annual composition.

the urban area would grow continuously outwards and the lit pixels would become brighter over time (Liu et al. 2012).

Twelve composites acquired from two different satellites in the same year, containing the data set from 1994 and the data sets from 1997 to 2007, were intra-calibrated in this study. The intra-annual composition results were the average values of the corresponding pixels in each data set that have different DN values from the two satellites:

$$D_n = (D_{n_1} + D_{n_2})/2, \tag{2}$$

where

D_n is the DN value of the data in the nth year after intra-annual composition; and

D_{n_1} and D_{n_2} are the DN values acquired in the nth year from satellites 1 and 2, respectively.

The intra-annual composition equation is applicable to the year n = 1994, 1997, 1998...,2007. For the other years in which there is only one satellite data available, D_n remains unchanged.

After intra-annual composition, series correction of the data was performed by the following equation (Liu et al. 2012):

$$D_{(n,i)} = \begin{cases} 0 & D_{(n+1,i)} = 0 \\ D_{(n-1,i)} & D_{(n+1,i)} > 0 \ D_{(n-1,i)} > D_{(n,i)} , \\ D_{(n,i)} & \text{otherwise} \end{cases} \tag{3}$$

where

$D_{(n,i)}, D_{(n-1,i)}$ and $D_{(n+1,i)}$ are the DN values of the i th pixel in the nth, $(n-1)$th, and $(n+1)$th year, respectively. In the series correction, the data of the first year remained unchanged, whereas all the data of other years were processed using the above equation.

3.3. Stable lit pixel extraction

The calibration procedure aims to make the different DN values under the same night-time light condition as close as possible. To evaluate the effectiveness of RFM-based data calibration, we developed a method for generating stable lit pixels, whose DN value mapping relationship is less influenced by noises and real light condition changes.

Owing to the differences of measurement equipment and procedure, the DN values from different satellites under the same night-time light condition (acquired at the same position and the same time) may not be the same. The general forms of the mapping functions from the reference image to the calibrated images and from the calibrated images to the reference image can be represented by the following equations, respectively:

$$D_R = F_{(n,R)}(D_n) + \delta_{(n,R)}, \tag{4}$$

$$D_n = F_{(R,n)}(D_R) + \delta_{(R,n)},$$

where

D_R is the DN value of the reference image,

D_n is the DN value of the calibrated image in the nth year,

$F_{(n,R)}$ and $F_{(R,n)}$ are the mapping functions from the calibrated image to the reference image and the mapping function from the reference image to the calibrated image, respectively, and

$\delta_{(n,R)}$ and $\delta_{(R,n)}$ are the noise or error existing in the relationship.

By counting the DN value correspondences of the pixels at the same position between the reference image and the calibrated image, we considered the maximum correspondence numbers of all the DN values (1–62) as the candidates. The candidate sets are named as $O_{(n,R)}$ and $O_{(R,n)}$, respectively. We defined the pixels that exist both in set $O_{(n,R)}$ and in set $O_{(R,n)}$ as stable lit pixels. For example, if (53, 48) is in $O_{(n,R)}$ and (48, 53) is in $O_{(R,n)}$, then 48 and 53 are a pair of stable lit pixels. Stable lit pixels are pixels with the stable DN value mapping relationship on the same position of the reference image and the calibrated images. As the distribution of the stable lit pixels is not always homogeneous on the DN value dimension, these points are only used to verify the effectiveness of RFM. The closer the fitted model is to the actual values of the stable lit pixels, the better the model is.

4. Results and discussion

4.1. *DMSP/OLS data calibration results*

Thirty-four DMSP/OLS data sets were calibrated using RFM. The calibration results (Figure 2) show that the calibration method in this study improved the comparability of DMSP/OLS data from 1992 to 2013. In Figure 2(c), the total number of lit pixels did not change after inter-calibration, because the RFM method can only change the DN value of the pixels but cannot change the lit pixels into non-lit pixels. We can find that after inter-calibration, the average DN values (Figure 2(d)) increase more smoothly than those in an un-calibrated image (Figure 2(b)). In most situations, inter-calibration reduced the discrepancies of the DN values between two satellites in the same year (Figure 2(d)). However, there are some abnormal fluctuations in the DN values from time-series data. The total number of lit pixels (Figure 2(e)) and the average DN values (Figure 2(f)) after intra-annual composition and series correction show that the discrepancies between the data were reduced. Overall, the fluctuations in time-series data were generally removed and the continuity and comparability of the time-series data were improved after the proposed calibration method.

To compare the performances of the RFM-based calibration and the traditional calibration, the same DMSP/OLS data sets were also calibrated using quadratic polynomials with the same reference satellite year (F162005) in Wuhan. Results from the two calibration methods are evaluated and compared based on the stable lit pixels. The mean squared error (MSE) of each fitted curve based on stable lit pixels was calculated based on the following equations:

$$M_{R_n} = \sum_{i=1}^{N_n} \left(Dc_{(n,i)} - I_{(n,i)} \right)^2 / N_n, \tag{5}$$

$$M_{Q_n} = \sum_{i=1}^{N_n} \left(Qc_{(n,i)} - I_{(n,i)} \right)^2 / N_n, \tag{6}$$

where
M_{R_n} is the MSE of RFM,
M_{Q_n} is the MSE of quadratic polynomials,
$Dc_{(n,i)}$ is the inter-calibrated DN value using the RFM method,
$Qc_{(n,i)}$ is the calibrated DN value using quadratic polynomials,

$I_{(n,i)}$ is the observed DN value of the extracted stable lit pixels in the nth data set,
N_n is the number of stable lit pixels, and
n represents the serial number of each data set.

We calculated the difference of mean square error (DMSE) between M_{R_n} and M_{Q_n} as a comparison as follows:

$$E_n = M_{Q_n} - M_{R_n},$$

(7)

where E_n is the DMSE of the nth year. Positive DMSE means RFM is better than the quadratic polynomial. The larger the DMSE, the better the performance of the RFM compared with the quadratic polynomial.

Figure 3 shows the estimated RFM curves and the quadratic polynomials with all the lit pixels in each data set. From this figure, we can directly find that the RFM performs better than the quadratic polynomials. The RFM can fit the stable lit pixels well, especially for pixel DN values close to 63. Figure 4 shows the quantitative analysis results of each data set. The DMSE shows that RFM is better than the quadratic polynomials for 76% of the data set. Based on the above analysis, the RFM was chosen in this study to calibrate the night-time light data.

4.2. Urban area expansion and construction policy analysis in Wuhan from 1992 to 2013

DMSP/OLS night-time light data, which have more observation years, are feasible for long-term and macro-scale analysis. The weighted light urban areas in Wuhan in the DMSP/OLS data set from 1992 to 2013 were extracted based on the weighted light area, which was calculated by the sum of lit pixels multiplied by the normalized DN values (Equation 8). Pixels with DN values larger than 11 and less than or equal to 63 are considered in calculating the weighted light area to reduce the effects of dim lights (Ma et al. 2012). As shown in Figure 5, urban expansion was tremendous in Wuhan from 1992 to 2013. Wuhan has been the core city among the cities in Wuhan city circle; according to the city circle planning policy, the development of Wuhan was supposed to promote the development of the other cities in the city circle. Figure 5 shows that the development of the cities in Wuhan city circle had the same trend as that of Wuhan.

$$A = \sum S_i \times \left(d_i / \sum d_i \right), \ (d_i = 12, 13, 14 \ldots 63),$$

(8)

where
  A is the weighted light area,
  S_i is the pixel number of each DN value, and
  d_i is the DN value.

The increment of weighted light area in the time-series data is shown in Figure 6. The increment value in each year was calculated based on the following equation:

$$I(n) = A(n) - A(n-1), \ n = 1993, 1994, 1995 \ldots 2013,$$

(9)

where
  I is the increment value,
  A is the weighted light area, and
  n represents the year.

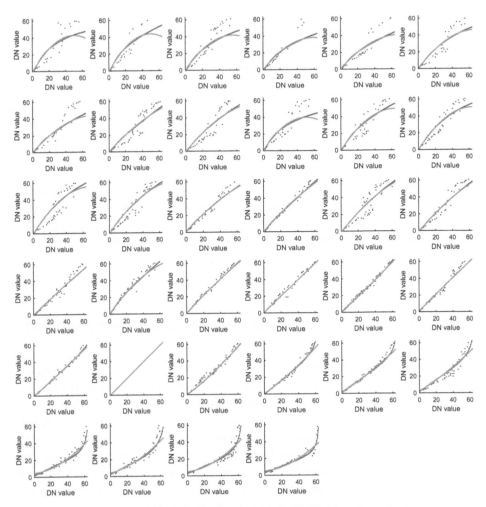

Figure 3. Fitted curves of RFM and the quadratic polynomials for the data sets. Each graph corresponds to one different data set as listed in Table 1. The blue points are the extracted stable lit pixels in each data set. The red lines are the RFM fitted curves with all the lit pixels and the green lines are the quadratic polynomials fitted curves with all the lit pixels in each data set.

According to Wuhan's construction policies from 1988 to 2010, there were four expansion stages in three Wuhan economic zones (Donghu New Technology Development Zone, Wuhan Economic and Technological Development Zone, and Wujiashan Economic Development Zone), including the start-up stage (1988–1993), the first expansion stage (2000–2001), the second expansion stage (2005–2007), and the third expansion stage (2010). Figure 6 shows five maximum values in five different years, and four of them reflect the four stages of expansion. It is also interesting to see that the maximum values in Figure 6 appear one year later than the end years of the expansion stages. This time lag is reasonable because the annual average data cannot reveal the constructions of buildings and roads that are finished and utilized in the end of the year.

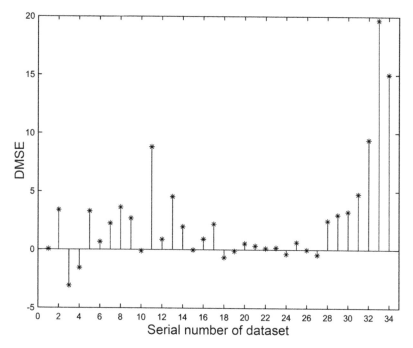

Figure 4. The scatter plot of DMSE in each data set.

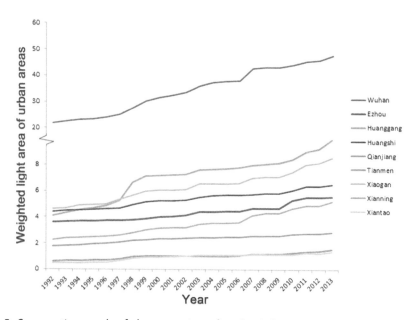

Figure 5. Comparative graph of the expansion of each of the cities in Wuhan city circle from 1992 to 2013.

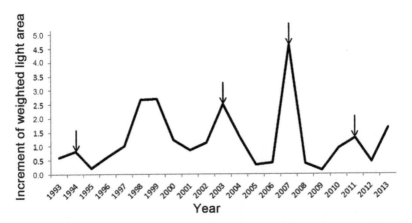

Figure 6. Increment of weighted light area per year in the time-series data in Wuhan. The arrows represent the expansion stages reflected in DMSP/OLS data.

4.3. *Socio-economic analysis in Wuhan and a comparative city Ordos*

The relationships between the weighted light area and socio-economic factors, GDP (X1), urban area (X2), electricity consumption (X3), population (X4), and population density (X5), were calculated through linear correlation analysis. The correlation coefficients (r, Equation 10) between the weighted light area and the socio-economic factors are shown in Figure 7. All the factors have a strong correlation with the weighted light area with a correlation coefficient (r) higher than 0.75. The results show that night-time light data have the highest correlation with population (X4).

$$r = \frac{\sum (x_i - \bar{x})(y_i - \bar{y})}{\sqrt{\sum (x_i - \bar{x})^2 \sum (y_i - \bar{y})^2}}, \tag{10}$$

where
   r is the correlation coefficient,
   x_i and y_i are the weighted light area and the socio-economic factor, respectively, and
   $\bar{x}$ and $\bar{y}$ are the mean values of the weighted light area and the socio-economic factor, respectively.

   We also performed linear regression analysis for all the five factors (X1, X2, X3, X4, and X5). Table 3 shows the results of unary and multiple linear regression analysis. It is identical to the correlation analysis results that the adjusted coefficient of determination (R^2, equation 11) of population (X4) is higher than that of any other factor in the unary linear regression. In the two-factor regression analysis, we can see that the values of R^2 of the combinations of urban area (X2) and other factors are always higher than those of the single factors (i.e. R^2 of XnX2 is higher than that of Xn and X2, n = 1, 3, 4, 5), and population (X4) has the same enhancement effect on the other factors. Thus, we can conclude that population and urban area are the enhancement factors in Wuhan. We can also find that some combinations containing population density (X5) resulted in a lower R^2 than that of the single factors.

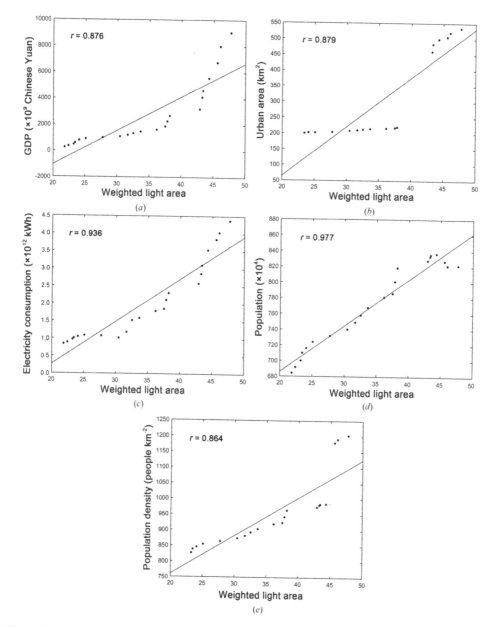

Figure 7. Linear correlation analysis results. (*a*) GDP versus weighted lighted area, (*b*) urban area versus weighted light area, (*c*) electricity consumption versus weighted light area, (*d*) population versus weighted lighted area, and (*e*) population density versus weighted light area. The points are the socio-economic factor values in different weighted light areas from 1992 to 2013. The lines are the regression curves.

$$R^2 = 1 - \left(\frac{m-1}{m-e}\right)\frac{\sum\left(Y_i - \widehat{Y}_i\right)^2}{\sum\left(Y_i - \overline{Y}_i\right)^2}, \tag{11}$$

Table 3. The R^2 of unary and multiple linear regression analysis in Wuhan.

One factor	R^2	Two factors	R^2
X1	0.775	X1X2	0.776
X2	0.757	X1X3	0.911
X3	0.870	X1X4	0.981
X4	0.951	X1X5	0.738
X5	0.733	X2X3	0.880
		X2X4	0.955
		X2X5	0.791
		X3X4	0.977
		X3X5	0.864
		X4X5	0.979

where

R^2 is the adjusted coefficient of determination,

Y_i is the observed value,

$\widehat{Y}_i$ is the estimated value,

$\overline{Y}_i$ is the mean value,

m is the number of observations, and

e is the number of regression coefficients.

To illustrate the significance of multiple linear regression analysis, the same analysis was performed for Ordos, a city in Inner Mongolia, China, as a comparative experiment. Table 4 shows the multiple linear regression result in Ordos. It shows that GDP (X1) has the highest R^2 among all the factors, urban area has the lowest R^2 among all the factors, and population density is the enhancement factor in Ordos.

There was a good linear relation between electricity consumption (X3) and weighted light area in Wuhan and Ordos, because electric lighting has a direct causal relationship with electricity consumption. By comparison, we find that population (X4) and weighted light area had the highest correlation ($R^2 = 0.951$) in Wuhan. Population in Wuhan has the highest R^2 because the increasing population has similar growth trends as the urban area. However, Wuhan has been developed around the Yangtze River, and there is an urban expansion limit that made the R^2 of the combination containing population density (X5) lower than that of single factors in the corresponding combinations. However, in Ordos, GDP (X1) and weighted light area have the highest correlation ($R^2 = 0.908$). Ordos is a typical resource-driven city in China. The exploitation of the resources resulted in the rapid growth of GDP and urban expansion, but there was no growth in population accordingly. Therefore, night-time light data in Ordos has a larger

Table 4. The R^2 of unary and multiple linear regression analysis in Ordos.

One factor	R^2	Two factors	R^2
X1	0.908	X1X2	0.907
X2	0.761	X1X3	0.904
X3	0.890	X1X4	0.899
X4	0.849	X1X5	0.914
X5	0.839	X2X3	0.881
		X2X4	0.863
		X2X5	0.874
		X3X4	0.890
		X3X5	0.928
		X4X5	0.880

correlation with GDP than population. Compared with analysing the relation between a single socio-economic factor and night-time light data, multiple linear regression analysis revealed more information about the city. The development driving forces of the cities could be reflected according to our research.

4.4. Dynamic analysis between urban expansion and construction policy using NPP

NPP/VIIRS data, which have a higher resolution than DMSP data, were employed for a more detailed study in the Wuhan City Administrative Region. The lit area was extracted through an optimal threshold (Milesi et al. 2003), which was selected through the correspondence between the lit area from the December 2014 and January 2015 NPP/VIIRS data and the urban area in Statistic Yearbook of Wuhan (Wuhan Bureau of Statistics 1992-2013). Through comparison between the lit area and the urban area in Wuhan, we used 24.72 as the threshold of lit area extraction. The extracted urban and urban change information are shown in Figure 8. The stable area (area that did not change in the different time phases), the expansion area, and the shrinkage area from January 2014 to August 2015 were all extracted by this method. By combining information on the construction policies, we extracted five typical change regions, including A) the Gusaoshu expressway (open to traffic on 28 April 2014), B) the Hanyang Binjiang Avenue (the second-stage project in 2015), C) the Yingwuzhou Yangtze River Bridge (open to traffic on 28 December 2014), D) the Lizhi Road (the first stage of reconstruction was completed at the end of 2014), and E) the Second Yangtze River Bridge of Wuhan (traffic control was cancelled from 22 July 2015).

The NPP/VIIRS night-time light data not only reflected the urban construction policy in the long term (blue rectangle in Figure 8), but also manifested the urban transport policy in the short term (green rectangle in Figure 8). The micro-scale human activities, e.g. those caused by traffic restriction on the Wuhan Yangtze River Bridge and the opening of new roads, can be revealed in the NPP/VIIRS night-time light data.

5. Conclusions

This article presented a new data calibration method for DMSP/OLS based on RFM. The proposed method achieved lower MSE than the conventional method (the quadratic polynomial model). The urban areas of Wuhan were extracted based on DMSP/OLS data after the calibration, and the extracted urban areas reflected the socio-economic activities and the construction policy. Correlation analysis was applied between the socio-economic data and the DMSP/OLS night-time light data from 1992 to 2013. Our results showed a strong correlation between the night-time light data and the socio-economic factors, especially for the population ($r = 0.977$). Multiple linear regression analysis showed that the R^2 of the combination results that contained the urban area and the population was higher than that of any other single factor in the corresponding combination result in Wuhan. A comparative experiment of multiple linear regression analysis was also carried out in Ordos, and the results show GDP has the highest R^2 among all the factors, and population density is the enhancement factor in Ordos. The comparative results indicated that the development driving forces of the cities could be reflected in multiple linear regression analysis of the socio-economic data and the night-time light data. A detailed study was carried out in Wuhan based on NPP/VIIRS night-time light data to illustrate the capability of revealing short-term construction policy.

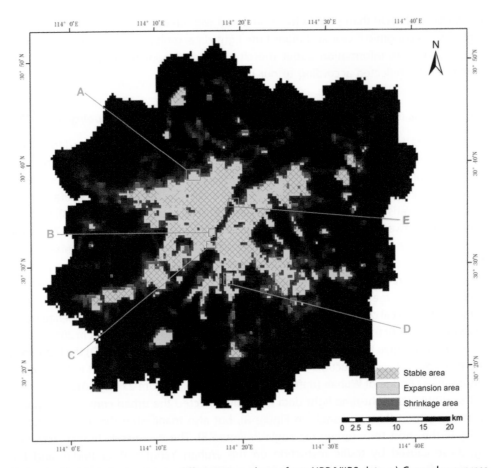

Figure 8. Changing results among different time phases from NPP/VIIRS data. a) Gusaoshu expressway, b) Hanyang Binjiang Avenue, c) Yingwuzhou Yangtze River Bridge, d) Lizhi Road, and e) Second Yangtze River Bridge of Wuhan. The green grid with yellow background covers the stable area (area that did not change in the different time phases), the yellow background covers the expansion area (area that increased in the next time phases), and the red background covered the shrinkage area (area that reduced in the next time phases) The NPP/VIIRS data were acquired in July 2015.

The results showed that high-resolution night-time light data can be used in monitoring micro-scale human activities caused by traffic restriction on the Wuhan Yangtze River Bridge and by the opening of new roads.

This research provides a case study in Wuhan and a comparative study in Ordos, which may not be representative of many other cities. Future research may focus on applying our methods to other major cities in different parts of the world, and evaluating whether our methods can be well generalized.

Acknowledgements

The authors are grateful to the editor Professor Warner and the three reviewers for their careful proofreading and insightful comments. This work was supported in part by National Natural Science Foundation of China (Grant Nos. 91538106 and 41501503). The authors would like to thank the NGDC/NOAA for providing the DMSP/OLS and NPP/VIIRS night-time light data, the

Wuhan Bureau of Statistics for the statistical data, the Wuhan Academy of Surveying and Mapping for the policy data and the information of expansion stages in three Wuhan economic zones.

Disclosure statement

No potential conflict of interest was reported by the authors.

Funding

This work was supported in part by National Natural Science Foundation of China (Grant Nos. 91538106 and 41501503).

References

Amaral, S., G. Câmara, A. M. V. Monteiro, J. A. Quintanilha, and C. D. Elvidge. 2005. "Estimating Population and Energy Consumption in Brazilian Amazonia Using DMSP Night-Time Satellite Data." *Computers, Environment and Urban Systems* 29: 179–195. doi:10.1016/j.compenvurbsys.2003.09.004.

Elvidge, C. D., K. E. Baugh, E. A. Kihn, H. W. Korehl, and E. R. Davis. 1997. "Mapping City Lights with Nighttime Data from the DMSP Operational Linescan System." *Photogrammetric Engineering & Remote Sensing* 63 (6): 727–734.

Elvidge, C. D., P. Cinzano, D. R. Pettit, J. Arvesen, P. Sutton, C. Small, R. Nemani, et al. 2007. "The Nightsat Mission Concept." *International Journal of Remote Sensing* 28 (12): 2645–2670. doi:10.1080/01431160600981525.

Elvidge, C. D., H. W. Kroehl, E. A. Kihn, K. E. Baugh, E. R. Davis, and W. M. Hao. 1995. "Algorithm for the Retrieval of Fire Pixels from DMSP Operational Linescan System Data." In *Global Biomass Burning*, edited by J. S. Levine, 73–85. Cambridge, Massachusetts: MIT Press.

Elvidge, C. D., D. Ziskin, K. E. Baugh, B. T. Tuttle, T. Ghosh, D. W. Pack, E. H. Erwin, and M. Zhizhin. 2009. "A Fifteen Year Record of Global Natural Gas Flaring Derived from Satellite Data." *Energies* 2: 595–622. doi:10.3390/en20300595.

Imhoff, M. L., W. T. Lawrence, D. C. Stutzer, and C. D. Elvidge. 1997. "A Technique for Using Composite DMSP/OLS "City Lights" Satellite Data to Map Urban Area." *Remote Sensing of Environment* 61: 361–370. doi:10.1016/s0034-4257(97)00046-1.

Johnson, D. B., P. Flament, and R. L. Bernstein. 1994. "High-Resolution Satellite Imagery for Mesoscale Meteorological Studies." *Bulletin of the American Meteorological Society* 75 (1): 5–33. doi:10.1175/1520-0477(1994)075<0005:HRSIFM>2.0.CO;2.

Lee, T. E., S. D. Miller, F. J. Turk, C. Schueler, R. Julian, S. Deyo, P. Dills, and S. Wang. 2006. "The NPOESS VIIRS Day/Night Visible Sensor." *Bulletin of the American Meteorological Society* 87 (2): 191–199. doi:10.1175/BAMS-87-2-191.

Levin, N., and Y. Duke. 2012. "High Spatial Resolution Night-Time Light Images for Demographic and Socio-Economic Studies." *Remote Sensing of Environment* 119: 1–10. doi:10.1016/j.rse.2011.12.005.

Li, B. 2008. "Urban Development Policies of China and the Influence to Chinese Urbanization." *(In Chinese) Urban Studies* 14 (2): 26–37. doi:10.3969/j.issn.1006-3862.2008.02.005.

Li, X., and D. Li. 2014. "Can Night-Time Light Images Play a Role in Evaluating the Syrian Crisis." *International Journal of Remote Sensing* 35 (18): 6648–6661. doi:10.1080/01431161.2014.971469.

Li, X., R. Zhang, C. Huang, and D. Li. 2015. "Detecting 2014 Northern Iraq Insurgency Using Night-Time Light Imagery." *International Journal of Remote Sensing* 36 (13): 3446–3458. doi:10.1080/01431161.2015.1059968.

Liu, Z. F., C. Y. He, Q. F. Zhang, Q. X. Huang, and Y. Yang. 2012. "Extracting the Dynamics of Urban Expansion in China Using DMSP-OLS Nighttime Light Data from 1992 to 2008." *Landscape and Urban Planning* 106: 62–72. doi:10.1016/j.landurbplan.2012.02.013.

Ma, T., C. Zhou, T. Pei, S. Haynie, and J. Fan. 2012. "Quantitative Estimation of Urbanization Dynamics Using Time Series of DMSP/OLS Nighttime Light Data: A Comparative Case Study from China's Cities." *Remote Sensing of Environment* 124: 99–107. doi:10.1016/j.rse.2012.04.018.

Milesi, C., C. D. Elvidge, R. R. Nemani, and S. W. Running. 2003. "Assessing the Impact of Urban Land Development on Net Primary Productivity in the Southeastern United States." *Remote Sensing of Environment* 86: 401–410. doi:10.1016/S0034-4257(03)00081-6.

Mills, S., S. Weiss, and C. Liang. 2013. "VIIRS Day/Night Band (DNB) Stray Light Characterization and Correction." *Proceedings of SPIE* 8866 (1P). doi:10.1117/12.2023107.

Ou, J., X. Liu, X. Li, M. Li, and W. Li. 2015. "Evaluation of NPP-VIIRS Nighttime Light Data for Mapping Global Fossil Fuel Combustion CO_2 Emissions: A Comparison with DMSP-OLS Nighttime Light Data." *PLoS One* 10 (9): e0138310. doi:10.1371/journal.pone.0138310.

Press, W. H., S. A. Teukolsky, W. T. Vetterling, and B. P. Flannery, (2007). "Rational Function Interpolation and Extrapolation". Chap. 3.4. In *Numerical Recipes: The Art of Scientific Computing*. 3rd. New York, NY: Cambridge University Press.

Shi, K., C. Huang, B. Yu, B. Yin, Y. Huang, and J. Wu. 2014a. "Evaluation of NPP-VIIRS Night-Time Light Composite Data for Extracting Built-Up Urban Areas." *Remote Sensing Letters* 5 (4): 358–366. doi:10.1080/2150704X.2014.905728.

Shi, K., B. Yu, Y. Huang, Y. Hu, B. Yin, Z. Chen, L. Chen, and J. Wu. 2014b. "Evaluating the Ability of NPP-VIIRS Nighttime Light Data to Estimate the Gross Domestic Product and the Electric Power Consumption of China at Multiple Scales: A Comparison with DMSP-OLS Data." *Remote Sensing* 6: 1705–1724. doi:10.3390/rs6021705.

Small, C., F. Pozzi, and C. D. Elvidge. 2005. "Spatial Analysis of Global Urban Extent from DMSP-OLS Night Lights." *Remote Sensing of Environment* 96: 277–291. doi:10.1016/j.rse.2005.02.002.

Snyder, J. P. 1987. *Map Projections-A Working Manual*. Washington, DC: U.S. Government Printing Office.

Sutton, P., D. Roberts, C. Elvidge, and K. Baugh. 2001. "Census from Heaven: An Estimate of the Global Human Population Using Night-Time Satellite Imagery." *International Journal of Remote Sensing* 22 (16): 3061–3076. doi:10.1080/01431160010007015.

Wuhan Bureau of Statistics. 1992-2013. *Statistic Yearbook of Wuhan*. Beijing: China Statistics Press.

Xia, L., K. Mao, Z. Sun, Y. Ma, and F. Zhao. 2014. "Method for Detecting Cloud at Night from VIIRS Data Based on DNB." *Remote Sensing for Land & Resources* 26 (3): 74–79. doi:10.6046/gtzyyg.2014.03.12.

Zhang, Q., and K. C. Seto. 2011. "Mapping Urbanization Dynamics at Regional and Global Scales Using Multi-Temporal DMSP/OLS Nighttime Light Data." *Remote Sensing of Environment* 115: 2320–2329. doi:10.1016/j.rse.2011.04.032.

Zhao, N., Y. Zhou, and E. L. Samson. 2015. "Correcting Incompatible DN Values and Geometric Errors in Nighttime Lights Time-Series Images." *IEEE Transactions on Geoscience and Remote Sensing* 53 (4): 2039–2049. doi:10.1109/TGRS.2014.2352598.

Zhou, Y., S. J. Smith, C. D. Elvidge, K. Zhao, A. Thomson, and M. Imhoff. 2014. "A Cluster-Based Method to Map Urban Area from DMSP/OLS Nightlights." *Remote Sensing of Environment* 147: 173–185. doi:10.1016/j.rse.2014.03.004.

Zhu, Z., Y. Fu, C. E. Woodcock, P. Olofsson, J. E. Vogelmann, C. Holden, M. Wang, S. Dai, and Y. Yu. 2016. "Including Land Cover Change in Analysis of Greenness Trends Using All Available Landsat 5, 7, and 8 Images: A Case Study from Guangzhou, China (2000–2014)." *Remote Sensing of Environment* 185: 243–257. doi:10.1016/j.rse.2016.03.036.

Predicting potential fishing zones of Japanese common squid (*Todarodes pacificus*) using remotely sensed images in coastal waters of south-western Hokkaido, Japan

Xun Zhang, Sei-Ichi Saitoh and Toru Hirawake

ABSTRACT
The present study used nighttime visible satellite images to identify the daily presence and absence of the fishing vessel aggregations, targeting the Japanese common squid (*Todarodes pacificus*) in the coastal waters of south-western Hokkaido, Japan. Here, statistical (generalized additive model (GAM) and generalized linear model (GLM)) and machine learning models (boosted regression tree (BRT)) were developed using a 3-year (2000–2002) presence/absence information from squid fishing aggregations and environmental variables (night-time sea surface temperature (SST), chlorophyll-*a* (Chl-*a*) concentration, K_d(490) (diffuse attenuation coefficients of downwelling irradiance at 490 nm), and bathymetry). Our findings showed that BRT outperformed the regression-based models in predicting the potential squid fishing zones during the validation period (2003). Results from BRT indicated that potential fishing zones were closely associated with water depth. Both SST and Chl-*a* concentration were also found highly influential to squid occurrence, while K_d(490), which is related to the water transparency, showed relatively less impact on the squid distribution. The spatial predictions using daily data from 2000 to 2003 revealed the gradual eastward movement of potential fishing zones between June and December, consistent with the pattern of squid fishing activities. Four experimental fishing surveys were further conducted to validate and improve our model predictions against experience-based fishing surveys. The results showed that the squid catches using our model predictions in 2012 substantially exceeded the average catches of experience-based fishing in 2011.

1. Introduction

Japanese common squid is of large commercial importance and increasing conservation concern in the recent decades. In 2012, catches of this species from Japanese fisheries reached 166,400 tons and account for about 78% of Japan's total squid landing (Ministry of Agriculture, Forestry and Fisheries 2014). This squid species is commonly divided into two main cohorts according to their peak-spawning season: autumn cohort and winter cohort (Murata 1990). Earlier studies showed that the autumn cohort generally performs a south–north feeding migration, assisted by the warm Tsushima Current in the Sea of Japan. In winter, it migrates south for spawning using almost the same route after reaching Hokkaido. The winter cohort, however, is mainly transported by the Kuroshio Current as its spawning areas are located further south relative to the autumn cohort. The juveniles of this cohort undergo annual feeding migration along the eastern coast-lines of Japan towards Hokkaido and retreat to the south via the Sea of Japan (Kasahara and Ito 1968; Sakurai et al. 2000; Choi et al. 2008; Rosa, Yamamoto, and Sakurai 2011).

As a result of the cephalopod's high sensitivities to ambient environment, annual Japanese landing records of Japanese common squid showed wide fluctuations during the past two regime shifts: late 1970s and late 1980s (Sakurai et al. 2000). Reasons for these conspicuous catch variations have been thoroughly discussed from perspective of environmental impacts on squid behavior. There is evidence that the water temperature variation altered spawning regions of winter cohort to the region which was unfavour-able to the survival of early stages and induced dramatic decrease of the stock size in the late 1970s (Sakurai et al. 2000; Rosa, Yamamoto, and Sakurai 2011). The warming environment during the late 1980s improved suitable spawning ground formation that facilitated feeding migration and thus markedly increased catches (Sakurai et al. 2000; Tian et al. 2008; Kidokoro et al. 2010). The increase in water temperature coupled with the long-term increase in zooplankton biomass, triggered the increase in catches during the 1990s (Kang et al. 2002). Moreover, currents are also thought to play an important role during the life cycles of Japanese common squid, as paralarvae off southwest the coastal waters the position of the Kuroshio Front (Bower et al. 1999). The importance of water turbidity in determining the Japanese common squid distribution is poorly known, however, the relationship between turbidity and fish distributions has been documented in some piscivorous species (Robertis et al. 2003). The spawning behaviour and catches varied with turbidity in other squid species, such as the California market squid (*Loligo opalescens*) and chokka squid (*Loligo reynaudii*) (Maxwell et al. 2004; Downey, Roberts, and Baird 2010).

Previous studies have shown a close relationship between squid distributions and the ambient environment. However, few studies on the impact of multiple environmental factors on spatial and temporal distributions of Japanese common squid are available. Until recently, it remains a challenge to have the local fisheries benefit from the findings of scientific studies of this species. Habitat suitability index (HSI) modelling has been recognized as a robust tool to explore environmental associations and is frequently used in the prediction of potential fishing zones for fish species to facilitate fishing activities (Bigelow, Boggs, and He 1999; Agenbag et al. 2003; Zainuddin, Saitoh, and Saitoh 2008; Sanchez et al. 2008; Chen et al. 2010; Mugo et al. 2010; Wang et al. 2015). However, due to the lack of fisheries data on actual squid catches in the coastal waters of south-western Hokkaido, no related work has been applied to improve the local fishery of Japanese common squid in the coastal waters.

Fishing activities in coastal waters of south-western Hokkaido primarily consist of small-scale jigging vessels (<20 GT). Local fishermen keep their records of catches and fishing positions in private. They also refuse to open the data for scientific use, so we cannot obtain appropriate fisheries data for our modelling or validation process.

Squid jigging fisheries have existed for over a hundred years in this area and most of the local fishermen often make good catches based on long years of fishing experience. Since all fishing vessels are equipped with powerful lights to attract squid aggregations (Choi et al. 2008; Yamashita, Matsushita, and Azuno 2012), the positions of nighttime sea light can be detected by the Defense Meteorological Satellite Program/Operational Linescan System (DMSP/OLS). This information can be used to represent the location of the daily presence and absence of the squid aggregations (Rodhouse, Elvidge, and Trathan 2001; Kiyofuji and Saitoh 2004; Waluda, Griffiths, and Rodhouse 2008; Choi et al. 2008). After matching with the feasible short-term environmental datasets, it is then possible to characterize the impacts of changes in oceanographic environment to squid potential fishing zones. Following this process, our goal is to generate the daily prediction of the potential fishing zones for the Japanese common squid to assist the local squid fisheries, especially under a highly variable climate.

In this research, to achieve more accurate predictions of potential fishing zones, we compared three of the extensively used models in ecological applications including, the machine-learning based boosted regression tree (BRT; Elith, Leathwick, and Hastie 2008; Leathwick et al. 2006, 2008) and regression-based statistical models: the generalized additive model (GAM) and generalized linear model (GLM) (Bigelow, Boggs, and He 1999; Agenbag et al. 2003; Zainuddin, Saitoh, and Saitoh 2008; Sanchez et al. 2008; Mugo et al. 2010). The environmental variables used as indicators for this study were selected based on their reported effects on squid behaviour and distributions. In addition, we conducted four fishery-independent squid jigging surveys using the T/S Ushio-Maru (179 GT, Hokkaido University) in 2012, to confirm the feasibility of our daily prediction.

The objectives of this research were (1) to identify the environmental associations of Japanese common squid in coastal waters of south-western Hokkaido, Japan, and (2) to develop an appropriate model to generate reliable daily predictions of potential fishing zones.

2. Materials and methods

2.1. Study area

This research covered the coastal waters of south-western Hokkaido, Japan (138°–143° E and 40°–43° N), where the predominant current systems are composed of the warm Tsushima Current, warm Tsugaru Current (one branch of the Tsushima Current), and the cold Oyashio Current (Figure 1). These current features result in the massive growth of phytoplankton in the water, consequently forming the rich foraging grounds for many marine species and create productive fishing areas for local fisheries (Kono et al. 2004; Sakurai 2007). Our study area is the main feeding ground of Japanese common squid, both autumn and winter cohorts undertake long-distance migrations to here, and the arrivals are mainly comprised of squid with more than 10 cm mantle length (Kawabata et al. 2006). Squid jigging vessels equipped with powerful lights gather in our study area to fish, starting from June until the end of December. We focused on coastal squid fishing activities, thus the information on the offshore fishing activities was discarded, and only fishing activities within the 30 nautical miles from the land were considered (Kiyofuji et al. 2005).

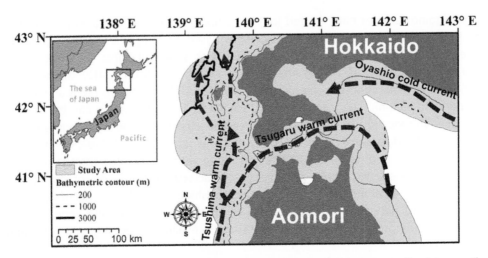

Figure 1. Map of the study area showing the key oceanographic structures. Overlain are the bathymetric contours and shaded region corresponds to the model domain.

2.2. DMSP/OLS nighttime visible data

Since there were very few other fishing activities using powerful lights in this coastal area, all strong light signals were considered to be emitted by the squid fishing vessels. Data on squid fishing vessels' position were obtained from cloud-free nighttime visible images of the DMSP/OLS through a threshold-based filtration method (Table 1), using the digital number (DN) values of the image ranging from 0 to 63 (Kiyofuji and Saitoh 2004). The extracted positions were then used to represent the presence and absence of the Japanese common squid aggregations.

A total of 954 single-pass images between June and December during a 3-year period (2000–2002) were used. We resampled these data into a daily 2.7 km resolution for establishing the habitat suitability model. The nighttime visible data for 2003 were used for comparison of the model performance (Table 2). Comparing with other years, the

Table 1. The threshold of identifying fishing locations from the DMSP/OLS nighttime visible data (Kiyofuji and Saitoh 2004).

Month	June	July	August	September	October	November	December
Threshold	39	42	39	37	38	36	36

Table 2. Data used in the model analyses.

Data	Source	Period	Original resolution	
			Spatial	Temporal
Fishing location	DMSP/OLS	2000–2003	2.7 km	Daily
Nighttime SST	AVHRR	2000–2003	4 km	Daily
Chl-*a* concentration	SeaWiFS	2000–2003	1 km	Daily
$K_d(490)$	SeaWiFS	2000–2003	1 km	Daily
Nighttime SST	MODIS	2012	1 km	Daily
Chl-*a* concentration	MODIS	2012	1 km	Daily
$K_d(490)$	MODIS	2012	1 km	Daily
Depth	JODC		500 m	

period from 2000 to 2003 had the highest numbers of available cloud-free DMSP/OLS data in summer. Thus, we selected this study period (2000–2003) to prevent potential sampling bias in the model building and validation.

2.3. Satellite-derived environment predictors

Satellite-derived environment factors were nighttime SST (sea surface temperature), SSC (sea surface Chl-a concentration), and $K_d(490)$ (diffuse attenuation coefficients of downwelling irradiance at 490 nm) which is an indicator of the turbidity of the water column (Udy et al. 2005) (Table 2). These available data were reported to influence squid behaviour and were also provided on finer spatial and temporal resolutions. Other environmental factors, such as salinity and sea surface height (SSH), were not included in the models due to their low spatial and temporal resolutions in the coastal areas that may cause large predictive uncertainties.

The daily nighttime SST data between 2000 and 2003 were downloaded from the Advanced Very High Resolution Radiometer (AVHRR) Pathfinder 5.0 dataset with an original 4 km spatial resolution. The ocean–colour data, Level-2 SSC and $K_d(490)$ of the Sea-viewing Wide Field-of-view Sensor (SeaWiFS), were obtained from the Ocean Color website (http://oceancolor.gsfc.nasa.gov/). To have the resolution of environment data consistent with the nighttime visible images (2.7 km), SeaDAS version 6.4 was utilized to resample and reprocess these data. These daily environmental data were also used to generate the prediction of potential fishing zones, to examine the patterns of squid distribution. During the cruises on T/S *Ushio-Maru*, the daily potential fishing zones were predicted using the daily environmental parameters derived from moderate resolution imaging spectroradiometer (MODIS) level-3 data at 1 km spatial resolution.

The change of the satellite database was mainly due to the data availability at much higher spatial resolution. MODIS was launched in 2002, therefore, only AVHRR (SST) and SeaWiFS (Chl-a and $K_d(490)$) data between 2000 and 2003 were then available for our study. In 2010, SeaWiFS stopped collecting data, thus, we used MODIS to derive the Chl-a and $K_d(490)$ in 2012. In addition, the SST from MODIS (1 km) has higher spatial resolution than AVHRR (4 km). Improvement on the spatial resolution is significant for local-scale studies, especially in application and forecasting.

2.4. Bathymetric data

The bathymetric data were obtained from the website of the Japan Oceanographic Data Center (http://www.jodc.go.jp/data_set/jodc/jegg_intro.html). The data were named J-EGG500 (JODC-Expert Grid Data for Geographic −500 m) and collected by several ocean research institutions using depth-sounding survey around Japan. We resampled the obtained bathymetric data into same spatial resolution with nighttime visible data (2.7 km) and MODIS environmental layers (1 km) for model building and predictions.

2.5. Model fitting

Investigation of squid distribution and prediction of potential fishing zones in this study were carried out by using the habitat suitability model, based on the presence/absence of fishing locations and corresponding environmental factors. The squid in this study area are a mix of autumn cohort and winter cohort, and currently it is not possible to differentiate the cohort here. Local squid jigging vessels target large juvenile and adult squid only, and this greatly reduced the effect of cohort on habitat preference in the model. BRT is regarded as a relatively new statistical technique of machine learning approach (Elith, Leathwick, and Hastie 2008; Leathwick et al. 2006, 2008). It combines the advantages of two statistical techniques: the decision trees and boosting. Decision trees model has no limitation on the shape and type of the predictor variables due to the growth of the tree is from repeated binary splits until reaching the specific criterion instead of fitting a particular function (Elith, Leathwick, and Hastie 2008). Since the uncorrelated predictor variables are rarely selected to accomplish the model in the process of decision trees, decision trees model is insensitive to the outliers and missing data, but decision trees model can be hardly used alone on account of its poor performance in their interpretation and prediction due to the low accuracy comparing to other methods. The boosting technique is a method to optimize the model accuracy through the combination of several weak simple models into a reliable model. It helps to cover the accuracy shortage of the decision trees and makes the BRT models fairly suitable for analyzing ecological datasets. Because the BRT model largely ignores non-informative predictors when fitting trees, it is not necessary to select the combination of variables before fitting the models (Elith, Leathwick, and Hastie 2008). A double-way interaction BRT (tree complexity = 2) using all environmental predictor variables with the 0.01 learning rate and 0.7 bag fraction was fitted to the presence/absence database. The other frequently used models in ecological researches, GAM (Equation (1)) and GLM (Equation (2)), were also applied to generate an appropriate prediction of potential fishing zones for Japanese common squid. Combinations of the variables in GAM and GLM were decided by sequentially removing the factors from a full model until a minimum Akaike information criterion (AIC) is reached (Alabia et al. 2016; Oppel et al. 2012).

$$Y = a + \sum_{i=1}^{n} f_i(X_i) + \varepsilon \tag{1}$$

$$Y = a + \sum_{i=1}^{n} \beta X_i + \varepsilon \tag{2}$$

where Y is the response (possibility of occurrence), X_i is the predictor (SST, Chl-a concentration, $K_d(490)$, and bathymetry), a and β are constants, ε is the error, and the f_i is the smoothing function for each predictor of the model.

The final model selection was dependent on comparing the relative performance in predicting the fishing location in 2003. All models were fitted with binomial response type in R version 2.15.1.

2.6. Model comparison

The predictive capability of each model approach was evaluated by comparing the true presence/absence and predicted values using three indices of performance: area under the curve (AUC) of the receiver operating characteristics (ROC), maximum kappa value, and point–biserial correlation. AUC is a common statistical test of ROC which is expressed as a proportion of the total area of the unit square defined by the false positive and true positive axes, AUC values range from 0.5 for no better than random guessing to 1 for a perfect fit (Pearce and Ferrier 2000). AUC has several superiorities over other performance indices as it measures the overall model performance without considering the species prevalence. Cohen's kappa measures the proportion of agreement beyond the chance effects in the predictive model, and maximum kappa value is widely used for optimizing the probability thresholds of models (Manel, Williams, and Ormerod 2001; Moisen et al. 2006; Leathwick et al. 2006; Freeman and Moisen 2008). The point–biserial correlation between the true presence/absence and the predictive probability of occurrence is sensitive to both discrimination and calibration (Oppel et al. 2012). This indicates whether the models are predicting the presence or absence more successfully or not. The suitable model from the criteria above was used to interpret the relationship between squid distribution and environmental factors. The inferred relationships from the selected models were used as the basis for the daily prediction of potential fishing zones for Japanese common squid.

2.7. Validation of the potential fishing zones prediction by field application

The performance of our model prediction was validated with four squid experimental fishery surveys on board the T/S *Ushio-Maru* in 2012. An automatic averaging process of previous data was developed to reduce the impact of low data coverage for the Chl-*a* concentration and $K_d(490)$ from the cloud contamination. The previous-day data were averaged when the coverage rate is less than 0.4. In order to maintain the accuracy, this averaging process was only conducted within a 15-day period. The captain of the T/S *Ushio-Maru* receives the prediction map on board through the communication satellite and then conducts fishing at places with higher prediction value (Figure 2). The catches per hour of Japanese common squid were recorded and compared with the average catches in 2011. The squid jigging sites in 2011 were all decided based on the captain's fishing experience. Due to the long history of commercial squid fishing in our study area, those experiences actually reflected the features of squid distributions from the perspective of statistics. It stands to reason that an experience-based squid jigging will have higher catches than a random sampling.

3. Results

3.1. Assessment of the model performance

Both GAM and GLM had the smallest AIC value when all environmental factors were included, hence, all three models (GAM, GLM, and BRT) used a full set environmental data (bathymetry, SST, Chl-*a* concentration, and $K_d(490)$) as explanatory variables. The predictive performance of GAM, GLM, and BRT based on the validation dataset (2003) for

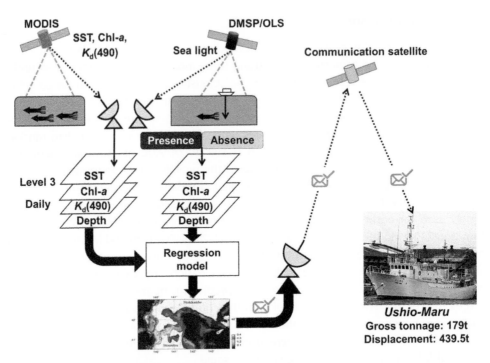

Figure 2. The schematic diagram summarizing the work flow involved during the model development, generation, and subsequent validation of predicted habitat.

Table 3. Predictive performances computed using independent validation data (2003 nighttime visible data).

Model	AUC	AUC standard deviation	Maximum kappa	Kappa standard deviation	Point–biserial correlation	Threshold
BRT	0.72	0.0034	0.09	0.0037	0.13	0.15
GAM	0.71	0.0035	0.08	0.0024	0.13	0.08
GLM	0.64	0.0036	0.04	0.0012	0.07	0.05

each model is shown in Table 3. The results of the model comparison showed that the BRT had slightly better predictive performance compared with the GAM in terms of AUC, maximum kappa, and the biserial correlation coefficient. The advantage of using BRT became larger when the practical applications of these models were applied in the prediction. Due to the low occurrence rate (0.04) of the fishing vessel in this whole area, prediction values of 2003 from all three models were very small, and BRT predictions can better extract the potential fishing zones as it gave more distinct boundary value (threshold = 0.15) between presence and absence than GAM (threshold = 0.08) and GLM (threshold = 0.05). The threshold derived for each model was based on highest maximum kappa value, which is commonly used to detect thresholds. The histogram of presence and absence with the optimized threshold (Figure 3) showed that the threshold of BRT had better ability to detect prediction boundaries. The predictive performance of the GLM was the worst among the three models because all indices were much lower than GAM and BRT. Therefore, BRT model was selected to further explore the squid habitat and predict the potential fishing zones in this study.

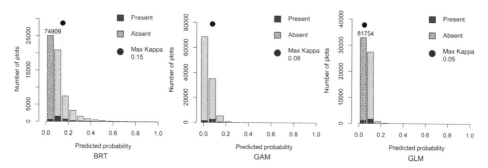

Figure 3. Presence–absence histogram for each model with optimized threshold based on the maximum kappa.

3.2. Environmental relationships of squid distributions

The fitted functions (Figure 4) indicated that squid assembles most frequently in the shallow waters, where the sea-surface temperature ranges from 8°C to 24°C with low Chl-*a* concentration and less turbid waters. The average percentage contribution of each variable is shown in Figure 4. The water depth contributed most to squid assemblages (53.1%), followed by sea-surface temperature (24.6%), and Chl-*a* concentration (16.8%). The $K_d(490)$ had the least model contribution (5.5%). The interaction between SST and

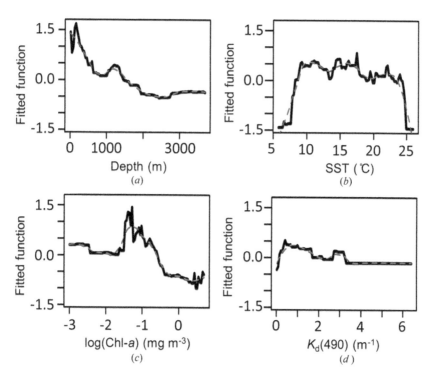

Figure 4. Fitted function for each variable in the BRT main effect model, indicating the dependence of squid presence on each predictor variable. The relative individual contributions of variables to the BRT model are as follows: 53.1% (water depth), 24.6% (SST), 16.8% (log(Chl-*a*)), and 5.6% ($K_d(490)$).

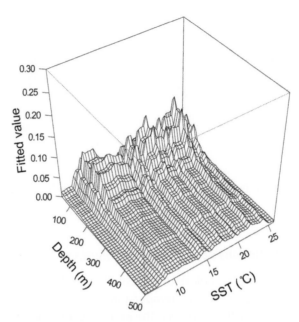

Figure 5. Plot of the 3D partial dependence for the strongest variable interaction in the model. The *x*, *y*, and *z* axes correspond to SST, depth, and fitted value, respectively.

water depth (Figure 5) had the highest median contribution to the BRT. This interaction captures the interdependence of environmental variables to one another. In the case of our study, the high potential fishing zones occur in the shallow waters only when the surface temperature reached the optimal ranges. Daily simulated predictions were combined into monthly average predictions of potential fishing zones between 2000 and 2003 (Figure 6). This revealed the specific location patterns of the potential fishing zones during these periods. High probability zone occurred near to the Sea of Japan in the early fishing season and gradually moved to the Tsugaru Strait and Pacific waters in the next several months. After October, there were very few potential fishing zones in the waters off the Sea of Japan.

3.3. Field application

In 2011, the T/S *Ushio-Maru* (Hokkaido University) survey designed its squid jigging sites based on the experiences of the captain. The mean record catch of 2011 was 42 catches per hour, and the maximum squid catch was at 92 catches per hour (Figure 7). The main purpose of the 2011 fishing survey was for on-board crew consumption, thus the captain always adjusted the fishing locations to find good fishing points and did not stay at a low-catch jigging site for a long time. Daily catches in 2011 (Figure 7) showed the fishing results at certain regions, and represented the squid catches of those areas.

In 2012, four jigging sites were chosen based on potential fishing zones predicted by the BRT model (Figures 8(*a*)–(*c*)). These four jigging points, located at the three sub-areas with very different water environment, were largely influenced by the Oyashio Current

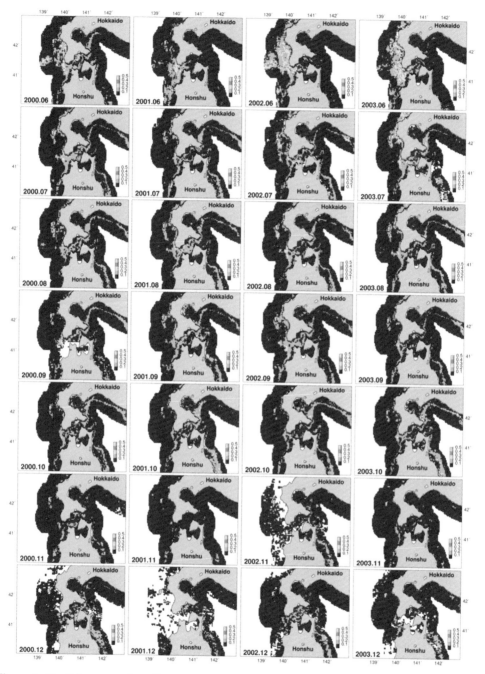

Figure 6. Monthly averaged prediction maps of potential fishing zones between 2000 and 2003. The prediction value shows the occurrence probability of squid assemblage (maximum probability = 0.5, minimum probability = 0; blank area was due to cloud).

(Figure 8(*a*)), Tsushima Current (Figure 8(*b*)), and Tsugaru Current (Figure 8(*c*)). Fishing catches reached an average of 76 catches per hour, when squid jigging was performed in area with high predicted catch. This largely exceeded the average fishing results of 2011 (42 catches per hour). On 2 October 2012, the squid catches also exceeded the

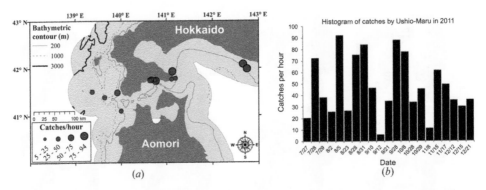

Figure 7. Map showing the squid jigging positions and corresponding catch information, scaled proportional to the amount of catch, on-board the T/S Ushio-Maru in 2011 (*a*). The frequency distribution of squid catches for the entire survey period is shown in (*b*).

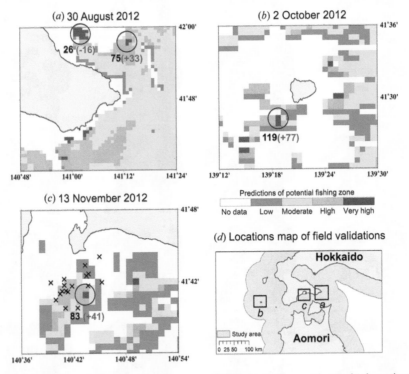

Figure 8. Prediction maps of potential fishing zones binned into low, moderate, high, and very high, for selected days of validation: (*a*) 30 August 2012, (*b*) 2 October 2012, and (*c*) 13 November 2012. The geographic sites for each validation period are shown in (*d*). The squid jigging region is encircled and corresponding differences between the current and mean squid catches in 2011 are shown in blue (decrease) and red (increase) numbers. Black crosses correspond to the actual fishing locations detected by the ship radar system.

maximum catch of 2011 (Figure 8(*b*)). During three out of the four fishing observations, the surveys were carried out in areas where no other fishing vessels were around (Figures 8(*a*) and (*b*)). In addition, the two fishing surveys at the low predicted area were as bad as expected and only caught a total of 16 squid in 11.5 jigging hours.

Figure 8(c) shows the map of the real fishing vessels' sites (black cross) overlaid on the prediction map of the potential fishing zones on 13 November 2012. The vessels' locations were obtained using radar plotting system on board. During this time, most of the fishermen conducted their fishing activities in areas, which the model predicted as high potential fishing zones.

4. Discussion

Recent developments of modelling technique have greatly promoted the understanding of fish behaviour and distribution. However, the interpretations of how these predictors affect the response variables will be very different between models. Consequently, the prediction of species distributions can be very different. Thus, the models have to be carefully selected for various purposes (Hirzel et al. 2006), especially in cases when the practical applications are addressed. In our development of habitat suitability model to predict daily potential fishing zones of Japanese common squid, the BRT model is proven to be the most suitable model according to the criteria of AUC, maximum kappa, and biserial correlation coefficient (Table 3). The advantages of using BRT model in this research relative to conventional methods (GAM and GLM) include (1) the insensitivity of the model to outliers of the predictors, (2) no pre-assumption of the relationship between response variable and predictor variables (the model completely learned from training data), (3) capability to handle the complex interactions among predictors which avoids the excessive prediction deriving from one single variable (Elith, Leathwick, and Hastie 2008; Leathwick et al. 2006, 2008). This is especially important when there are some invariant parameters such as bathymetry or distance from the port. These advances in the predictive ability allow a more reliable comprehension of squid habitat, and assist the local fisheries.

The relative importance of predictor variables which were derived from the BRT model revealed that water depth had the most significant effect on squid assemblage, and this is presumably due to vertical diel movement behaviour of the squid (Murata and Nakamura 1998). They generally dived into the deep waters and rested on continental shelf and slope before sunset. Thus, the frequency of occurrence at the water within specific depth ranges was much higher than in other places. Additionally, fishermen's fishing behaviour may also cause the strong influence on the depth preference. The SST is another relevant factor next to water depth. This result points to the close relationship between water temperature and squid behaviour (Sakurai et al. 2000; Mokrin, Novikov, and Zuenko 2002; Kidokoro and Sakurai 2008). In our study area, high surface Chl-a concentration area mainly concentrated on the place very close to shore with shallow depth, particularly inside of the Funka Bay and Mutsu Bays where few fishing activity happens throughout the year. Even though the SSC reflects high nutrient waters that may affect prey availability, the preference of squid to low SSC waters in our results does not mean that the food availability is not important for squid assemblages. In fact, Tsugaru Current, Oyashio Current, and their confluence create numbers of eddies and rings, and result in the sub-surface nutrient enhancement that supports feeding for squid (Itoh and Yasuda 2010; Yoda, Shiomi, and Sato 2014). The $K_d(490)$ only had 5.3% contribution to the final BRT model. Most of the extremely clear waters are distributed far from the coast where the fishing vessels were seldom located, and extremely low turbid environments would contain little food. Turbid water also has the negative effect due to the obstacle to forage in the sea for squid (Sanchez et al. 2008), and it further influences on the catch efficiency as reducing fishing light penetration and visibility of jig in the water. In fact, high $K_d(490)$ values (higher than 2 m^{-1}) accounted for a very small amount in the data. The second smallest peak of $K_d(490)$ in Figure 4 can be hardly detected, and the very high $K_d(490)$ areas in satellite data are probably contaminated by cloud.

The seasonal variation of favourable habitats for Japanese common squid was identified in the predictive simulation between 2000 and 2003. The potential fishing zones in the Sea of Japan side predominantly occurred between June and September. In the Pacific side, it occurred between July and December. These predictions were in agreement with the major seasonal changes in fishing grounds (Nakamura 2003). This agreement can also be seen as another evidence of the good model reliability. In June, almost all the potential fishing zones were only located in the Sea of Japan. It indicated the early arrival of squid off the Sea of Japan area were mainly composed of the autumn cohort, as most of the winter cohort arrived at the Hokkaido area in July (Murata 1989; Sakurai et al. 2000; Rosa, Yamamoto, and Sakurai 2011). One part of autumn cohort entered into the Pacific side through the Tsugaru Strait. Another part stayed at the Sea of Japan for a short period or moved farther north, and then returned to the southern spawning regions through the Sea of Japan to finish their life cycle. Most of the winter cohort came from the Pacific side, and comprised the main component of the eastern fishing grounds in the study area. This group then reached the Sea of Japan through the Tsugaru Strait during its spawning migration. These migration patterns also determined the formation of the potential squid fishing zones, which coincided with our predictions.

The process of deriving presence/absence location of the squid assemblages from DMSP/OLS nighttime visible data was based on the assumption that squid were caught in areas where fishing vessels were located (Kiyofuji and Saitoh 2004). This means that the prediction was related to the frequency of fishing vessels' occurrence, not the actual squid catches. Therefore, our validation using the fishing locations may not be enough to demonstrate the exact accuracy of predicting the potential squid fishing zones. Moreover, the presence/absence model only reflects the occurrence rate rather than the amount of catches. The incorporation of the actual catches from the local fishermen can improve the predictive performance of our model, and should therefore be considered in future studies, should they become available. The increased fishing catches from our cruises based on the prediction maps highlighted the high potential of our method when applied into the actual fishery forecasts. This result indicates the good feasibility and accuracy of the BRT model to predict potential squid fishing zones.

5. Conclusions

This study examined and compared the performances of three modelling techniques to predict the daily potential fishing zones of Japanese common squid in coastal waters of Hokkaido, Japan. The analysis also focused on identifying the relationships between environment and squid distributions. The use of the nighttime visible data resolved the difficulties of accessing actual fisheries data. This is the first time to relate the nighttime visible data with the environmental data to establish the habitat suitability model for

Japanese common squid in coastal waters. Such approach showed high feasibility based on both statistical and practical perspectives. In summary, our work has shown that the squid distribution is highly associated with the bathymetry, SST, and Chl-*a* concentration, with least influence from $K_d(490)$.

BRT outperformed other conventional statistical models in this research. The field validation of the daily predictions showed good model performance and the high feasibility of our prediction into the actual fishery forecasting. As such, we recommend the use of habitat suitability models to improve fishing efficiency and productivity in the future.

Acknowledgments

This study was supported by 'Hakodate Marine Bio Cluster Project' from the Ministry of Education, Culture, Sports, Science and Technology (MEXT), Japan. It was also supported by the Japan Aerospace Exploration Agency (JAXA) SGLI/GCOM-C Project. We are very grateful to the captain, officers, and crews of the T/S *Ushio-Maru* for their kind assistance and helpful suggestions during cruises.

Disclosure statement

No potential conflict of interest was reported by the authors.

Funding

This work was supported by the Ministry of Education, Culture, Sports, Science and Technology (MEXT), Japan; Japan Aerospace Exploration Agency (JAXA).

References

Agenbag, J. J., A. J. Richardson, H. Demarcq, P. Fréon, S. Weeks, and F. A. Shillington. 2003. "Estimating Environmental Preferences of South African Pelagic Fish Species Using Catch Size- and Remote Sensing Data." *Progress in Oceanography* 59 (2–3): 275–300. doi:10.1016/j.pocean.2003.07.004.

Alabia, I. D., S. Saitoh, H. Igarashi, Y. Ishikawa, N. Usui, M. Kamachi, T. Awaji, and M. Seito. 2016. "Ensemble Squid Habitat Model Using Three-Dimensional Ocean Data." *ICES Journal of Marine Science* 73 (7): 1863–1874. doi:10.1093/icesjms/fsw075.

Bigelow, K. A., C. H. Boggs, and X. He. 1999. "Environmental Effects on Swordfish and Blue Shark Catch Rates in the US North Pacific Longline Fishery." *Fisheries Oceanography* 8 (3): 178–198. doi:10.1046/j.1365-2419.1999.00105.x.

Bower, J. R., Y. Nakamura, K. Mori, J. Yamamoto, Y. Isoda, and Y. Sakurai. 1999. "Distribution of *Todarodes Pacificus* (Cephalopoda: Ommastrephidae) Paralarvae near the Kuroshio off Southern Kyushu, Japan." *Marine Biology* 135 (1): 99–106. doi:10.1007/s002270050606.

Chen, X., S. Tian, Y. Chen, and B. Liu. 2010. "A Modeling Approach to Identify Optimal Habitat and Suitable Fishing Grounds for Neon Flying Squid (*Ommastrephes Bartramii*) in the Northwest Pacific Ocean." *Fishery Bulletin* 108 (1): 1–14. http://agris.fao.org/agris-search/search/display.do?f=2013/AV/AV2012089600896.xml;AV2012089645.

Choi, K., C. I. Lee, K. Hwang, S.-W. Kim, J.-H. Park, and Y. Gong. 2008. "Distribution and Migration of Japanese Common Squid, *Todarodes Pacificus*, in the South-western Part of the East (Japan) Sea." *Fisheries Research* 91 (2–3): 281–290. doi:10.1016/j.fishres.2007.12.009.

Downey, N. J., M. J. Roberts, and D. Baird. 2010. "An Investigation of the Spawning Behaviour of the Chokka Squid *Loligo Reynaudii* and the Potential Effects of Temperature Using Acoustic Telemetry." *ICES Journal of Marine Science* 67 (2): 231–243. doi:10.1093/icesjms/fsp237.

Elith, J., J. R. Leathwick, and T. Hastie. 2008. "A Working Guide to Boosted Regression Trees." *Journal of Animal Ecology* 77 (4): 802–813. doi:10.1111/j.1365-2656.2008.01390.x.

Freeman, E. A., and G. Moisen. 2008. "Presenceabsence: an R Package for Presence Absence Analysis." *Journal of Statistical Software* 23 (11): 1–31. doi:10.18637/jss.v023.i11.

Hirzel, A. H., G. L. Lay, V. Helfer, C. Randin, and A. Guisan. 2006. "Evaluating the Ability of Habitat Suitability Models to Predict Species Presences." *Ecological Modelling* 199 (2): 142–152. doi:10.1016/j.ecolmodel.2006.05.017.

Itoh, S., and I. Yasuda. 2010. "Characteristics of Mesoscale Eddies in the Kuroshio–Oyashio Extension Region Detected from the Distribution of the Sea Surface Height Anomaly." *Journal of Physical Oceanography* 40 (5): 1018–1034. doi:10.1175/2009JPO4265.1.

Kang, Y. S., J. Y. Kim, H. G. Kim, and J. H. Park. 2002. "Long-Term Changes in Zooplankton and Its Relationship with Squid, *Todarodes Pacificus*, Catch in Japan/East Sea." *Fisheries Oceanography* 11 (6): 337–346. doi:10.1046/j.1365-2419.2002.00211.x.

Kasahara, S., and S. Ito. 1968. "Studies on the Migration of Common Squids in the Japan Sea. 2. Migrations and Some Biological Aspects of Common Squids Having Occurred in the Offshore Region of the Japan Sea during the Autumn Season of 1966 and 1967." *Bulletin of the Japan Sea Regional Fisheries Research Laboratory* 20: 49–70.

Kawabata, A., A. Yatsu, Y. Ueno, S. Suyama, and Y. Kurita. 2006. "Spatial Distribution of the Japanese Common Squid, Todarodes Pacificus, during Its Northward Migration in the Western North Pacific Ocean." *Fisheries Oceanography* 15 (2): 113–124. doi:10.1111/j.1365-2419.2006.00356.x.

Kidokoro, H., T. Goto, T. Nagasawa, H. Nishida, T. Akamine, and Y. Sakurai. 2010. "Impact of a Climate Regime Shift on the Migration of Japanese Common Squid (*Todarodes Pacificus*) in the Sea of Japan." *ICES Journal of Marine Science* 67 (7): 1314–1322. doi:10.1093/icesjms/fsq043.

Kidokoro, H., and Y. Sakurai. 2008. "Effect of Water Temperature on Gonadal Development and Emaciation of Japanese Common Squid *Todarodes Pacificus* (Ommastrephidae)." *Fisheries Science* 74 (3): 553–561. doi:10.1111/j.1444-2906.2008.01558.x.

Kiyofuji, H., K. Kumagai, S. Saitoh, Y. Arai, and K. Sakai. 2005. "Spatial Relationship between Japanese Common Squid (*Todarodes Pacificus*) Fishing Ground Formation and Fishing Ports: an Analysis Using Remote Sensing and Geographical Information Systems." *GIS/Spatial Analyses in Fishery and Aquatic Sciences* 2: 339–352.

Kiyofuji, H., and S. Saitoh. 2004. "Use of Nighttime Visible Images to Detect Japanese Common Squid *Todarodes Pacificus* Fishing Areas and Potential Migration Routes in the Sea of Japan." *Marine Ecology Progress Series* 276 (1): 173–186. doi:10.3354/meps276173.

Kono, T., M. Foreman, P. Chandler, and M. Kashiwai. 2004. "Coastal Oyashio South of Hokkaido, Japan." *Journal of Physical Oceanography* 34: 1477–1494. doi:10.1175/1520-0485(2004)034<1477:COSOHJ>2.0.CO;2.

Leathwick, J. R., J. Elith, W. L. Chadderton, D. Rowe, and T. Hastie. 2008. "Dispersal, Disturbance and the Contrasting Biogeographies of New Zealand's Diadromous and Non-Diadromous Fish Species." *Journal of Biogeography* 35 (8): 1481–1497. doi:10.1111/j.1365-2699.2008.01887.x.

Leathwick, J. R., J. Elith, M. P. Francis, T. Hastie, and P. Taylor. 2006. "Variation in Demersal Fish Species Richness in the Oceans Surrounding New Zealand: an Analysis Using Boosted Regression Trees." *Marine Ecology Progress Series* 321: 267–281. doi:10.3354/meps321267.

Manel, S., H. C. Williams, and S. J. Ormerod. 2001. "Evaluating Presence-Absence Models in Ecology: the Need to Account for Prevalence." *Journal of Applied Ecology* 38 (5): 921–931. doi:10.1046/j.1365-2664.2001.00647.x.

Maxwell, M. R., A. Henry, C. D. Elvidge, J. Safran, V. R. Hobson, I. Nelson, B. T. Tuttle, J. B. Dietz, and J. R. Hunter. 2004. "Fishery Dynamics of the California Market Squid (*Loligo Opalescens*), as Measured by Satellite Remote Sensing." *Fishery Bulletin* 102 (4): 661–670. http://www.biomedsearch.com/article/Fishery-dynamics-California-market-squid/125227470.html.

Ministry of Agriculture, Forestry and Fisheries. 2014. "*2012 Annual Statistics on the Fishery and Culture Production.*" Accessed September 2014. http://www.maff.go.jp/j/tokei/kouhyou/kaimen_gyosei.

Moisen, G. G., E. A. Freeman, J. A. Blackard, T. S. Frescino, N. E. Zimmermann, and T. C. Edwards. 2006. "Predicting Tree Species Presence and Basal Area in Utah: A Comparison of Stochastic Gradient Boosting, Generalized Additive Models, and Tree-Based Methods." *Ecological Modelling* 199 (2): 176–187. doi:10.1016/j.ecolmodel.2006.05.021.

Mokrin, N. M., Y. V. Novikov, and Y. I. Zuenko. 2002. "Seasonal Migrations and Oceanographic Conditions for Concentration of the Japanese Flying Squid (*Todarodes Pacificus* Steenstrup, 1880) in the Northwestern Japan Sea." *Bulletin of Marine Science* 71 (1): 487–499.

Mugo, R., S. Saitoh, A. Nihira, and T. Kuroyama. 2010. "Habitat Characteristics of Skipjack Tuna (*Katsuwonus Pelamis*) in the Western North Pacific: A Remote Sensing Perspective." *Fisheries Oceanography* 19 (5): 382–396. doi:10.1111/j.1365-2419.2010.00552.x.

Murata, M. 1989. "Population Assessment, Management and Fishery Forecasting for the Japanese Common Squid, *Todarodes Pacificus.*" *Marine Invertebrate Fisheries: Their Assessment and Management* 27: 613–636.

Murata, M. 1990. "Oceanic Resources of Squids." *Marine Behaviour and Physiology* 18 (1): 19–71. doi:10.1080/10236249009378779.

Murata, M., and Y. Nakamura. 1998. "Seasonal Migration and Diel Vertical Migration of the Neon Flying Squid, Ommastrephes Bartramii in the North Pacific." In *Contributed Papers to International Symposium on Large Pelagic Squids*, 1996 July 18-19, for JAMARC's 25th Anniversary of Its Foundation, 13–30. http://www.vliz.be/en/imis?refid=8998.

Nakamura, Y. 2003. "In the World of Japanese Common Squid" edited by T. Arimoto and H. Inada, 1st ed., 134–148 (In Japanese). Tokyo: Seizando Press.

Oppel, S., A. Meirinho, I. Ramirez, B. Gardner, A. F. O'Connell, P. I. Miller, and M. Louzao. 2012. "Comparison of Five Modelling Techniques to Predict the Spatial Distribution and Abundance of Seabirds." *Biological Conservation* 156: 94–104. doi:10.1016/j.biocon.2011.11.013.

Pearce, J., and S. Ferrier. 2000. "Evaluating the Predictive Performance of Habitat Models Developed Using Logistic Regression." *Ecological Modelling* 133 (3): 225–245. doi:10.1016/S0304-3800(00)00322-7.

Robertis, A. D., C. H. Ryer, A. Veloza, and R. D. Brodeur. 2003. "Differential Effects of Turbidity on Prey Consumption of Piscivorous and Planktivorous Fish." *Canadian Journal of Fisheries and Aquatic Sciences* 60 (12): 1517–1526. doi:10.1139/f03-123.

Rodhouse, P. G., C. D. Elvidge, and P. N. Trathan. 2001. "Remote Sensing of the Global Light-Fishing Fleet: an Analysis of Interactions with Oceanography, Other Fisheries and Predators." *Advances in Marine Biology* 39: 262–303. doi:10.1016/S0065-2881(01)39010-7.

Rosa, A. L., J. Yamamoto, and Y. Sakurai. 2011. "Effects of Environmental Variability on the Spawning Areas, Catch, and Recruitment of the Japanese Common Squid, *Todarodes Pacificus* (Cephalopoda: Ommastrephidae), from the 1970s to the 2000s." *ICES Journal of Marine Science* 68 (6): 1114–1121. doi:10.1093/icesjms/fsr037.

Sakurai, Y. 2007. "An Overview of the Oyashio Ecosystem." *Deep-Sea Research Part II: Topical Studies in Oceanography* 54 (23–26): 2526–2542. doi:10.1016/j.dsr2.2007.02.007.

Sakurai, Y., H. Kiyofuji, S. Saitoh, T. Goto, and Y. Hiyama. 2000. "Changes in Inferred Spawning Areas of *Todarodes Pacificus*(Cephalopoda: Ommastrephidae) Due to Changing Environmental Conditions." *ICES Journal of Marine Science* 57 (1): 24–30. doi:10.1006/jmsc.2000.0667.

Sanchez, P., M. Demestre, L. Recasens, F. Maynou, and P. Martin. 2008. "Combining GIS and Gams to Identify Potential Habitats of Squid *Loligo Vulgaris* in the Northwestern Mediterranean." *Hydrobiologia* 612 (1): 91–98. doi:10.1007/s10750-008-9487-9.

Tian, Y., H. Kidokoro, T. Watanabe, and N. Iguchi. 2008. "The Late 1980s Regime Shift in the Ecosystem of Tsushima Warm Current in the Japan/East Sea: Evidence from Historical Data and Possible Mechanisms." *Progress in Oceanography* 77 (2–3): 127–145. doi:10.1016/j.pocean.2008.03.007.

Udy, J., M. Gall, B. Longstaff, K. Moore, C. Roelfsema, D. R. Spooner, and S. Albert. 2005. "Water Quality Monitoring: A Combined Approach to Investigate Gradients of Change in the Great

Barrier Reef, Australia." *Marine Pollution Bulletin* 51 (1–4): 224–238. doi:10.1016/j.marpolbul.2004.10.048.

Waluda, C. M., H. J. Griffiths, and P. G. Rodhouse. 2008. "Remotely Sensed Spatial Dynamics of the *Illex Argentinus* Fishery, Southwest Atlantic." *Fisheries Research* 91 (2–3): 196–202. doi:10.1016/j.fishres.2007.11.027.

Wang, J., W. Yu, X. Chen, L. Lei, and Y. Chen. 2015. "Detection of Potential Fishing Zones for Neon Flying Squid Based on Remote-Sensing Data in the Northwest Pacific Ocean Using an Artificial Neural Network." *International Journal of Remote Sensing* 36 (13): 3317–3330. doi:10.1080/01431161.2015.1042121.

Yamashita, Y., Y. Matsushita, and T. Azuno. 2012. "Catch Performance of Coastal Squid Jigging Boats Using LED Panels in Combination with Metal Halide Lamps." *Fisheries Research* 113 (1): 182–189. doi:10.1016/j.fishres.2011.10.011.

Yoda, K., K. Shiomi, and K. Sato. 2014. "Foraging Spots of Streaked Shearwaters in Relation to Ocean Surface Currents as Identified Using Their Drift Movements." *Progress in Oceanography* 122: 54–64. doi:10.1016/j.pocean.2013.12.002.

Zainuddin, M., K. Saitoh, and S. Saitoh. 2008. "Albacore (*Thunnus Alalunga*) Fishing Ground in Relation to Oceanographic Conditions in the Western North Pacific Ocean Using Remotely Sensed Satellite Data." *Fisheries Oceanography* 17 (2): 61–73. doi:10.1111/j.1365-2419.2008.00461.x.

Index

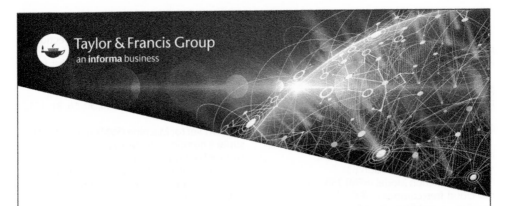

Printed and bound by CPI Group (UK) Ltd, Croydon, CR0 4YY

24/10/2024

01778292-0002